Fritz London was one of the twentieth century's key figures in the development of theoretical physics and chemistry. A quiet and self-effacing man, he was one of the founders of quantum chemistry, and was also the first to suggest that superconductivity and superfluidity could be viewed as macroscopic quantum phenomena. This thoroughly researched biography gives a detailed account of London's life and work, and, by following his correspondence with other leading physicists and chemists (such as Erwin Schrödinger, Walter Heitler, Max Born, John Bardeen, Max von Laue and Brian Pippard), examines the process by which scientific theories become legitimized.

Unusually, London first worked in philosophy, but then in 1925 he moved to Munich to study physics with Sommerfeld. Moving in 1927 to Zürich, he met Walter Heitler, with whom he published the classic paper on the homopolar bond. He then worked for a time in Berlin with Schrödinger, but, with the advent of Nazi rule, London was forced to resign his post. Leaving Germany, he went first to Oxford and together with his brother, Heinz, proposed the first successful theory of superconductivity. He then moved to Paris, where he worked on the problem of superfluidity, and continued to develop his views on macroscopic quantum phenomena. Before the outbreak of World War II, London took up a position at Duke University in the USA, where he died in 1954, at the age of just 54.

Covering a fascinating period in the development of theoretical physics, and containing an appraisal of London's work by the late John Bardeen, this book will be of great interest to physicists, chemists, and to anyone interested in the history of science.

Fritz London

a scientific biography

Fritz London

a scientific biography

Kostas Gavroglu
University of Athens, Greece

CAMBRIDGE UNIVERSITY PRESS
Cambridge, New York, Melbourne, Madrid, Cape Town, Singapore, São Paulo

Cambridge University Press
The Edinburgh Building, Cambridge CB2 2RU, UK

Published in the United States of America by Cambridge University Press, New York

www.cambridge.org
Information on this title: www.cambridge.org/9780521432733

First published 1995
This digitally printed first paperback version 2005

A catalogue record for this publication is available from the British Library

Library of Congress Cataloguing in Publication data

Gavroglu, Kostas.
Fritz London: a scientific biography/Kostas Gavroglu.
p. cm.
Includes index.
ISBN 0 521 43273 1
1. London, Fritz, 1900–1954. 2. Physicists – Germany – Biography. 3. Chemists – Germany – Biography. I. Title.
QC16.L645G38 1995
530'.092–dc20 [B] 94-691 CIP

ISBN-13 978-0-521-43273-3 hardback
ISBN-10 0-521-43273-1 hardback

ISBN-13 978-0-521-02319-1 paperback
ISBN-10 0-521-02319-X paperback

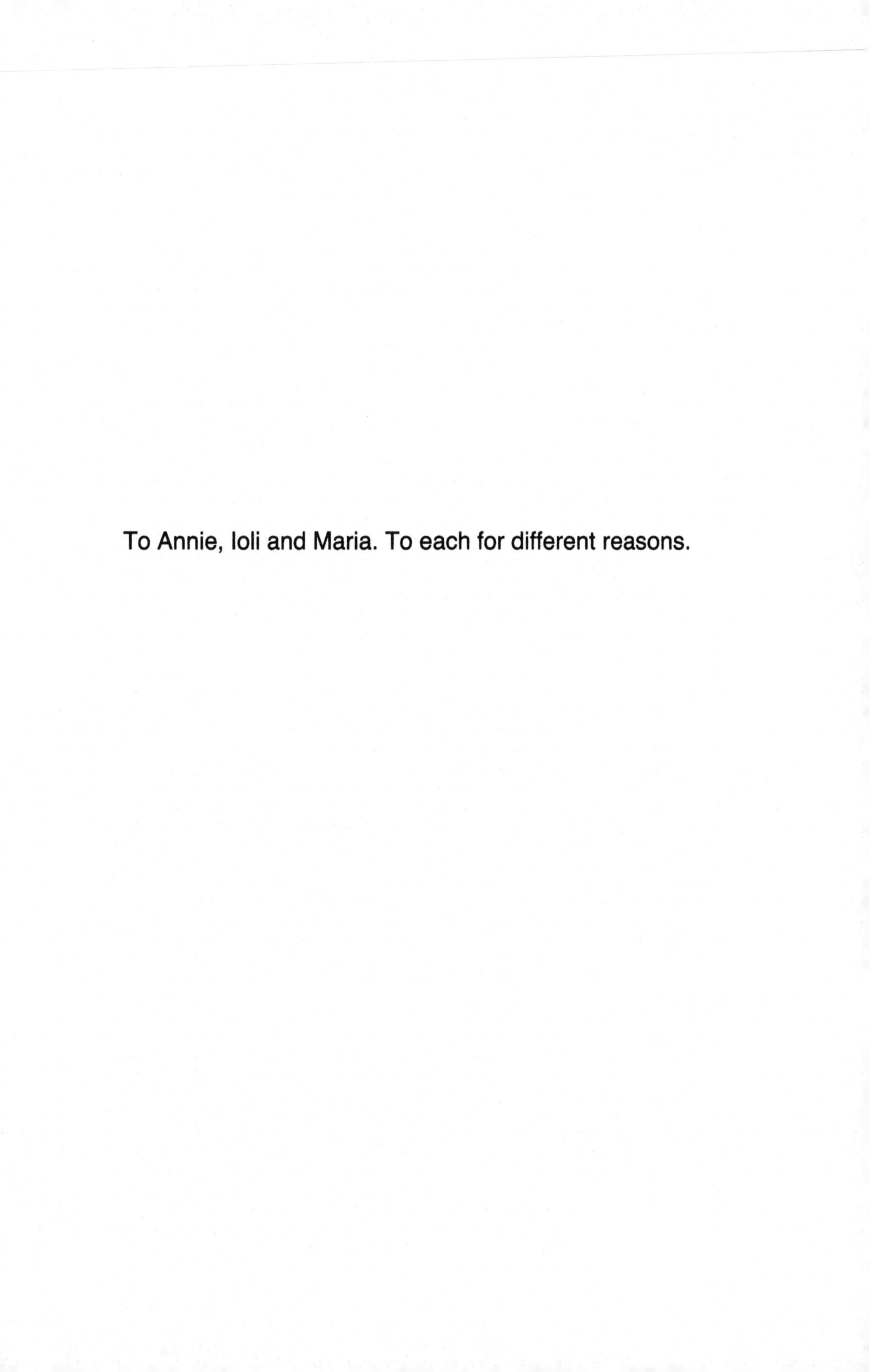

To Annie, Ioli and Maria. To each for different reasons.

Contents

Preface

In March 1953, the Royal Dutch Academy of Sciences informed Fritz London that it had awarded him the prestigious Lorentz Medal. The only other recipients were Max Planck, Peter Debye, Arnold Sommerfeld and Hendrick Kramers. In July of that year, London attended a ceremony at Leiden to receive the Medal. At the end of his short speech, Fritz London talked for the first and last time about himself publicly. 'During most of my life I have been so fortunate that I could do the things which my own nature drove me to do. It is embarassing to earn so much respect for just doing this. Yet, it is a great satisfaction for me to receive this particular sign of recognition, because it tells me that the work which was done, apparently by an internal necessity, has been found to be of some objective value'.

Some?

Surely it was a show of humility demanded under the circumstances, I thought when I first read that speech a few years ago. But, after reading and rereading his papers and books, going through his notebooks, doing most of the calculations to understand the missing steps, examining more than 3000 letters in his own and his correspondents' archives, talking to his family, friends and colleagues, I came to realize that Fritz London was not being humble that day at Leiden. He truly meant *some*. As I slowly realized that, the man whose work so deeply intrigued me, shrank into much smaller dimensions. And happily he become a very real person to me.

I have decided to write a scientific biography of Fritz London for many reasons. He was quite unique among the physicists of his generation. He started his career as a professional philosopher, writing a thesis in philosophy under one of the most prominent figures in the phenomenological movement, Alexander Pfänder. He never succumbed to the temptations of popular philosophizing. After a couple of years spent teaching at the Gymnasium, he started doing physics and made some insightful contributions to the development of the edifice of quantum mechanics. Soon after, with Walter Heitler, they solved the problem of homopolar bonding in chemistry and their joint paper has been regarded as the formal beginning of a new subdiscipline: that of quantum chemistry. After developing a theory of valence, and finding the correct form of the long range interatomic forces, he started working in low temperature physics and, in 1935 together with his brother Heinz, formulated the first successful theory of superconductivity. A few years later he daringly proposed that a hitherto neglected mechanism, first pointed out by Bose and Einstein, could be extended to account for superfluidity. His work in low temperature physics led to the possibility of extending quantum mechanics to large scale systems and Fritz London tried to legitimize a new possibility provided by quantum mechanics by defining macroscopic quantum phenomena. And, although he and Edmond Bauer wrote what is still regarded as one of the standard accounts for the quantum theory of measurement and despite his philosophical training, he did not feel that he could elaborate on any of the philosophical issues of quantum theory. He was a baptised German citizen of Jewish descent, he was very highly regarded by all his peers but was forced by the rise of the Nazis to leave Germany. After six years in Oxford and Paris, he was offered, at the age of 39 years, his first tenured position at Duke University in North Carolina. During the last eight years of his life his health worsened and he died in 1954 at the age of 54 years. He never held an administrative position at a university, he was never a member of an editorial board of a journal nor of an organizing committee for a conference, and he never did consultancy work for any length of time. Strangely, he never wrote a popular article.

To write about a person who started as a philosopher and who made such significant contributions to quantum chemistry, atomic physics, superconductivity and superfluidity, was bound to be not only interesting and intriguing, but also very challenging. It was a journey from philosophy to superfluidity, from Berlin to North Carolina, from the twenties of modernism to the fifties of McCarthyism. This was, at least, an eventful journey, but I slowly came to realize that it was also a painfully lonely journey.

The very scant biographical evidence from his first twenty years does not permit speculations about the possible effects of his early experiences upon his later life. I am strongly committed to attempts to map the influences of the early years upon a person's mature work and to use these influences to explicate a person's underlying agenda. But this seemingly commendable need for coherence in the narrative is often undermined by the ever present danger whereby what was intended as suggestive correlations turns into strictly causal pronouncements. I hope to have avoided such pitfalls. I am convinced that in the case of Fritz London's agenda the only viable way to chart these correlations meaningfully is to be guided by the ideas he expressed in his early philosophical writings.

In 1925, London published his first paper with Helmut Hönl on the intensity of band spectra. He lived through the exciting years of the founding of the new quantum mechanics. He did important work in transformation theory showing, among other things, quite impressive mathematical abilities. Would London have been among the founders of quantum mechanics had he not started with philosophy and had he not spent some time as a Gymnasium teacher? It is an irresistible question to pose, but to try to answer it is a thoroughly meaningless enterprise – at least for the way I would like to do biography. I do not think that history of science is about understanding connections between the past and what could have happened – however tempting it is for a scientist/biographer to chart such connections. This is not history, although it may be an interesting, if at times rather bizarre, pastime. It is much more interesting to explain why one path, rather than another, was followed. Here may be one of the few liberating lessons one draws from history of science: that science and its history are as much about successes as about failures.

In this book, I am also trying to formulate an approach to a problem in historiography. Ever since reading Thomas Kuhn's *The Structure of Scientific Revolutions* (1962) and following the ensuing discussions, I had become impressed with what appeared to be the easy part: normal science. The name, for better or worse, remained with us and, independent of whether we agree or disagree with Kuhn's schema, most of us agree on what does not constitute normal science. Writing the history of the revolutionary periods was undoubtedly an exceedingly complex task and there have been many excellent works. But what about the 'other' periods? Were they periods of harmonious collaboration, relatively uneventful developments, imaginative applications of theories already formulated and further clarification of the new conceptual issues involved? To a certain extent that is what they were. But they were also periods of diverging trends and periods of confluence. They were periods when disciplinary politics became a significant factor in scientific developments and when the cultural aspects of the scientific enterprise acquired a pronounced profile. I always felt that what has been regarded by many as Kuhn's deceptively unproblematic presentation of normal science, is, indeed, very rich in 'structure'.

In writing this biography and in trying to deal with the historiographical problem I mentioned above, I organized the available material and my arguments into five groups, all appearing in every chapter.

First, there is the biographical material and the attempt to bring in whatever was relevant from the wider social issues. Fritz London lived through the devastating, but creative, years of the Weimar Republic. Like almost all his colleagues, he never thought that the beer-hall rowdies could ever govern Germany – until 1933 when everything collapsed. Lindemann offered him a position at Oxford with funding from the Imperial Chemical Industries (ICI) and when there was no more funding in 1936, he was offered a position in Paris, at the Institut Henri Poincaré. He and his wife Edith arrived in Paris during the promising years of the Popular Front, the years of experimenting with cultural policies for the masses and the dominant role of the Left, where most of his friends – Joliot, Langevin, Perrin – belonged. Sensing the coming war, he decided to accept a job in the United States, becoming Professor of Theoretical Chemistry at Duke University in North Carolina. A few months earlier he had rejected an offer for the Chair of Mathematical Physics at the Hebrew University in Jerusalem for which he was strongly recommended by Einstein, among others. After the unhappy years in Oxford and the hopeful years in Paris, North Carolina was a melancholy place. The very private London was lonely and scientifically isolated. He became one of the most treasured acquisitions of Duke University. He was not associated, even peripherally, with the bomb project, and he was far away from where the action took place after the end of the war. His health was getting progressively worse. He made several trips to Europe and he was the main speaker at the first International Conference of Low Temperature Physics in 1946 at Cambridge, UK. At the end of the forties and the early fifties, he found himself in a surreal state, reminiscent of the way tragedies – and comedies – have a surreal sense about them. A German of Jewish descent who had been baptised, a man of deep culture, a humanist with left leanings, an internationalist, was in the American South at the beginning of the Cold War. As if the race relations and racism which shocked him so much in the beginning were not sufficiently taxing, he could now feel how fast his adopted country was being shrouded in the web so masterfully woven by McCarthyism. The pointing of fingers again brought frightening memories. He admired the democratic institutions of America and it was for that reason that he was so worried about what could happen, rather than what had happened to him. He felt that this excessively irrational situation could very easily lead to things hauntingly similar to what he had lived through in Germany twenty years before.

The second group concerns the issues related to his work: the writings in philosophy and the phenomenological movement; the work in quantum chemistry and the trends among the chemists in the late twenties and early thirties; the peculiar phenomena of very cold temperatures and superconductivity; the defiance of classical hydrodynamics and superfluidity; questions of quantum theory of measurement and his short but very influential booklet on the theory of measurement written with Bauer; and his views on biology. London's scientific biography is not intended as a popular account of his work nor as an attempt to make his work easily accessible. This is why I do not discuss other parts of his work or certain details of his papers. Most importantly, I have not attempted to assess London's work and to speculate on its significance today. I felt that too much detail is always detrimental to the story as a whole. To decide when one stops researching; to decide what not to include; to decide when a biography stops is a highly personal decision. Necessarily and naturally, a lot has been left out.

The third group has to do with the discussions in particular meetings and conferences whose role has been quite decisive in articulating, clarifying and legitimizing the rules of the game and for being meetings where a consensus was reached about the kinds of concepts and techniques to be used for the study of the particular problems. They were the meetings which, in a way, formulated the framework of mainstream research for the particular subdiscipline. Such meetings were convened by the Faraday Society in England and by the American Chemical Society in the USA in 1929, by the Royal Society in London in 1935. The 1946 Cambridge Conference on Low Temperature Physics played a similar role.

The fourth group is about the cognitive significance of disagreements. The use of disputes as a probe into normal science can become absolutely crucial in understanding subtle undercurrents involving, among other things, the process of legitimizing the new theories. In the case of Fritz London, I study this process by following the extended correspondence with key individuals. In his correspondence with Walter Heitler he expressed his antithesis to the schemata proposed by Linus Pauling, John Slater and Robert Mulliken concerning the chemical bond and it is possible to discern a rather strong dislike of the Americans' pragmatic approach to chemical problems. He disagreed strongly with Max von Laue about superconductivity. London's rather dogmatic rejection of Landau's treatment of superfluidity is not only in his published works, but also in his correspondence with Laszlo Tisza. Interestingly, in both the discussions with Heitler and Tisza, although there were many direct references to his disagreements, the emphasis was on the improvement of their mutual work and on the ways of communicating it to the community in a convincing manner. There were also shorter incidents like those with Herbert Fröhlich about the structure of solid helium and with Franz Simon about superfluidity.

The fifth group is about London's theoretical agenda. After formulating the theory of superconductivity, London proceeded to elaborate a new concept which would constitute his most insightful contribution to modern physics. It was the idea of a 'macroscopic quantum phenomenon'. He proposed that quantum mechanics provided the framework to discuss a new kind of order not in coordinate space, but, because of the uncertainty principle, in momentum space. I argue that there is a continuity from his thesis in philosophy to his work in superfluidity. The context for such a concept can be traced in his early school essays as well as in his philosophy thesis, and I am convinced that there is an undercurrent unifying his early readings of Kant, his avowed dislike of du Bois-Reymond's reductionism, his defense of Goethe the scientist, his adoption of Verrword's conditionalism, his discussion of the set theoretic methods, his perceived rigidity of the wave function of an electron in a superconductor and the Bose–Einstein condensation phenomenon which he used to explain the transition of liquid helium to the superfluid state. The 'macro' attitude had a long period of gestation and a difficult period before becoming accepted by the community. In all of London's writings, and especially in his later ones, there is a systematic attempt to explore the possibilities of a different theoretical approach, to be able to articulate the differences between a macroscopic theory on the one hand, and the phenomenological and fundamental theories on the other. London's aim, which I believe was largely fulfilled, was to formulate those theoretical, empirical, methodological criteria that would delineate the autonomous status of the macroscopic approach and to differentiate it from those approaches which aimed at constructing phenomenological theories.

One of the truly difficult parts in writing a biography of a scientist concerns decisions about the extent of the technical details to be included. It is one of those 'standard' problems, which, nevertheless, needs to be clarified and specified every time. The problem becomes even more difficult when the interpretation of the technical parts of the work does not have any 'grand' implications and, hence, cannot be intelligibly put into plain language. Time dilation, length contraction, the curvature of space, the discrete atomic orbits, the uncertainty principle and the reduction of the wave function are exceedingly complex notions which, nevertheless, could be reasonably well described and discussed without, in a first approximation, having to resort to the mathematical details behind them. It is obviously the case that I do not imply that whoever decides to write about these subjects without the heavy use of mathematics is guaranteed to do a good job. Quite the opposite is the case and the misunderstandings and myths around these subjects are mostly due to such popular writings.

Nevertheless, there have been excellent popular accounts of these developments, and what is more important, there have been superb scholarly works where the use of the technical background was the optimum for the comprehension of the implications of the theory. So, how does one begin to explain the work of scientists whose extremely significant contributions are inextricably tied up with the understandings of the technical details? If one knows nothing about the subject and does not have any training in the general area of the subject matter, then it is impossible to learn the subject just by reading the biography, no matter how conscientiously the biographer presents the technical details. On the other hand, for those readers who either know the subject or can follow the technical details because of their training, what is included may appear to be a rather watered down version which does not do much justice to the depth of a particular formulation. I do not think that a scientific biography should be aimed only towards a readership with scientific background and I did my best to make the overall story not depend on the technical details. The sisyphean pattern between a popular account of the scientific work and a highly technical presentation of it, is almost an inherent feature of a scientific biography. My background in physics and my involvement for some time in theoretical high energy physics, did not leave me indifferent when encountering the ingenuity involved in the intermediate steps, the imaginative interpretation of the mathematics and the beauty of the final product. I always tried to disperse this intimidating feeling of awe by looking again and again at London's overall agenda and understanding its overall coherence.

I avoid all problems related to the history of the birth of quantum mechanics. Part of our recent understanding of very many issues concerning the birth of quantum mechanics has also been due to the biographies of almost all the founders of quantum mechanics. Biographies and autobiographies about the pioneers of quantum mechanics have brought us additional insights into the first stages of a very complicated period in the history of science. Although Fritz London was not among the founders of quantum mechanics, his work, like that of many others, was exceedingly significant for the development of quantum mechanics and for the understanding of the complex patterns of normal science. For nearly two years, London worked on problems of quantum theory. He started his work in the months right after the publication of Heisenberg's matrix mechanics and stopped working on these kinds of problems at the time when Schrödinger completed his wave mechanics. I have tried very hard to avoid discussing London's work in relation to the host of issues about the history of quantum mechanics and especially those of the period 1925–1927. Though London's work in quantum theory was quite interesting, I feel that it was neither among the important contributions for the development of quantum mechanics nor did it play a decisive role in the development of his own later work. Alternatively, I am convinced that his early work and development of ideas in philosophy are found in various forms in his later more mature writings in physics.

Unlike some other biographers, I never came to know Fritz London. When he died I was 7 years old. Often, though, I tried to imagine what it would be like to have met him *after* writing this biography. I am sure he would have been gracious and kind, but rather apprehensive and somewhat suspicious. I think I would have answered – quite satisfactorily – any questions he posed about his work, his colleagues, and the situations he found himself in. But what about if he asked why I wrote so much about such a private man? It is something I have, still, not come fully to terms with and it is the question I would be afraid he would ask, but, just to make things worse, he would never have asked.

K. G.

Acknowledgements

My most deep-felt thanks go to two persons in particular – Edith London and Horst Meyer.

In many respects, this book is about Edith London as well. I am very fortunate to have met this remarkable lady and I am very sorry to have asked her to resurrect sad and painful memories. There was not a single instance when I felt that she wanted to impose her own version of events, and she never attempted to discourage me from pursuing all kinds of leads. When I first met Edith London, I had all kinds of prejudices, since I was convinced that feelings of admiration and bitterness would dominate our conversations. Perhaps my strongest shock was when I immediately realized that she did not want someone to write a hagiographic account of Fritz London's life. But there was something more than that. What I treasure most from the many discussions with Edith London is her ability to communicate her deep love for Fritz London. She has been an exemplary counterexample to all those cases who so egotistically want a biography to be the written account of their own narrative about a particular person. It was my request to have a portrait she had painted of Fritz on the jacket of the book, and even then she said that if I changed my mind it would be fine with her as well. But hers is a life of extraordinary perseverance. She took advantage of every conceivable opportunity to study and to further her passion for art, under the incredibly taxing conditions of their life after leaving Germany in 1933. In 1954, when Fritz died at the age of 54 years, Edith was 50 years old and she continued to paint and to experiment with new techniques. She was a member of the Art Department of Duke University from 1955 to 1969, taking part in numerous exhibitions. In 1988, she was awarded the North Carolina Governor's Medal of highest merit.

The completion of the biography of Fritz London, I owe above all to Horst Meyer, Professor of Physics at Duke University and one of the authorities on experimental low temperature physics. He was the first to suggest, in 1988, that I undertake the task and he has continuously been the person to whom I turned to solve all kinds of technical problems – including the arduous task of copying papers and correspondence on a slow copier and transcribing some of Heinz London's impossible handwriting! Horst also raised the necessary funds to complete the book by writing proposals, bargaining with Department Chairmen and making personal phone calls to various Duke alumni. He always reminded me of deadlines I promised and did not keep, of various chores I had at some point told him I would need be done and then had forgotten. But as was the case with Edith London, from Horst I also had a glimpse of the tortured soul of so many Europeans of Jewish descent and their determination to beat the odds. Though he had never met London, he had a deep insight into London's abstruse agenda. I am thankful to Horst Meyer not only for his multifarious help, but also in furthering my understanding of some of the trickier aspects of the low temperature experiments. I was truly elated when I was informed that Horst Meyer was one of the recipients of the fifteenth Fritz London Memorial Award in 1993 for his many significant experimental contributions in the study of solid hydrogen and helium mixtures.

Peter Harman, from the Department of History, University of Lancaster, Gerald Holton from the Departments of Physics and History of Science, Harvard University, Sy Mauskopf from the Department of History, Duke University, and Sam Schweber from the Department of Physics, Bradeis University, and the Department of History of Science, Harvard University, have urged and encouraged me from the very beginning to accept the task and have helped me enormously throughout. I thank them all.

I was very fortunate to meet two close, personal friends of Fritz London: Ernst Manasse, Emeritus Professor of Classics of North Carolina Central University, and Judah Goldin, Emeritus Professor of Theology of the University of Pennsylvania, have provided me with valuable insights into many aspects of Fritz London's personality. The lengthy conversations with Fritz London's son Frank have been very useful for me and I enjoyed his refreshing humor very much. In the book, there is an appreciable amount of material about Heinz London and I am sorry that I could not include more. I am obliged to Lucie London, Heinz London's wife, who has helped me tremendously by sending photographs and letters and many details about Heinz.

I was also fortunate to have had many discussions with Laszlo Tisza, emeritus Professor of Physics at MIT, and Sir Brian Pippard, formerly Cavendish Professor at the University of Cambridge. They both knew London and both had extensive discussions and correspondence with him on a number of scientific issues. They were very patient in answering all my queries and in clarifying a series of scientific and historical issues and, also, in making available to me all the material they had in their possession which pertained to London.

The late John Bardeen is the only person to have received the Nobel Prize twice for the same discipline. He shared the Prize in 1956 for his work in semiconductors and in 1972 for his work in superconductivity. In 1973, he provided the funds for an endowment to honour the 'distinguished scientist' Fritz London. In the ceremony announcing the endowment, Bardeen stated that 'Fritz London more than anyone else pointed to the path that eventually led to the theory of superconductivity'. Professor Bardeen was enthusiastically supportive of my project. He sent me copies of his correspondence with London and when I asked him to write a short note as an afterword to the book he immediately and gladly responded to the request.

I have sent parts of the manuscript to various colleagues for criticism. I am deeply thankful to Theodore Arabatzis, Cathy Carson, Allan Griffin, Ed Jurkowitz, Thomas Mormann, Mary Jo Nye, Richard A. Palmer, Skuli Sigurdsson, Ana Simoes, Thomas Soderqvist, and Norton Wise for taking the time to read what I sent them and to make detailed comments. I received the most detailed comments, criticisms and suggestions for the whole manuscript from an anonymous referee whose comments forced me to make many revisions. I thank him for his kind words about the book and his incisive suggestions which I am convinced have contributed to a better final manuscript.

I have also received many comments and suggestions from Aristidis Baltas, Panayiotis Benetatos, Jed Buchwald, Bob Cohen, Vasilis Drolias, Yorgos Goudaroulis, Ian Hacking, Steve Heims, Erwin Hiebert, Vasilis Karakostas, Thomas Kuhn, Rosie Martin (née London), Meir Hemmo, Jurgen Renn, Nikos Sarlis, Abner Shimony, John Stachel and Tassos Tsiadoulas. I had useful discussions with Diana Barkan, Theodor Benfey, James Bohning, Jean Christianidis, Kostas Christodoulidis, Kostas Crimbas, Bascom Deaver, Dimitri Dialetis, Hans van Duyneveldt, Larry Friedman, Marcus Hobbs, Dieter Hoffmann, Robert Kohler, Henry Linschitz, Sir Neville Mott, Everett Mendelssohn, E. Merzbacher, Efthimios Nikolaidis, Rudolf de Bruyn Ouboter, Linus Pauling, Michael Redhead, Barbara Rosencrantz, Bashi Sabra, Simon Schaffer, David Shoenberg, Arnold Thackray, Toni Travis, Spencer Weart, Victor Weisskopf and Paul Zilsel. I thank them all. I regret that I never received an answer from many of the researchers in low temperature physics in Russia to whom I repeatedly wrote.

All translations into English have been done by myself, unless stated otherwise. For the latter, I have had help from Skuli Sigurdsson, Arthur Daemmrich, Stamatis Gerogiogakis, Kerina Kordela, Alekos Tsitsovits and Tassos Tsiadoulas who have helped me with some of the translations.

I have visited and made extensive use of many collections and archives. My main source, of course, was the Fritz London Papers deposited in the archives of the Library of Duke University, USA. I thank the archivists Thomas Harkins and William King for their help. I also visited and used the collections of the American Institute of Physics, mainly for the material in the Archive for the History of Quantum Physics and van Vleck's papers; of Nuffield College in Oxford, UK, for the Chrewell (Lindemann) papers; of the Society for the Protection of Science and Learning 1933–1987, Royal Commission on Historical Manuscripts deposited at the Bodleian Library, Oxford, UK; of the Library of the University of Zürich, Switzerland, for the papers of Walter Heitler; of the Deutsches Museum in Munich, Germany, for the archives of Max von Laue; of the Oregon State University Library, USA, where the Linus Pauling Papers are; of the Bancroft Library at the University of California at Berkeley, USA, for the papers of Gilbert Newton Lewis; of the Willis Memorial Library of the University of Bristol, UK, for the Archive of Heinz London; of the Library of the University of Chicago, USA, for the correspondence of Robert Mulliken; of the American Philosophical Society in Pennsylvania, USA, for the papers of John Slater; of the Kamerlingh Onnes Laboratory of the University of Leiden, Holland, for the correspondence of Cornelius Gorter and of the Boerhaave Museum in Leiden for the papers of de Haas and Keesom.

I have received partial financial support from the Mary Duke Biddle Foundation, the Physics and Chemistry Departments of Duke University, the Dean of the Graduate School of Duke University, and the American Institute of Physics. Part of the work was completed during my stay as Edelstein Fellow at the Beckman Center for the History of Chemistry, University of Pennsylvania, USA.

Finally I wish to express my many thanks to Simon Capelin, senior physics editor of Cambridge University Press, UK, whose help and interest about the project has often transcended the bounds of professional obligations. Philip Meyler, physics editor of Cambridge University Press, had the misfortune of dealing with all kinds of technical details in the manuscript. I was very lucky to have had such a competent and helpful copy editor: Jane Barrett's thoughtful interventions helped me greatly to improve the text. I thank them both for their patience and help.

K. G.

1

From philosophy to physics

Fritz London's first published paper in a professional journal was in philosophy. In 1921, the year he graduated from the University of Munich, whilst supervised by one of the most well-known phenomenologists, Alexander Pfänder, he wrote a thesis that dealt with deductive systems. It was among the very first attempts to investigate ideas about philosophy of science expressed by the founder of the phenomenological movement in philosophy, Edmund Husserl. It was a remarkable piece of work for someone who was 21 years old. In this work, London developed an antipositivist and antireductionist view. This is all the more surprising, given London's knowledge of and interest in science, and the appeal of positivism to many scientists. London also intervened in the controversy between Richard Tolman and Tatiana Ehrenfest-Afanassjewa about the possibility of finding physical laws by dimensional considerations alone. Many of the ideas elaborated by London in his later researches, including his insightful suggestions and discussions of macroscopic quantum phenomena, can, indeed, be traced back to these early philosophical wanderings.

Right after his graduation from the University he started teaching in the Gymnasium and, when he was ready to matriculate as a Gymnasium teacher, he resigned and went to Max Born, who was at the University of Göttingen, to work in philosophy. Born did his utmost to discourage him, but to no avail. Born's only hope to persuade the young London to do an actual calculation, like all others beginning a career in physics, was to convince him to go to Munich and to study with Arnold Sommerfeld. There he did his first calculations in spectroscopy and, in 1925, he published his first paper in physics with Helmut Hönl on the intensity of band spectra. Soon afterwards he became Peter Ewald's assistant in Stuttgart and there he did his work on quantum mechanics.

His researches in transformation theory showed, among other things, London's quite impressive mathematical abilities. London also tried to express Hermann Weyl's theory – unifying gravitation and electromagnetism – within the quantum mechanical framework and he was among the first to introduce complex phases in quantum mechanics.

The years that left nothing unaffected

Though greatness and war are incommensurate notions, historians insist on calling one of the most catastrophic events in human history the 'Great War'. When in the fall of 1918 the war ended, there were nearly two million dead and four million wounded. The end of the war did formally bring peace. But it was also the start of a long protracted disagreement between those who thought they had won the war and Germany, whom no one doubted had lost the war. It was not peace with honour but peace with defeat, and soon the whole of German society was torn apart by those stormy passions which nurture only on defeat.

The mutiny of the sailors at Kiel, the proclamation of a Soviet Republic in Bavaria by Kurt Eisner, and the huge demonstrations by workers in Berlin were events which took place in less than ten days. Prince Max of Baden announced the abdication of the Emperor and appointed as Chancellor the leader of the Social Democrats, Friedrich Ebert. Another of the leaders of the Social Democrats, Philipp Scheidemann, ostensibly fearing that unless he proceeded quickly, the Spartacists would proclaim a Soviet Republic, proclaimed the Republic in the city of Weimar. November 9, 1918, became the formal beginning of a period in Germany whose significance both for Germany and Europe can hardly be overstated. The Weimar Republic was born in defeat and it was tragically defeated in less than fifteen years. The elections of January 1919 brought what appeared to be a viable coalition. On January 15, 1919, the two leaders of the Spartacists, Karl Liebknecht and Rosa Luxemburg were murdered in cold blood by right-wing thugs. The hopes that with an elected government the anarchy of the preceding months would come to an end were soon gone. At the end of February, Eisner was murdered. As a result, there was a general strike in Bavaria and the proclamation of the Socialist Council's Republic. There ensured bloody clashes, when two months later the government troops, intent on crushing the insurgence, used all their might on guilty and innocent alike. The Allies with their recent successes in the battlefields fresh in their memories, failing to realize their Pyrrhic victory, dictated their humiliating terms to Germany. By the end of the summer of 1919, Germany had signed the Treaty of Versailles and had a new constitution. Germany was to become a democratic republic. 'The next four years stood under the signs of domestic violence and foreign intransigence, the two interacting, and to Germany's misfortune, reinforcing one another.'[1]

For a long time, Germany resembled a helpless and desperate person in quicksand. Every move to escape the paralyzing humiliation of defeat in the Great War got the country in a greater mess. As Germany faced its past and tried to transcend it, the Left faced its own past and tried to assert it. Both would soon fail. That the Right would be against the Left was, of course, expected. But ironically as the war was the unmaking of the Right, peace brought the downfall of the Left. There was much bitterness, hate, violence and political maneuvering by the Left, about the Left and, most importantly, against the leftists. Utopias that appeared to be so immaculate and, hence, so resistant against all kinds of 'evil forces' were reduced to unrecognizable amateurish exercises on paper, after the realities and contingencies of governmental and administrative responsibilities set in. None of the old ideas and ideologies was left unaffected. None of the institutions could claim itself impregnable against the onslaught of 'modernity'. The years of the Weimar Republic left their mark on everyone and everything. And, then, the whole thing (and the exact determination of the 'thing' for many members of a generation, among them Fritz London, was rather immaterial) started crumbling like an old newspaper, into many small pieces, each carrying part of the story.

Fritz Wolfgang London was born in Breslau (now Wroclaw, Poland) on March 7, 1900. His mother, Louise née Hamburger, was the daughter of Heinrich Hamburger, one of the co-owners of the I. Z. Hamburger cotton mill in Liegnitz with business offices in Breslau. His father, Franz London, whose father was Meyer London, was born in 1863 in Liegnitz. He studied at the Universities of Marburg, Leipzig, Giessen, Zürich, Breslau and Berlin and in 1886 he received his doctorate in mathematics from the University of Breslau. In 1889, he became a *Privatdozent* in Breslau and also worked for an insurance company. His research field was constructive geometry. In 1896, he was appointed as titular Professor at the University of Breslau and when he died on February 27, 1917, he was Professor of Mathematics at the University of Bonn. Both parents were Jews. Franz had to wait many years to have a 'proper' academic appointment and this difficulty must have been one of the factors in his decision to have both his sons baptised. Fritz was baptised when he was 7 years old.

Fritz London's last four years at the Gymnasium in Bonn were the years of the Great War. In fact, he graduated from the Gymnasium earlier to be able to do volunteer work for the army, but was not accepted because of the problems he had with his spine. His spine was a little bent and needed massaging, and for that reason he, often, could not freely play with other children. Nevertheless, this ailment did not prevent him from learning mountain climbing, which was his second hobby after piano playing. After his father's death, from angina when he was 54 years old, the 17-year-old Fritz, the oldest male in the family, felt a deep sense of responsibility. Naturally Franz's death was a strong blow to Louise London, and Fritz felt that he should do all he could so that she would recover. Later, Fritz often talked about what his father's loss meant to his mother, but never discussed what the loss did to him.

Figure 1. Fritz London's parents, Louise (née Hamburger) and Franz London, at their home in Bonn in 1915. (Courtesy of Lucie London.)

Figure 2. Fritz London, seven years old, at the gate of his house in Bonn. His father is at the window on the left, the grandfather Alfred Hamburger is at the window on the right. The woman is unidentified. (Courtesy of Lucie London.)

The appeal of ideas

Between his last year in the Gymnasium in 1917–1918 and the final manuscript of his graduating thesis in philosophy from the University of Munich in 1921, London had written a number of essays which were quite suggestive of his early interests and influences. The only other document from his school days is a personal notebook from 1915 to 1916. Among its contents are several drawings: of a young man in what appear to be school surroundings titled *Morning Sickness*; figures in various positions; stars of David; Gothic buildings; geometrical figures and some geometrical proofs; bored and snoozing students in a class with a teacher; a drawing of a fiendish figure with rather strong features, with, scribbled in Greek, διαβολος the word for devil; maps and travel itineraries. There are a few drawings with graveyards with crosses and with the buried person's soul emanating from the grave towards the sky. One such grave has his name, his date of birth and the date of death with the last two digits erased. On the same page there is also a vase with (his) ashes and the title *In the Crematorium*. All this is followed by his will, dated March 20, 1915, where he stipulates that should God in his infinite wisdom decide that he should die, his bedroom should be used as a lodging for poor women and his money should be distributed to the poor. There is also the beginning of a play written in 1916 by Fritz and titled *The Comedy About Nakedness*. The characters have Greek names, the girls are blond with blue eyes and 'often naked' as he stipulates, as is Philo the heroine's love; their parents are 'never naked'. The only text of the play is detailed stage directions about the opening scene in a busy street as the sun sets and the two lovers enter.

The first full essay written by London when he was still in school was an essay about Goethe the scientist. There is also an earlier essay on the history and principles of telegraphy. Two other essays were either written in his last year at school, or, more probably, in his first years while he attended university. One of them is about the absoluteness of knowledge and the other is a critique of the methods of cognition. Furthermore, there is an essay titled *Relative/absolute/dynamical*, written most probably in 1919, and, finally, an untitled essay which almost certainly was the essay he showed to Alexander Pfänder (at the University of Munich in 1921) who, then, told him to go ahead and expand it for a thesis, being one of the necessary prerequisites for receiving one's degree from the university. Interestingly, London added various comments to these school essays during his later years, as is evident from his notes in the adjacent pages and margins.

London's excursions in philosophy should not be regarded merely as expressions of someone with a cultivated and inquiring mind who tried his luck by letting himself be lured by the appeal of the philosophical tradition of Germany. Themes and issues first raised in these school essays found a more mature expression in his philosophy thesis. They also appeared in his later publications in physical chemistry and molecular physics, and took a dramatic expression in his low temperature researches and the conception of the macroscopic quantum phenomena. It is not, of course, the case that everything London did can be traced back to his philosophical papers, nor that he did not have any novel and radical suggestions afterwards. Quite the opposite is the case: his lasting contributions were proposed without an explicit reference to any of his philosophical thoughts, yet there are unmistakable signs of links that can be discerned between his early writings in philosophy and his later polemics defending his unique approach to the interpretation of superconductivity and superfluidity as macroscopic quantum phenomena.

Concerning his approach to philosophy, London did not follow the practice of many physicists who were either among the founders of quantum mechanics or among its first practitioners. Most of these physicists wrote some kind of a philosophical piece after having made those contributions by which they established their reputation in the community. Some of these pieces are texts for a rather sophisticated audience, some are explanations of the implications of quantum mechanics and relativity, some are historico-philosophical accounts of the development of what is called 'modern physics', and others are attempts to present, in a systematic manner, a series of metaphysical issues within the context of the new developments. London followed a different path. His work in philosophy, never mentioned by others when there is reference to the philosophical writings of this generation, was of the professional kind, and was impressively ambitious: he wanted to discuss the issue of a deductive theory and the conditions for the existence of such a theory. In a thoughtful essay examining Husserl's philosophy of science, Thomas Mormann (1991) considers London's thesis, together with Husserl's ideas concerning philosophy of science, as having anticipated the semantic approach to the philosophy of science.

Goethe as a scientist

Goethe als Naturforscher was completed on July 1, 1917, and there we find the first traces of ideas that were to be fully expanded in his philosophical thesis. It was well received by his teacher. The language of the essay is heroic and, in places, almost poetic.

London's youthful admiration for Goethe was almost uncontrollable. He complained that the German people did not think highly of Goethe as a scientist because of his incorrect conjectures in optics. 'The foreigners' he noted 'surpass us in objectivity', since Goethe's works had been extensively translated and respectfully read in France and England. He felt that it was wrong to assess the importance of Goethe's work solely by his views on optics. He thought that Goethe's researches in osteology and botany were serious contributions to science. And these contributions resulted from Goethe's mystic conception of nature and, more specifically, from his notion of *Urphanomen* – that of the primal, archetypal phenomenon.

Even though London's assessment of Goethe's color theory was that of a physicist (the theory was wrong, whereas Newton's was right), London wanted to change the focus of criticism and to argue that Goethe should be considered as being the person who developed an aesthetic theory of the harmony of colors. Goethe had claimed that he was actually doing physics, and the reason for such a claim was because he generalized his psychological experiences. For Goethe the experience of seeing was the basis of knowledge, and he thought that whatever was being unravelled through seeing was also the unravelling of the real world. London considered Goethe to be in conflict with Kant. Nevertheless, one had to be lenient. A genius should be excused for attempting to turn such a method into a general rule. After all, Goethe, as London pointed out, was in a position to grasp intuitively the dependencies among various entities through his unconditional and absolute 'surrender to the whole'.

At the end, London wondered whether it might have been Goethe's idea to identify the *Urphanomen* with God, since the creation regarded as the *Urphanomen* cannot be explained by any science and the philosophers choose the way of speculative research. In the investigations for the *Urphanomen*, one witnessed the dark struggle of one who sought God as being the 'nucleus of the physical research of Goethe'. Attracted by Goethe's force of intellect, London, nevertheless, was unwilling to adopt Goethe's methodology, not because its results rendered it questionable, but because it was problematic on epistemological grounds. And at that level his sympathies and preferences were with Kant.

How absolute is our knowledge?

The second essay from his early years is titled *Speculations about the absoluteness of our knowledge*. London wrote this essay in November 1918, having already graduated from the Gymnasium, and there are no indications that he wrote it for any kind of course requirements. It is a serious and disciplined expression of his attempts to clarify for himself a series of issues that he was thinking about. The practice of writing a long piece to clarify the conceptual issues of a particular problem characterized London to the end of his life. Another important characteristic of this essay is London's explicit commitment to Kantianism.

Science seeks the absolute. Whichever way one looks at the absolute it does not change. Everything depends on the absolute, but it does not depend on anything else and there is no explanation for that. The relative is to be understood as a function. It is the 'flickering form of Proteus, which takes on a different appearance whenever one changes one's vantage point'. Thinking depends on the notion of function. The search for the absolute can be realized by answering the following question: in these function-like subordinations, is there a member which is not subordinate to any other? Interestingly, London referred to the work of scientists for the answers to such philosophical questions, and he started a systematic presentation of the methodologies of various disciplines to see the extent to which the researchers in these disciplines had reached any conclusions concerning the absolute. There was no possibility of finding the absolute in theology, since the miracle of the world could not be understood rationally. Neither in geology nor in mineralogy could one find absolutes. Astronomy did not have an absolute, since it became possible to provide an equally satisfactory explanation of the phenomena, independent of whether the earth was in the center of the universe or not. It may acquire an absolute if it was verified that the universe is closed, but what about if there are many universes? It may be the case that the elements in chemistry could be considered as absolutes, but this is not so, since they are differentiated from each other with respect to their specific weight, and, more importantly, they have structure. It is also not possible to consider either mass or energy as being absolutes, since one can be transformed into the other and it is not known which one is primary. Mechanics is the only discipline where such a discussion is rendered unnecessary because of the special theory of relativity. Geometry, also, could not provide absolutes: not only can one start from the axioms to derive a number of relations, but also one can start from the properties and these can lead to the axioms. Therefore, geometry is not a strictly deductive system, but it resembles a spiral.

At this point there is an addition to the text and the handwriting indicates that this intervention was made sometime after the mid-1920s. In the additional text, London discusses the impossibility of determining an absolute through systems of coordinates, since there cannot be a privileged system. He must have (re)read his essay after becoming conversant with the general theory of relativity, most probably during 1926 when he worked on Weyl's ideas to unify the gravitational and the electromagnetic fields. His conclusion was that there are no absolutes and that the tendency of the sciences to find absolutes was wrong and was an expression of their anthropocentrism – since the human spirit demands limitations.

London, then, proceeded to a discussion on the nature of reality. He considered it to be like the interior of a reflecting sphere where all points are equivalent: every point reflects itself upon itself *ad infinitum*. If it were possible to enter this sphere, then this may lead us to the belief that there is something absolute in the sphere by looking at our 'indeterminately disfigured' image. But this absolute changes in a function-like manner when we change our position in the sphere or our method of observation. For this reason we delimit ourselves to the view that if we leave the inside of the sphere then the absolute disappears. And though we think the absolute disappears, we do not believe that reality disappears.

> Everything is a dream, it is the great whole in its admirable aimless attempts, in its causal regularities. All this is the work of our spirit. All our knowledge is a function of ourselves. The only absolute is the thinking spirit.
>
> *Cogito ergo sum.*

Acquiring knowledge

The *Critique of the methods of cognition* is of special interest, since it contains, in embryonic form, some of the ideas that Fritz London developed analytically in his thesis. London took a very strong anti-reductionist stand, expressing his abhorrence for any approach that had as its strategy the formulation of the equations of motion for the minutest constituents as a necessary step for deriving the behavior of the whole. Apart from this pronounced antireductionism, there are three thoughts in this essay which he would develop further in his philosophy thesis. Firstly, the relation between an object and a condition is not one of subordination, but one of coordination. Secondly, concepts are not hierarchically arranged, but acquire their meaning only with respect to other 'similar' concepts within the same framework. Thus, the meaning of concepts is primarily contextual. Thirdly, deduction is a process that produces truths and, hence, the construction of a deductive theory should transcend the possibilities provided by formal logic.

London also pointed out an intrinsic incoherence in du Bois-Reymond's scheme. On the one hand, du Bois-Reymond's scepticism was expressed by *ignoramus et ignorabimus*, on the other, he stressed the necessity of an approach to nature that conceived the world as being describable by means of a single mathematical formula determining all that happens. London found refuge in 'Verworrn's conditionalism': to explain is to describe all the conditions of an event or of a state of affairs. Instead of trying to understand nature as an object, one attempts to examine the kinds of functions that are present in nature 'ignoring the question about objective reality'. And in his sympathy to conditionalism we witness the first signs of what eventually became his commitment to phenomenology.

London's teachers in philosophy: Alexander Pfänder and Erich Becher

London, in his discussions with his dear friend and scholar of ancient Greek philosophy, Ernst Moritz Manasse,[2] always mentioned that he proceeded to write a thesis in philosophy as a result of mere coincidence. London and Manasse met in Durham in 1939 when they had both come to the USA from Europe, and they stayed close friends until London's death in 1954. London had become deeply interested in problems of epistemology after reading Kant's *Critique of Pure Reason*, but he never intended to become a philosopher and he had gone to Munich to study physics with Sommerfeld. One day he had a long discussion with Alexander Pfänder, the best known and most senior member of the Munich group of phenomenologists. London explained to Pfänder his objections to the logical interpretation of deductive theory. He showed Pfänder something he had written about the matter. Pfänder was highly impressed with what he heard and read, and promised London that if he wrote all this as a scholarly essay, he would accept it as a dissertation. Manasse, with whom London discussed his days in Munich, remembered that 'according to London's account this had not been planned by him and that he never wavered in his determination to become a physicist, even before he became excited by the developments which led to quantum mechanics'.[3] Still, it was not until 1925, four years after completing his thesis in philosophy, that he published his first paper in physics.

Alexander Pfänder (1870–1941) was at the time the second most prominent member of the new phenomenological movement. By all accounts, he was an imaginative teacher. In the words of a student of his, Herbert Spiegelberg, who has so meticulously chronicled the early developments in phenomenology, 'the most characteristic thing about Pfänder was his intellectual and moral integrity, his frankness to the point of bluntness'.[4] His style has been described as Socratic and in the *Festschrift* commemorating his sixtieth birthday there are many instances where Pfänder was portrayed as a meticulous thinker who would spend a lot of time with students and ask all kinds of questions to ensure that the subject under discussion had been fully clarified.[5] He did not encourage the enunciation of grand programs which were so fashionable at the time. All these characteristics must have been very appealing to the young London, and, in fact, some of them, characterized his own later work. Pfänder's style was catalytic to Fritz London who was overflowing with philosophical ideas and who was so bent on clarifying ideas. London, as he told his wife on many occasions, felt a deep respect for Pfänder both as a teacher and a thinker. There are no records that London actually took a course by Pfänder, even though, during 1921, he took a course in the history of philosophy and the philosophy of mathematics by Moritz Geiger (1880–1937), who was another important member of the Munich phenomenologists and whose philosophical interests and contributions ranged from mathematics to aesthetics to experimental psychology. He also took courses by Erich Becher on logic and epistemology and by Clemens Baeumker on Aristotle.[6] In 1906, Husserl had asked Pfänder to write a textbook on logic and *Logik* was first published in 1921, the same year as London's thesis was completed.[7] Therefore, when Pfänder and London had their discussions, questions of logic were very much in Pfänder's mind. Pfänder's initial education in engineering and London's interest and knowledge of physics and mathematics made the rapport between the two even stronger. Pfänder's undertaking in his *Logik* was to argue that logic cannot be considered to be a branch of psychology. And many aspects of London's thoughts about logic, and London's explicit antipsychologism resonated with Pfänder's agenda, who had himself vigorously rejected the psychologism of his teacher, Theodor Lipps.

Becher (1885–1929) together with Hans Driesch, Max Frischeisen-Kohler and Traugott Konstantin Oesterreich expressed the 'critical realist' trend. Despite their various objections, all were strongly influenced by the neo-Kantianism of the Marburg school. The critical realists argued that, since the neo-Kantians denied the possibility of a substantive knowledge of the external world and they emphasized methodology rather than metaphysics, they had turned philosophy into a set of formal statements. Becher had studied with Erdman in Bonn and his doctorate was about Spinoza's doctrine of attributes.[8] He was more attracted to the German philosophers of the nineteenth century who did discuss metaphysics, but were drawn to *Natuphilosopie* or to any other theory of reality. He considered sense impressions to be complete images of external objects and advocated that there were spiritual as well as mechanistic types of causality. In his studies on the mind–body problem he argued against psycho-physical parallelism, occasionalism and epiphenomenalism and defended interactionalism. The critical realists acknowledged their debt to Husserl and the other phenomenologists for helping them to articulate their new epistemology.

Husserl's teachings

The essential parts of Husserl's philosophy of science[9] were formulated during the last years of the nineteenth century and they were first published in his *Logische Untersuchungen* in 1900/1901.[10] His proposals for a philosophy of science did not have any effect on logical positivism and they were even ignored among the phenomenologists. The only discussions of these ideas appear to be in London's thesis and in the writings of Oscar Becker.[11]

Husserl argued that the logical reconstruction of a theory of, say, physics necessitated the use of mathematics rather than logic. The task of the philosopher of science was the mathematical description of a theory's domain, i.e. the 'world' the theory intends to talk about. It is not sufficient for a philosophically adequate description of a theory to describe only its linguistic features. What is also needed is a mathematical description of its formal ontology, i.e. of its models. This is because an empirical theory T can be identified with its class of models $M(T)$, the models being specifically structured sets. Thus the first task of philosophy of science, according to the semantic approach, is to describe the models of a theory in set theoretic or mathematical terms. The aim of Husserl's philosophy of science was to reconstruct a global structure of empirical theories and may 'therefore be characterized as a *macrological* approach in contrast to the *micrological* one favored by the members of the Vienna Circle in the early decades of the century'.[12] This macrological analysis which took into account the global structures of empirical theories as structured wholes would become one of London's reference points in developing his own ideas about deductive systems.

Phenomenology's antireductionism must have been one of its more appealing aspects for London, who had already been thinking in terms of an antireductionist approach during his later school years. The logical positivists, and Rudolf Carnap in particular, had a strong reductionist program. The phenomenologists asserted that if there are implications of certain assumptions that are in conflict with the facts, the phenomenologists' first task was to reexamine the initial assumptions. This was in contrast to the approach of the reductionists who would, instead, redefine the terms used to describe the facts about the world in such a way that the contradictions between these descriptions of facts and the implications of the original assumptions disappeared. For the positivists, the role of mathematics was never autonomous, nor was it independent of logic, whose progress was perceived in its growing reductive and unifying powers. Mathematics and syllogistics were absorbed by an all embracing theory of formal logical syntax – which was to become the basis of any scientific philosophy. But for Husserl, London and Becker, the appropriate tools for the formal description of scientific theories was to be mathematics, not logic. Husserl maintained an independent position for mathematics. What London was thinking programmatically in 1921 was very close to

Husserl's thoughts. In this sense London's *problematique* was not marginal at all.

It also seems that London was influenced by Heinrich Rickert's views as well. There are many notes by London in the margins of his copy of Rickert's *The Limits of Concept Formation in Natural Science*. Rickert had argued against the received view that there is one scientific method of analyzing and conceptualizing the empirical material in the natural as well as the 'historical sciences'. Fritz London appeared to be in agreement with most of the arguments supporting this suggestion, but disagreed with Rickert when the latter proposed that in their development, the natural sciences become progressively more and more quantitative and less and less qualitative. London scribbled that this was a 'prejudice' on the part of Rickert.[13]

Abhorrence of reductionist schemata

In Alexander Pfänder's work, phenomenology was first and foremost a method. Furthermore, he was an outspoken realist and one of his avowed aims in *Logik* was to 'save the fundamental logical laws from triviality'.[14] The student did well, enriching the already novel approach and going a step further. If anything, London's elaborations in his thesis did advance the argument that it was possible to transcend that 'old prejudice' whereby deduction meant a process leading from the general to the more particular. As he pointed out, the emphasis on logic was not an emphasis on reductionism, but an attempt to have a holistic approach to science. This view was substantiated years later, in the mid-1930s, in London's first elaborations of his remarkable notion of macroscopic quantum phenomena.

London's early writings provide enough hints to be able to trace additional and, perhaps, equally significant influences upon him. There was his abhorrence of Emil du Bois-Reymond's reductionist program and London's argument that du Bois-Reymond's 'recipe' for a mechanical explanation of everything was inherently contradictory. His adoption of Verworrn's conditionalism, whereby there was no subordination between events and that all their conditions of existence provided a convincing alternative to du Bois-Reymond's schema. Then, there was Goethe: not his scientific achievements, but his view about the sciences that endeared him to the *Natuphilosophen*, his romantic view of all encompassing knowledge. Last, and most certainly not least, there was the all encompassing Kant himself. His distinction between the form of experience supplied by the human mind and the content of experience coming from the material world impressed the young London. Hence, even before starting his thesis, London was quite sensitive and receptive to ideas related to the semantic approach to the philosophy of science. Perhaps it was this quite unique blend of his early intellectual wanderings which attracted him to those esoteric aspects of Husserl's work.

The philosophy thesis

London's philosophy thesis was published in 1923 in the *Jahrbuch für Philosophie und phaenomenologische Forschung*. Alexander Pfänder, together with M. Geiger and M. Scheler, was one of the coeditors of the *Jahrbuch*, whose editor-in-chief was the founder of phenomenology, Edmund Husserl. During the early 1920s, Pfänder had overall editorial responsibility for the *Jahrbuch*. The dominant features of Fritz London's thesis place it within the phenomenological movement, itself striving to acquire a profile and a niche within the unsettled scene of philosophy during the first quarter of the twentieth century.

Among London's essays there is an untitled manuscript which ends rather abruptly. In all probability, it must be the essay he showed to Pfänder, who told him to go ahead and write a dissertation along the same lines.

London argued that there is an *a priori* tendency of the mind to consider all real contents as being dynamical. This, in a way, means that whatever comes from experience does not contain contradictions – a condition which is necessary for actually doing science. Another necessary condition is induction, without which it is not possible to do science. Therefore, in quite general terms there are two conditions which have to be satisfied by the transformations through which reality becomes an object of our consciousness. The first is that the character of these transformations is selective, since they transform only dynamical contents. The second is that reality is mainly made up of non-dynamical contents.

London, then, defined two kinds of formal minimal systems – systems which are comprised of axioms based on experience and everything that is derived from them. In these systems one attempts to choose axioms which are independent of each other. Two (or more) minimal systems are equivalent if both systems contain the same sentences – called by London 'relative contents' – which, however, have been derived by a different logical order. The preference of such a system over its equivalent(s) is mainly due to psychological reasons and, usually, the criterion of the intelligibility of the axioms with respect to experience plays a rather decisive role in the final choice.

The relative contents are derived from experience through a transformation T_2 and experience comes from reality through a transformation T_1.

$$\text{Reality} \underset{T_1}{\rightarrow} \text{Experience} \underset{T_2}{\rightarrow} \text{Language}$$

There are always uncertainties in measurements and, therefore, there are no 'correct' minimal systems. Because of T_2, there are 'deformed' minimal systems. Einstein's relativity showed that the Newtonian system, which was thought of as being a minimal system was, in fact, a deformed system. Such uncertainty of the measurements introduces another notion. The 'boundary' content can be acquired whenever the extrapolation of the relative content to infinity gives finite results. Newton's first law is such an example of a boundary content. Therefore, every mathematical formulation of a regularity was a boundary content.

In his thesis, London undertook an exhaustive analysis of the relations of the objects of thought among themselves, clarifying old definitions and devising new notions. He rejected the almost exclusive emphasis that had been given to the analytical aspects of deduction and showed that, if properly carried out, the process of deduction can bring about significant syntheses and that one can go not only from the general to the particular, but also from the 'general to what is equally general'.[15] His rejection of earlier views that considered logic as being part of psychology, and his strong and repeated emphasis that logic is not the study of the act of thinking but the study of the products of thought and their laws, was nothing but the adoption of Husserl's views in *Prolegomena zur reinen Logik* and Pfänder's overall attitude about logic. Rejecting the arguments of the adherents of psychologism, London vowed to examine only the domain of judicious knowledge, whose terms about objects emerge from a concrete finite number of indefinable basic terms by pure nominal definitions. His purpose was to understand the logic of acquiring knowledge and, hence, what became important were the presuppositions of formal logic and not the principles of cognition. Hence, London was not interested in the question 'How is theoretically formed knowledge possible?'. What he asked was 'How is theoretical knowledge procured, in case it were possible?'. He felt that such a change of focus would enable him to study this problem free from the psychogenetic points of view.

London started his thesis by expressing his dissatisfaction with the various researches in logic. He did, of course, acknowledge that the works of Boole, Frege, Peirce, Schroeder, the Peano School, and, above all, Russell were, in fact, extraordinary and transformed the whole subject. Nevertheless, these investigations dealt exclusively with the analysis of the logical 'elementary processes'. Through these researches it became possible to know the structure and interrelationships of the processes, proofs, and definitions that are found in a theory. None of them, though, dealt with theories as wholes.[16] London's aim in the thesis was:

> [To] enter upon a regressive procedure, through which the essential features of theoretical knowledge come out of their shells and are revealed, but of course not proved. We strive for insight and not for proof, and consider any other sort of reasoning as nonsensical and impossible. The discussion here has nothing to do with consequences, such as in Mathematics, but with principles.[17]

London stated that not all presuppositions of the theorems of a deductive theory can be proved, nor can one define all the terms that appear in these theorems. What was important in order to build a theory was not the content of the basic terms, but the (relative) relations which correlated the axioms to the objects that are referred to by these basic terms. Hence, the axioms can be considered as being the definitions of the basic terms. Nevertheless, the axioms do not define the terms in an exhaustive manner and part of the content of the terms is determined by their relations to other terms. The implicit definition worked out by David Hilbert is being used as a useful vehicle for explicating the relative relations. An implicit definition fixes immediately a whole group of terms, not in any absolute way, but only relatively with respect to the other terms of the theory.

At this point, one is obliged to discuss 'relative relations'. Could these relations be perceived by direct knowledge, or is there something more primitive, by which the different relations can be satisfied and which should, therefore, be examined as having a logical priority? The 'law of connection' was a way to differentiate the specificity of relations. If R_n is a relation which subsists among n objects that are in a well-determined succession, then the volume (R_n) of this relation is the number of cases in which the particular relations subsist. By the volume of a term, one is to understand the number of the objects which come under the particular term. In an analogous manner, the volume of a relation means the number of the states of affairs which correspond to the particular relation.

Since relations are defined by the law of connection, and basic terms by relations, these 'original terms' make up the basis for the formation of every possible pure deductive theory. The original terms represent the inventory of those terms which cannot be reduced to the already known terms, by the formal process of a nominal definition or an implicit definition. These must be conceived through direct knowledge of the objects that have been perceived.

London, then, discussed the notion of the 'state of affairs of a relation' – 'the set of the states of affairs within which a specific relation exists'. London called this set the 'volume of the relation'. One further clarification was required by the law of connection, and that was about the relation between a set and a partial set. It will be incorrect to define implicitly this relation as a relation, that is by its law of connection ('the part of a part is part of the whole' as London mentioned in his thesis), since such a definition would already presuppose knowledge of this relation. The statement about identity and its negation cannot be defined as an equivalence relation by its law of connection (transitivity, symmetry), since the definition would for the same reasons be circular as well as insufficient, like the definition of the relation of subsumption. All these, together with the principles of logic, comprised London's inventory for his main argument.

A given group of judgements constitutes the system of axioms w of a theory. The terms and relations that appear in these axioms are determined only relatively to each other by this system of axioms. The system of axioms w consists both of existential and subsumption axioms, and entails the following. Firstly, a set of terms that is not generally finite. The objects of this set exist because of the existential axioms in the ideal world outlined by the system of axioms w together with the basic terms and, secondly, a set of syllogistical conclusions (usually not finite) of which the states of affairs exist in that ideal world simultaneously with the axiomatically determined relations.

The set of the propositions that can be derived from a group a in a deductive way is called the 'domain' of the group a, and this is written as $B(a)$. A domain conceived as a group of judgements cannot obviously lead to syllogisms, the results of which are not already included in the domain. The reason for this is that the deduction of a deduction is again a deduction and is already included as such in the original domain. A domain is, therefore, a group of judgements which possesses a characteristic quality of closure. This characteristic is expressed as:

$$B(B[a]) \equiv B(a)$$

The content of a relation is restricted by the following: in case it exists in an orderly multitude, it should at the same time exist in another specific multitude derived from the former by rearrangement. There is, then, a crucial difference between the subsumption axioms and the existential axioms. The subsumptions of volumes do not define all objects, but certain pure ordinal ideal objects, i.e. relations. The formal objects that comprise the manifold are introduced in 'implicit definitions' after having made as their basis the already known relations. These implicit definitions of the basic terms do not have the form of a subsumption of volumes, but rather they include the assertion of the (logical) existence of the particular category of objects. He called these axioms 'existential axioms'. Somehow, they single out from the space encompassing all the 'objects in general', a number of classes.

Having made these conventions, London proceeded to a particularly interesting classification of judgements reminiscent of the Kantian categories, but with an important difference concerning the synthetic judgements, whereby London did not divide them into *a priori* and *a posteriori*. A deductive manifold was considered to be a relative unit of connection of truth. The following two terms are used: 'content of judgements' – the set of the states of affairs which are signified by the judgement – and 'content of a group of judgements', which represents the content of the logical product of the single judgements. All judgements can be classified according to the kinds of relations that the content of judgements has with the content of w that signifies the system of axioms. He called a judgement 'analytic with respect to w' when the content of w is a partial set of the content of the judgement, 'synthetic with respect to w' when at least a partial set of its content is a proper partial set of the content of w. He did not consider as important the remaining judgements whose content has no common elements with that of w. The contradictory judgements were analytic with respect to w, while those judgements whose contradictory judgements were synthetic with respect to w, were themselves synthetic judgements with respect to w.[18] A judgement that is analytic with respect to w is always true whenever w is true and it is logically implied by the system of axioms w and belongs to the domain $B(w)$. The truth of the judgements that are synthetic with respect to w does not necessarily follow the truth of w.

> From the truth of w it is not possible to derive the necessity, but only the (formal) possibility of a judgement that is synthetic with respect to w. The other principle, from which the necessity of a synthetic judgement arises, is according to Kant experience for the *a posteriori* synthetic judgements and principles of pure knowledge for the *a priori* synthetic judgements.[19]

Interestingly and somewhat apologetically, London needed to add that he had not intended to criticize or to correct Kant's distinction between the two types of synthetic judgements.

Let me elaborate a little on an example London gave about physics, since the examples from physics are very few indeed in the thesis. In physical theories, the judgements that are analytic with respect to the system of axioms are the general theorems, while the synthetic ones are the 'peripheral statements', which are not substantiated by the theory, but by another principle, namely that of experience. This substantiation by experience must necessarily be compatible with the theoretical possibilities. In their turn, the general theorems of physics can derive the necessary connections among the unimportant statements, without substantiating the truth of the not so significant data. Therefore, no experience aiming at stating the truth of these data is necessary for building up theoretical physics, after establishing the general basic laws from experience in the system of axioms. The derived propositions represent relative connections among those peripheral data that are possible. The establishment of the truth of synthetic judgements through experience is required only when one deals with special problems.

A new notion was introduced at this point. This was the notion of 'potency'. When two groups of judgements a and b have identical domains, then they are called 'equipotent'. The set of groups of judgements which are equipotent to a group of judgements a, is a well defined set. This set defines a quality by its volume, which is common for all (equipotent) groups of judgements appearing in this set and belonging only to them. This well defined common quality is called 'potency', $P(a)$. The potency is a function of the group of judgements a; $P(a) = P(b)$ only if $B(a) \equiv B(b)$. There are just as many potencies as there are non-identical domains.

It is often not possible to decide which of the two groups of judgements has the larger potency, because a case may arise where neither of the two domains is fully included in the other one and that the partial sets are, at most, identical. Such groups were called 'incomparable'. The groups of judgements belonging to domains that do not have a common partial set are independent from each other. In this case none of the judgements in one of the domains can be deduced from the group of the other domain. One cannot define a relation between the potencies of two incomparable (therefore independent) groups. When a group a is such that none of its judgements is to be found in the domain of the rest of the judgements of the group a, then the group is irreducible. By introducing the term of potency, the concept could be used for a complex of judgements to signify that it says more, or that it is more significant than another.

London believed that the outstanding problem during the historical formation of theories was to be able to find the criteria or the conditions – for a particular quality – which are neither too specific nor too general, so that they are both necessary and sufficient. 'The theoretician's passion for research (in contrast to that of an empiricist) is satisfied only by the knowledge of these conditions.' One can, in such a situation, identify the presuppositions that are being fulfilled or the criteria of that quality so that these presuppositions or criteria could represent the quality or the quality could represent the criteria. The version that displays judgements for such knowledge has the form of a total judgement.

London discussed the way by which it became possible to proceed from a total judgement analytic with respect to *w*, to other total judgements – not, though, by the simple formation of syllogisms. Results obtained by such syllogisms were equipotent premises: they represented only a very trivial extension of knowledge. London claimed that there were no such deductions in any theory. One went directly to those judgements that were synthetic with respect to *w*, the states of affairs which represented objects of that total analytic judgement. The syllogistic process interfered with these synthetic judgements and transformed them (with the help of the subsumption axioms and of the analytic subsumption judgements derived from them) into different syntheses. These new syntheses existed if and only if the original ones existed, since the subsumption axioms represented total judgements. When the premises were analytic, these syllogisms did not seem to be conclusion processes, but substitutions of terms in the subject and the predicate by others, which have the same content with them, according to certain axiomatically fixed rules. 'Such a process is usually called a "calculation".' It was nothing but an abbreviated syllogistic process of the judgements that are synthetic with respect to *w*. This is so, as long as the second statement of the conclusion is omitted, and the 'law of connection' is always implicitly presupposed. 'That is the fundamental meaning of a calculation in a theory.' Hence it was now possible to see the way the dualism between the two kinds of judgements, i.e. between those analytic with respect to *w* and those synthetic with respect to *w*, corresponded to the dualism in the system of axioms. There were, the subsumption axioms that defined only connection laws and those were distinguished from the existential axioms which introduced objects that obeyed such relations. And with this, London, rested his case – 'we are [thus] able to perceive constructively the whole mechanism of deductive manifolds'. He felt that he had developed the necessary notions to examine that 'old prejudice whereby a deductive conclusion should necessarily lead from the general to the more particular ... [and show] that it is also possible to deduce from something general, something equally general'. London was convinced that the only way to comprehend the syllogistic processes involved in proofs was to devise a way of thinking which regarded theories as wholes. This could only be done with the application of the methods which had been first created in the manifold theory, and for this reason the problem is only indirectly of a logical origin.

Figure 3. Louise London with Heinz London at a picnic around 1920. (Courtesy of Lucie London.)

> The issues that have been raised by these methods, such as the ones about the potency, the independence, etc. of judgement systems, cannot be dealt with by subtle logic of the elementary judgement and of the elementary conclusion process. *For the whole of a deductive manifold is not the sum of its parts.*[20] (Emphasis added).

To the end of his thesis, London never stopped emphasizing that such an understanding could not be achieved by the more traditional researches on syllogisms, despite the comprehensive and definitive treatment of the theory of calculation by Russell and by Peano's school, so long as these researches were confined only to problems concerning judgements and conclusions.

Tolman's principle of similitude

Among Fritz London's philosophical works, there is a paper which basically deals with methodological questions and it is titled *The conditions for the possibility of determining the measure in a physical manifold and the principle of similitude*. The philosophical issues in the paper are discussed only indirectly. This paper, published in 1922,[21] was a reaction to some thoughts expressed by Richard Tolman in 1914 and to a discussion between Tolman and Ehrenfest-Afanassjewa in 1916.

Tolman's 1914 paper advanced some remarkable claims. He first formulated what he called the 'principle of similitude': 'The fundamental entities out of which the physical universe is constructed are of such a nature that from them a miniature universe could be constructed exactly similar in every respect to the present universe.'[22] Newton had formulated a similar principle for specific kinds of dynamic systems, but, as Tolman showed, a complete development of the consequences of the principle of similitude depended both on the electron theory and the special theory of relativity. He argued that this principle provided a simple method for finding the form of the functional relations connecting physical magnitudes. After completing such a program, he was confident that it would become possible to regard the principle of similitude at least as a 'temporary criterion for the correctness of physical theories which may be advanced'.[23]

Tolman went on to derive the validity of all the laws of physics, using the principle of similitude and the assumption that they were accepted as given in one universe. He considered two observers O and O′ to have instruments for making physical measurements. Observer O has the ordinary meter sticks, clocks and other instruments of the kind we now possess, and makes measurements in the present physical universe. Observer O′ has correspondingly altered measuring instruments, making measurements in the 'miniature' universe, but, according to the principle of similitude, he derives exactly the same numerical results as O for all the experiments they both do. He assumed the existence of a construct which appeared the same for both observers. Therefore, the properties of this construct obeyed the same laws, independent of the particular observer who did the measuring. By using the transformations developed for the physical magnitudes, Tolman derived a series of very interesting results. Assuming the product of pressure and volume (PV) to be equal to a function of temperature ($F(T)$), he found that $F(T) = k(T)$, k being a constant. He derived both Stephan's law and Wien's law, and showed that Maxwell's equations were in accord with the principle of similitude. Tolman used the constancy of light explicitly to derive the transformation for time, and he, himself, stated that the special theory of relativity was necessary for deriving his results. In a way, this principle became the equivalent of the relativity principle of the special theory, except for the crucial fact that the latter referred only to inertial frames, and, hence, there were no difficulties concerning gravitation.

A further, generalized discussion of the principle of similitude was inextricably related to a discussion about the possibilities of dimensional analysis for deriving laws of nature.[24] The principle of dimensional homogeneity stipulates that for all equations of the form:

$$A' + B' + \ldots + N' = 0$$

the A', B' ... N' have the same dimensions, and the equation itself remains unaltered if one multiplies every value of length by an arbitrary multiplier x, every time interval by another arbitrary multiplier y, every mass by z, etc. It was, indeed, possible to derive various laws, but only under certain conditions. First, the equations derived by dimensional considerations always involved numerical constants, which could be calculated by other methods or by experiments. Second, a law was derived only if one knew in advance the physical properties which were expected to be found to be connected by the equation. The third disadvantage of dimensional homogeneity was in those cases where it is impossible to construct a dimensionless product of the quantities involved, without introducing a dimensional constant. In order to get around these difficulties, Tolman advanced the principle of similitude, especially on the basis of the complete relativity of size, which assumes the possibility of a coordinate change in the magnitudes of all measurable properties (lengths, time intervals, masses, temperatures, etc.) of such a nature that no measurements of any description can lead to its detection.

The necessary clarifications

London's (1922) paper was a direct response to the proposals of Tolman and to some of the criticisms by Ehrenfest-Afanassjewa in 1916. Characteristically, the main thrust of London's arguments was to clarify many of the presuppositions made by Tolman, to reformulate many of his claims in a mathematically more satisfactory manner and to show that the scheme was interesting and useful, but by no means *ars inveniendi*. London's interest in such an approach, like Tolman's, was that, somehow, it provided the possibility of finding the laws of physics deductively – an issue about which London had thought a great deal, as we saw.

This is also the only text where London, in a paper which was primarily intended for an audience of scientists, expressed his thoughts on certain epistemological issues. The imperfection of the empirical sources of knowledge, despite the continuous perfection of the experimental techniques, made it impossible to acquire a knowledge of the physical laws by empirical means. But this was also not possible by reason alone, as Kant's critique destroyed 'the scholastic hopes for deducing the real from the possible'.[25] He thought that his thesis did provide a way out from such an impasse. Concerning the physical laws as such, it was possible to formulate certain general statements, most of which were not synthetic *a priori* judgements but appeared to be plausible and reasonable. Considering the fact that the requirements of general relativity placed constraints on the form of the physical laws, could further constraints be made to enable specific laws to be derived as analytic judgements from the notion of physical laws? London was convinced that this derivation was not possible and that the only way to do that was to find the interrelationship among certain physical magnitudes – and only among these – when some kind of a result about the kind of a functional relation could be possible.[26]

London agreed with Ehrenfest-Afanassjewa that the principle of similitude did not refer to an observation of two different worlds, but to the measurement of the same world by two observers with different conventions for the units of all the manifolds. The aim was to formulate Tolman's principle by taking into consideration Ehrenfest-Afanassjewa's objections that the system of new variables may not be finite. Such considerations, according to London, may raise questions about the possibility of a deductive theory of the physical world of phenomena. He pointed out that in his thesis he had developed the 'Conditions for the possibility of a deductive theory'. Ehrenfest-Afanassjewa's objection questioned a necessary condition whose content was self-evident: 'The number of basic terms of a deductive theory has to remain below a finite limit in each successive revelation of judgements of a particular branch of knowledge, in a process that continues indefinitely.'[27] No one can stop somebody doubting the validity of these necessary preconditions of a manifold of physical phenomena.

But Tolman's method, according to London, was of limited use. It took advantage of those empirical situations that have been considered by London to be the necessary prerequisites for the determination of a measure for a physical manifold. London concluded that Tolman's theory could not be considered to be a deductive theory and that its range of possibilities was not all that significant. The principle of similitude gave only equations of state, when it is known that most of the laws of physics are not in the form of equations of state, but they are differential equations of physical quantities. London's reaction did not leave much to be expected from Tolman's method. He wrote: 'To have this perfect knowledge is in many instances more difficult than to be able to get the law by different means.'

London's first serious excursion into the 'outside' world after he completed his thesis in philosophy was to intervene in a dispute between two physicists in the pages of a journal read mainly by physicists. London's paper did not receive any attention, except for a letter by Norman Campbell, who had adopted an approach similar to London's in his *Physics: The Elements* and who also thought that Tolman's principle was 'an utter nonsense'.[28] London had sent a copy of the paper to Pfänder who responded with no comments.[29] The only person who discussed London's ideas in a more detailed manner was G. Wallot, but nothing came of it.[30] London's wish to discuss these issues with the scientists would perpetrate his professional isolation. It was the time when such writings were becoming quite numerous. It was also the time when one of the leading figures among the physicists and astronomers, Arthur Eddington, suggested that a warning be posted over the entrance gate of the new physics saying 'structural alterations in progress – no admittance except on business. Under no circumstances would the doorkeeper admit prying philosophers'.[31] The indifference and, perhaps, hostility that London encountered among the scientists about his work did not leave him with many choices: he decided to work in mainstream physics.

Work on quantum theory

In November 1921, after a short stay at the Universities of Frankfurt-am-Main and Göttingen, London received his degree *summa cum laude* from the University of Munich with a thesis in philosophy. He took the civil service examinations to qualify as a Gymnasium teacher in May 1922 in Bonn, and passed them with distinction. He spent one year as an assistant teacher, teaching science. One day, when he was doing his practice teaching, he took the children into the woods to illustrate various aspects of natural science for which he was severely reprimanded. Apart from this incident, everything in school went smoothly and he appeared to be successfully preparing for the quite prestigious career of a Gymnasium teacher. But six months before he took the examinations to become a full-time teacher, he changed his mind. In the fall of 1923, he decided to go to Göttingen to continue his work with Max Born along the lines of his paper on dimensional analysis. Born, at the time, was not receptive to such projects, and tried to convince London to do some straightforward physics. Born's work on the interpretation of quantum mechanics was among the first papers which gave the impetus for many of the discussions about the foundations of modern physics. Ironically, although Born became interested in philosophy later on, at the time he was 'very much opposed to philosophizing ... we were not very philosophical and we disliked many of the letters from philosophers and theologians we got'.[32] Also, London was not attracted to Göttingen's highly formalized approach to physics and, after one year, left for Osnabruck where he started, again, to prepare for his examinations to matriculate as a Gymnasium teacher, and where he also started to study mathematics and physics systematically. At the end of the few soul-searching months in this calm mediaeval town, he decided to go to the University of Munich and to study with Arnold Sommerfeld. He worked in Sommerfeld's institute for about six months from April to October 1925, and made calculations on a problem related to the intensity of the band lines in spectra.

London's stay with Sommerfeld coincided with the heyday of the Weimar Republic. London was in his mid-twenties and he lived through the craving for modernity, its (re)definition in painting, music, theater, cinema, dance, literature and architecture. Though he was never a strong adherent of trends in the arts and the humanities which saw the radical break with the past as the necessary prerequisite for modernization, London was open to the new ideas. The traditional academic disciplines were changing as well. Among the sociologists, there were the debates concerning the nature of the discipline itself and what it meant for the social sciences to be value free. In the beginning, philosophers were fiercely arguing the legitimacy of the claims of phenomenology to present philosophy as a science. Later, in 1929, when the Vienna Circle in its first manifesto argued that the problems confronting its German colleagues were a matter of semantic confusion and that a totally new approach to philosophy was needed,

London was giving a talk to these 'German colleagues' at Berlin. Out of the work in phenomenology the first samples of existential writings made their presence felt. Gestalt psychology, which was itself boosted by the philosophical discussions, exerted, in turn, significant influence upon philosophy. And though the origins of the supporters of psychoanalysis could be traced before the Great War, it was the discussions during the Weimar period which set out many of the themes which would condition the future of the discipline. But, above all, those years witnessed a radical change in the classical views about physical processes, the reality they implied and the ensuing discussions about the assessment of their peculiar nature. London himself had felt gratitude for being a witness of and, to a certain extent, a participant in an episode in the history of science which he felt was comparable to Galileo's achievements. He thought that the development of quantum mechanics was a kind of team effort and, as he told Manasse, he hoped that such team work might become the model for future research.

London's first academic appointment, starting on October 1925, was as P. P. Ewald's assistant at the Technische Hochschule in Stuttgart. Ewald was the Director of the Institute for Theoretical Physics. He had worked for his doctorate under Sommerfeld, and was first his assistant and then became a *Privatdozent* at Munich. Though London was not very enthusiastic, he accepted the appointment after the intervention of Sommerfeld, who was very keen for London to do so. Sommerfeld was convinced that London's contact with the experimental work carried out at the Technische Hochschule would be beneficial for his intellectual growth. 'God will then help in what follows. If he should extend his small finger, one should not brush if off. As the poet says, in your breast are the stars of your destiny.'[33] It was in Stuttgart that London started working on quantum theory.

Transformation theory

In the very early stages of the developments of quantum mechanics, London was spending his last months with Sommerfeld's group in Munich. He had already completed his first paper in physics, with Hönl, on the intensity of band spectra and, instead of continuing to work in spectroscopy as the 'Sommerfeld culture' stipulated, London, as soon as he reached Stuttgart, plunged into matrix mechanics.

He first used Jacobi's classical transformation theory of periodic systems and adopted it for matrix mechanics, proving that energy conservation was independent of the combination principle of atomic theory. This he proved after showing that the two definitions of the matrix derivative in the famous 'three-man paper' of Born–Heisenberg–Jordan, followed from his proposal of a more general definition of the matrix derivative.

His next two papers were quite significant in what came to be known as the transformation theory of quantum mechanics, a theory which was developed, independently and in more detail, by Paul Dirac and Pascual Jordan, and which was completed by them in 1926–1927.[34] Eventually, transformation theory allowed quantum mechanics to be formulated in the language of Hilbert spaces. In this new framework, quantum mechanics could be treated in a mathematically more satisfactory way and its results could acquire a consistent physical interpretation, dependent less on visualizability and on a description in space–time, and giving more emphasis on underlining the novel foundational characteristics of quantum mechanics.[35] The analogy was with the work of Poisson, Jacobi, Hamilton and Poincaré in classical mechanics.

Work in transformation theory had started before the advent of wave mechanics. In his paper completed in January 1926, Dirac was able to formulate a method by which it was indeed possible to carry-over the Hamilton–Jacobi theory of classical mechanics into matrix mechanics. The significance of such a possibility could hardly be overstated, since by the Hamilton–Jacobi method the frequencies of multiply periodic systems could be calculated directly. It was shown that the solution of this problem was, in effect, the finding of a system of canonical variables which would satisfy the commutation rules. In this case, the Hamiltonian becomes a function of one of these variables, and the position and momenta used to describe the system are multiply periodic functions of the other variable. In the spring of 1926, Jordan proved an important result, which was assumed to be valid in the three-man paper, although it had never been rigorously proven. Every transformation which left the commutation relations invariant could be written as:

$$P = SpS^{-1}; \; Q = SqS^{-1}$$

It was also shown that a result well known in classical mechanics – that every point transformation is a canonical transformation – was also true for quantum mechanics. This was not particularly practical, since the calculation of S^{-1} is quite an involved procedure.

London's first paper on transformation theory solved a difficulty in the formal formulation of quantum mechanics. Angle and action variables in matrix mechanics could not be used smoothly because the quantum condition

$$pq - qp = h/2\pi i$$

was not satisfied with constant p. London first generalized the condition by substituting q by a matrix function, by taking the proper limits and with a diagonal matrix p, he derived the original quantum condition. There was now no difficulty in the presence of action and angle variables in matrix mechanics as a result of the Hamilton–Jacobi method. He confidently concluded that to be able to utilize matrix mechanics in a practical manner one had to find the transformation matrix S.

Most likely, when London prepared this paper he was not aware of Schrödinger's proof of the equivalence of wave mechanics with matrix mechanics. In his second paper on transformation theory, London took a rather bold step, expressing his preference for wave mechanics over matrix mechanics. Transformation theory was by no means a complete theory in the framework of matrix mechanics. He tried to see what it all meant in the context of wave mechanics and, in fact, he was the first to do it. Wave mechanics used representations in configuration space and Schrödinger had used point transformations to solve his equation. In analogy with the use of canonical transformations in matrix mechanics, London constructed generators of a transformation T which, as he proved, left the eigenvalues of the original Schrödinger equation invariant. Then he proceeded to prove an important property of T, which entered into the expansion of the orthonormal eigenfunctions and found that

$$\tilde{T}^* T = 1$$

Thus, it was shown that T was a unitary matrix. His insight was to consider Jordan's result as a rotation in the infinite dimensional space of the coordinate system spanned by the orthogonal eigenfunctions. A straightforward result was that matrix elements, such as those of the electric dipole moment, could be shown to be invariant under such rotations.

It was only after the completion of his paper that London fully appreciated the intimate connection between his new conceptions and the theory of linear operators in functional spaces or, as it was called at the time, the 'theory of distributive functional operations'. In a footnote added to the typeset manuscript, London referred to a paper by Cazzaniga, to a memoir by Pincherle and to the latter's article in the *Encyclopedie der mathematischen Wissenschaften* as presentations of the abstract mathematical theory that underlies this interpretation. As Max Jammer mentions in his pioneering work on the development of quantum mechanics, when London added this footnote 'he could hardly have been aware of its historical significance. It was the first reference to the future language of theoretical physics'.[36] London had undoubtedly shown his mathematical prowess, but he was not a mathematician. As he would, thirty years later, explain to his publishers, he was a theoretical, rather than a mathematical, physicist. The expansion of his footnote into the mathematics of abstract spaces and the development of his observation about the complex phases into gauge theory were tasks which demanded a different kind of an approach to physics from London's.

When London moved to Stuttgart he sought an opportunity to resuscitate his original agenda and to do the kind of physics he enjoyed – like his 1922 paper criticizing Tolman's work. He tried to study a problem that involved some kind of unification, some kind of generalization, some kind of further abstraction: this was his work in transformation theory. The work was technical, abstruse, and the physical meanings were not easily recoverable. But London's work in transformation theory was, also, plagued by a feeling of frustration. In the increasing pace of the overall developments in quantum theory, London found himself, on more than one occasion, reading results published by Jordan or Dirac that he had been very close to deriving himself.[37]

Unsuccessful attempts at unification

London decided to stop working on transformation theory and to examine issues which he could have more time to think about and which would be commensurate with his views about the kind of problems he wanted to examine in physics. He decided to consider the theories which had already claimed to have unified the only two known forces – gravitation and electromagnetism – and to examine them within the newly emerging framework of quantum mechanics. Such a program was also very close to his overarching philosophical attempts. Weyl's attempts at such unification provided an attractive opportunity. In fact, the grandeur of London's scheme was reflected by the grandeur of the title of his paper (1927*a*): *Quantum mechanical interpretation of the theory which unifies gravitation and electromagnetism*. His plan was to provide a geometrical interpretation of the wave function and to make the structure of quantum mechanics resemble that of general relativity. The project turned out to be a dead end.

In *Space, Time and Matter*, published in 1920, Hermann Weyl presented a way of unifying gravitation and electromagnetism. He felt that the program by Lorentz and Mie to explain gravitation in terms of electromagnetism had failed. After the deep insights on the nature of gravitation resulting from Einstein's general theory, Weyl's attitude was that such a unification could become possible by regarding 'electromagnetic phenomena as well as gravitation, as an outcome of the geometry of the universe'.[38] This he did by using the geometric concept of space–time-dependent scale changes. Starting from the way the scale changes in going from one point to a neighbouring point in space–time and extending it to the way a function changes when its argument changes, he applied the scale change to the function itself. Weyl formulated a geometrical theory of electromagnetism by identifying the electromagnetic vector potential A with the space–time-dependent scale change operator. But Einstein formulated an objection to this idea. If such a suggestion of a non-integrable scale factor was correct, then two clocks starting from the same point, and brought along different paths back to the same point, would have their scales continuously changed. Thus by the time they got back to the point they started from, since they had traced different histories, the clocks would keep time at different rates. Thus, if Weyl's approach was correct, a clock's measure of time depended on its history.

In his work on the quantum mechanical interpretation of Weyl's theory, London emphasized a difference between gravitation and electricity. Einstein's theory, and especially his geometric interpretation of gravitation, was based on the equivalence of the inertial and gravitational mass. There was no similar 'principle' in electricity. There was, for example, no reason to believe in a universal influence of the electromagnetic field on the rigid rods or the clocks. According to London, this was the gist of Einstein's objection to Weyl's theory. Weyl had ignored Einstein's objection and London talked of Weyl's metaphysical conviction which must have led him into believing that 'nature would have made use of this beautiful possibility that was offered to her'.[39] Furthermore, such a way of building a theory was in conflict with London's methodological criteria: to base one's theory on a characteristic of nature which could be either assumed or read from the experimental data, and not on metaphysical beliefs about nature.

By examining the weaknesses of a radical continuum theory of matter, such as the new wave mechanics, London realized that the Riemannian metric had to be replaced. Now there could be no reproducible measuring rods and clocks. One may have been able to do the job by using Weyl's generalization of Riemannian geometry where a space had been created 'in which the non-reproducibility of the tract unit is provided as the consistent postulate of a radical local geometry'.[40] He thought, in fact, that he had found the answer. It was the complex amplitude of the de Broglie wave. In the electromagnetic field, this amplitude experienced the influence that Weyl had postulated for his measuring rod and to 'which – as a free floating element in the physics of that time – he had to assign a metaphysical existence'.[41] London's thought was to make use of a new entity, the ψ-field of quantum mechanics, to overcome Einstein's objection.[42] In a footnote, London expressed the view that, after the work of Born, Schrödinger's wave mechanics should be reinterpreted statistically. But since such reading entailed the rejection of any interpretation in space and time, it was of slight interest.[43]

In 1926, London observed that Weyl's identification of the potential with the generator of the scale-change would be correct if one proceeded to the formal replacement

$$\frac{\partial}{\partial \rho_\mu} \rightarrow -i\frac{\partial}{\partial \rho_\mu}$$

Now the two would be the same except for the insertion of $-i$. In other words, the correct way to do electrodynamics in quantum mechanics would be to use the operator

$$(d_\mu - ieA_\mu)\phi.$$

This also answered Einstein's objection. The two clocks acquired different phases. But that did not affect their respective speeds, since the phase difference produced no observable effect. This was the gist of the gauge principle. 'This insertion, although trivial formally, has profound physical consequences, because it changes the meaning of [Weyl's original] non-integrable factor into an expression which is a non-integrable *phase* factor.'[44] Now one had no longer a scale change, but a phase change – which could be thought of as an imaginary scale change. London did not do anything with this result. Two years later, in 1929, Weyl – with no reference to London's paper – developed his theory of gauge transformations and proposed the two-component theory for a spin-$\frac{1}{2}$ particle. Pauli, in the notes he added to the later editions of his *Theory of Relativity*, remarked that London and Weyl recognized the significance of the phase factor 'immediately after the discovery of wave mechanics'.[45] Although Pauli did not elaborate on what he meant by 'recognized the significance', London soon afterwards left the subject altogether and never returned to the issues concerning gauge invariance.

There is, however, another aspect to this paper which provides clues for a deeper insight into London's agenda. This is London's use of complex quantities. London was not the first to use complex quantities. Born had just introduced them, and had legitimized their use through his suggestions about the physical interpretation of the wave function. In quantum mechanics, interference phenomena could be understood in terms of these phases, and this particular characteristic of the new mechanics was, in fact, regarded as being the 'really new idea of quantum mechanics' by one of its founders, Paul Dirac. Whilst developing his program, London found that in setting up the phase coherence (in the semiclassical limit based on the old quantum theory limit of geometrical optics) a certain quantum would have a value equal to an integral number multipled by Planck's constant. Nevertheless, it does not appear to be the case that London later reinterpreted this result to use it for his rather timid prediction of flux quantization through a superconducting ring, mentioned in a footnote in the first volume of *Superfluids* in 1950.[46] The significance of these elaborations for London's later work, were primarily conceptual. It was the first use of a purely quantum mechanical notion, with no classical analogue, as, in fact, would be the case, in the coming years, with the exclusion principle in his quantum chemistry, the uncertainty principle in his intermolecular forces, the zero point energy in his proposed structure of solid helium, diamagnetism and the rigidity of the electronic wave function in superconductors, quantum statistics and the transition to superfluid helium.

London's work concerning the quantum mechanical interpretation of Weyl's theory has been discussed in connection with issues related to Schrödinger's receptiveness of de Broglie's ideas and their subsequent development in the form of wave mechanics.[47] In 1922, in a paper titled *On a remarkable property of the quantized orbits of a single electron*, Schrödinger related the quantization of the orbit of a single electron to a periodicity of Weyl's 'tract factor'. He discovered that if one imposed quantum conditions on Weyl's tract factor, the latter became an integral multiple of an exponential expression where there was a constant involved with the dimension of action. Schrödinger noticed that if this constant were a complex number then the length of the tract that an electron carried around would reproduce itself periodically. He concluded that 'it would be difficult to believe that this result is solely an accidental mathematical consequence of the quantum conditions and without deeper physical significance'.[48] In 1926–1927, London also noticed this result and speculated upon its significance, concluding that Schrödinger had in 1922 'the characteristic wave-mechanical periodicities which he was obliged to encounter again later from entirely different viewpoints'.[49]

At the beginning of December 1926, London was informed that he had received a fellowship from the International Education Board to go to Zürich and – as he hoped – to work with Schrödinger. Around December 10, he wrote to Schrödinger informing him about the Fellowship and that he was 'very happy that it is now certain that I may work with you'. There is no evidence to suggest that Schrödinger had made any such commitments earlier. London's letter was an overture to Schrödinger to find out what specific question he would be working on. Schrödinger, London wrote in no uncertain terms, was into wave mechanics in 1922, before de Broglie – in fact, before anyone else. He informed Schrödinger that: 'the most important thing is yet to be done'. Schrödinger's remark of 1922 was a theorem of the old quantum theory. 'One can with certainty expect that it will show its whole significance when it is brought into meaningful connection with wave mechanics (I have not done this yet). I think that it is your duty, after you have mystified the world in such a manner, now to clarify everything.' London continued the letter saying that he had stopped working on Weyl's theory – meaning, in other words, that he was ready and willing to work on this problem. Of course, London did not know that Schrödinger himself had written to Einstein in November 1925 about the similarities between his earlier work and de Broglie's ideas.[50]

It was an enticing letter. A young assistant had discovered that the master had (perhaps unknowingly) unravelled the mystery about the dual nature of matter much earlier than anyone else had thought, that the most important thing with that work was still to be done and that he knew how to do it, had not done it yet, but asked Schrödinger to do it! He wrote the letter on the day after he had received the Rockefeller Fellowship. London was not too subtle in his request. The hoped-for answer, where Schrödinger would gladly acknowledge London's suggestions and promise to look into the matter, never arrived. Instead he received a less than encouraging letter from Schrödinger.

> Although I frequently had the feeling that in the work you are talking about, something connected with today's change in the direction of quantum theory seems to be interesting, I must seriously ask you for the following favor. What you find in the murky material which has been so far discarded, you should attribute to you and not to me. If you extract something interesting, I shall be very pleased, indeed.[51]

The letter did not leave any doubts about Schrödinger's intentions. He did not appear to think that the ideas in his 1922 paper foreshadowed de Broglie's proposals, he did not want to get involved in any priority dispute and he did not want to commit himself to any collaboration with London. Thus, London was to continue his researches in Zürich independently of Schrödinger.

Notes to Chapter 1

1. Gay (1968), 151. (In general, where there are several quotations from the same source, the source citation is given with the note for the final quotation.)
2. Ernst Moritz Manasse was born in Germany in 1908. He studied at the Universities of Heidelberg, Munich, Berlin and Paris. His doctorate from the University of Heidelberg in 1933 was on the philosophical problems of the Platonic dialogues *Sophist* and *Statesman*. He was a Jew and left Germany in 1934. In 1939, he became

Professor of German, Latin and Philosophy at North Carolina Central University (then called North Carolina College for Negroes) in Durham, USA. He continued teaching there until his retirement in 1973. He was a member of the Institute for Advanced Studies in Princeton, USA, 1958–1959. In a series of articles in *Philosophische Rundschau*, he surveyed Platonic studies in Germany, UK, USA and France.

3. Private communication (April 15, 1990).
4. Spiegelberg (1982). One cannot but marvel at Spiegelberg's determination to make the works of his former teacher more accessible to a wider audience. In this connection see also Alexander Pfänder (1967) and Spielgelberg (1973).
5. Heller & Low (1933). In his autobiography, Karl Alexander Müller, one of the better known of the Munich historians who was a student of Pfänder's soon after the latter's appointment as a *Privatdozent* in 1900, remembered that

   After a few lectures in which he stuck closely to his manuscript, he forced himself to speak freely and asked for indulgence and possible halts. This was something entirely new as compared with the smooth lectures of the big shots, and established at the same time an inner bond between him and us which reached far beyond the mere relation of listening.

   H. Spiegelberg (1967) *Introduction to Alexander Pfänder*, xviii.
6. This is also in Jammer (1974), 483, but he overrates Becher's influence on London and does not stress Pfander's vigorous antipsychologism.
7. Spiegelberg is not consistent about the date that Husserl asked Pfander to write the textbook, and has given the date as 1909 as well as 1906. The latter is more likely, however, since it was shortly after they had a vacation together.
8. Becher (1905).
9. I am indebted to Prof. Thomas Mormann for the analytic draft of Mormann (1991).
10. Husserl had developed his ideas in Husserl (1900, 1906, 1929).
11. Becker (1923, 1927). Becker's first essay was published in the same issue of the *Jahrbuch* as London's thesis.
12. Mormann (1991), 72.
13. This copy of Rickert's book in London's possession had been given to Prof. Manasse whom I thank for sending me London's comments. The following are among the other philosophical books in Fritz London's personal library: *The Collected Works of Leibniz*; Husserl's *Logical Investigations and the Ideas*; Cassirer's *Substance and Conception of Matter and the Philosophy of Symbolism*; various books by Russell and Hegel. Surprisingly there are no copies of books by Kant. Of course, many books were left behind when London left Germany in 1933.
14. Spiegelberg (1982), 42.
15. London (1923), 371.
16. Except Padoa (1903), who, according to London, was the only 'voice that approves as not unnecessary the contemplations which have as their subject theories as *wholes*'.
17. London (1923), 336.
18. Ibid., 357.
19. Ibid., 358.
20. Ibid., 371.
21. This publication is considered to follow his thesis, since the thesis is cited in the paper.
22. Tolman (1914), 244.
23. Ibid., 244.
24. Tolman (1915).
25. London (1922), 2.
26. Ibid., 3.
27. Ibid., 12.
28. Norman Campbell to London (August 14, 1922).
29. Pfänder to London (September 14, 1922).
30. Wallot to London (September 24, 1922; October 3, 1922); London to Wallot (September 30, 1922).
31. Eddington (1928), 86.
32. Interview with Born by Thomas S. Kuhn and Friedrich Hund (October 17, 1962), from Archive for History of Quantum Physics (AHQP). When Born was interviewed he was 80 years old, and there are some inaccuracies, at least concerning London. For example, London went to Munich about a year later than the date implied by Born. Furthermore, at the time London went to Göttingen, none of the papers on quantum mechanics had been published and there was no 'philosophy of quantum mechanics' as Born specified London's interests. London was not interested in problems of 'modern physics' like causality, discontinuity, relativism, etc., but in questions which were predominantly epistemological.
33. Sommerfeld to London (undated, but most probably October 1925).
34. Transformation theory is discussed in Jammer (1989), Hendry (1984*b*) and Kragh (1990).
35. Jurkowitz (1995) has argued that in order to understand London's notion of 'rigidity' one has to examine the ideas which de

Broglie, Einstein and Schrödinger were considering during the development of wave mechanics, and that one source of this notion may have been a concept of 'coherence' which was applied to the analysis of the statistics of quantum gases and was originally formulated in the theory of blackbody radiation.

36. Jammer (1989), 314.
37. Interview with P. P. Ewald by Thomas Kuhn (May 8, 1962, AHQP).
38. Weyl (1921), 800. See also Sigurdsson (1991).
39. London (1927*c*), 376–77.
40. Ibid., 378.
41. Ibid., 380.
42. In a private communication, Skuli Sigurdsson noted that London's remarks showed that the relationship between physics and geometry was still a very vexing problem in 1927.
43. London (1927*c*), 378.
44. Yang (1986), 16.
45. Pauli (1958), 223–24.
46. Jurkowitz (1995) has argued that gauge considerations similar to those London first used in his work in quantum theory did play a role in London's prediction of the flux quantization. Yang (1987) made the suggestion that London may have been helped in his prediction by such a background but does not analyze it any further. Kragh (1990) discussed Dirac's work on magnetic monopoles and magnetic flux, etc., and noted the formal similarity between Dirac's expressions for the strength of a magnetic monopole and London's expression for the unit of quantization of flux in a superconductor. I agree with Kragh's conjecture that in spite of the formal similarity 'there seems to be no generic connection between Dirac's monopoles and London's theory'.
47. See Raman & Forman (1969).
48. Schrödinger (1922), 22.
49. London (1927*c*), 381–82.
50. Hanle (1977, 1979) discussed the Schrödinger–Einstein correspondence. Schrödinger had written to Einstein on November 3, 1925 (more than a year before London's letter to Schrödinger), that 'the de Broglie interpretation of the quantum rules seem[ed] to be related in some ways to my note' of 1922.
51. The postcard has no date and the date stamp on the postage stamp has been removed. There is, however, no doubt that this is Schrödinger's response to London's letter. This is the letter sought by Raman & Forman (1969) and has been found among some letters in possession of Mrs Edith London.

2

The years in Berlin and the beginnings of quantum chemistry

Fritz London went to Berlin at the start of the winter semester of 1927–1928 and left for Oxford at the end of August 1933. He was Schrödinger's assistant and became a Privatdozent *in 1928. During this time he was away for a month in Leningrad and for six months in Rome. In the joint paper with Walter Heitler, written when both were at Zürich in 1927, it was shown that the formation of the hydrogen molecule was a purely quantum mechanical phenomenon, whose understanding depended on the Pauli exclusion principle. In his Berlin years, London exploited fully the possibilities provided by the notion of a purely quantum phenomenon, and developed the rather abstruse group theoretical method to deal with polyelectronic molecules and with the difficulties of calculations involving many bodies. He also developed a theory of chemical reactions as activation processes. In 1930, together with Eisenschitz, he started to investigate the characteristics of the molecular forces. He found out that they could be understood only as manifestations of the uncertainty principle – another purely quantum mechanical notion. In Berlin, he was in an intellectually stimulating environment and was highly appreciated by all the holy men there: Einstein, von Laue, Schrödinger and the retired Planck. He signed a contract with Springer to write a book about atomic and molecular forces. He met Edith Caspary and they were married in 1929. He placed his younger brother Heinz in good hands; but after a short stay with Gerlach in Berlin, Heinz went to Breslau to work with Franz Simon. Nevertheless, life in Berlin bored Fritz, and he yearned for Bonn. Though he was recognized by his peers as one of the most promising physicists of the younger generation, it was frustrating not to be offered any kind of professorship in Germany.*

In the theoretical approach initiated with Heitler, actual calculations for more complicated molecules were impossible and the group theoretical methods provided a rationale for the qualitative results already found experimentally. The semiempirical methods devised in the USA by Linus Pauling and Robert Mulliken proved to be immensely successful and were quite enthusiastically adopted by the chemical community. Though they stood on theoretically shaky ground, Pauling and Mulliken devised a methodology that gave correct results. Such a sacrilege infuriated London, who thought that the pragmatic Americans had violated the prerogatives of theory building. He could not complete the book for Springer, progressively realizing the deadlock of doing chemistry from first principles.

Then came the Nazis and the decree of April 7, 1933. Though pressured by Planck and von Laue not to resign and to stay in Berlin, in August 1933 he gratefully accepted an offer to go to Oxford, England. In 1935, when both Heitler and London were in England, they started writing frantically to each other. Heitler and London tried hard to understand what was happening among the American physical chemists and wanted to save the scheme they had jointly proposed in 1927, which they saw as being allotted to the status of an interesting, but out of fashion, curiosity. But it was too late to reassert themselves in the chemical community.

The mysterious bond

Among the first successful applications of quantum mechanics were the calculation of the energy levels of the hydrogen atom and of the hydrogen molecular ion. By the beginning of 1927, many physicists and chemists talked about the possibilities of applying quantum mechanics to deal with actual chemical problems. The simplest, but deeply intriguing, chemical problem was the formation of the hydrogen molecule. The mechanism responsible for a such a formation – the homopolar bond – was quite puzzling, since it joined two electrically neutral atoms to form a molecule. Although there had been several suggestions to explain the homopolar bond, it was not until after 1916 that some remarkably insightful proposals by Gilbert Newton Lewis, within the framework of the old quantum theory, provided a rather simple rule to deal with such a puzzling bond. Lewis (1923) proposed that chemical bonding – both the ionic type and the homopolar type – could be explained in terms of shared electron pairs. It was a semiempirical theory. He had started from the observation that almost all the molecules had an even number of electrons. By requiring there to be eight electrons in the outer orbits of each atom in a molecule, he argued that two neutral atoms could be joined together only by sharing pairs of electrons between them. The possibility of explaining the homopolar bonding by considering some kind of mechanism of sharing or exchanging one or two electrons had already been discussed by various scientists. By 1914, J. J. Thomson (1914) had abandoned his plum-pudding atom, and he proposed a model whereby the stability of an atom could be guaranteed if the electrical forces between the nucleus and the orbiting electrons were confined to narrow tubes. Extending his ideas to the formation of the chemical bond, he asserted that polar and non-polar bonds are two distinct types of bonds. The polar bonds were realized through the transfer of electrons, whereas the non-polar bonds were conceived as a tube of force which connected an electron of one atom with the nucleus of another. William Arsem (1914) proposed a mechanism for molecule formation where the non-polar bond was explained by the sharing of one electron. Hence, the electrically neutral units of matter were the molecules and the electronic charge should have been twice the value used for the ionic mechanism. Alfred Parson (1915) explored the possibilities provided by magnetism for the formation of molecules. He proposed a model of the atom where the magnetons were arranged at the corners of a cube. Each magneton was a circular band of electricity in rapid revolution, and the chemical affinity was the result of the magnetic moment generated by this revolution. Walther Kossel (1916) proposed a series of ideas about valence which bore some similarity to those of Lewis. Kossel was a physicist whose proposals were by no means discussed by the community to the same extent as those of Lewis and Langmuir. Kossel was at best non-committal on the crucial point of the shared electron pair and the mechanism he proposed for chemical bonding depended on electron transfer – except for a special non-polar bond he postulated, whose existence was the result not of electron transfer, but of rings of two to five electrons whose orbits were perpendicular to the bond axis. The only aspect of his theory common to Lewis' schema was his insistence on the 'rule of eight' (the demand that there should be eight electrons in the outer orbit), and, thus, Kossel provided a model for the electropolar bond, and the similarities, if any, were with Lewis' early work.[1]

In 1923, after the publication of Lewis's book, the Faraday Society organized a meeting at the University of Cambridge, UK, titled the *Electronic theory of valency*. J. J. Thomson was the chairman of the meeting and in his opening statement he referred to the tasks awaiting the chemists. He felt that the problem of the chemical bond dominated chemistry and that the problem itself found its most suggestive mode of expression in terms of electrons.[2] Lewis gave the introductory talk and, though he thought that the aim of the meeting was to secure a better mutual understanding of divergent points of view, the expressions he used in his presentation failed to convince the audience of his tolerance about the kind of pluralism he advocated. He asserted that the cardinal phenomenon of all chemistry was the formation of electron pairs. For Lewis, this was an actual pairing and not a convenient mode of explanation. He could not offer an explanation of the mechanism itself, but he had no doubt that, sooner or later, quantum theory would provide an explanation of electron pairing.

At the same meeting, R. H. Fowler stressed that the great advantage of Lewis's work was that it was model independent: the only necessary assumption was that the atoms be composed of a positively charged nucleus and negatively charged electrons orbiting around it. Then, he systematically examined the various chemical problems, and found that the Bohr theory together with Lewis' proposal of shared electron pairs could provide qualitative explanations for many chemical problems, except for the problem of the tetrahedral symmetry for carbon. N. V. Sidgwick went a step further and attempted to explain why there should be a shared pair of electrons. He proposed that the orbit of each electron in the shared pair includes both nuclei. When the two electrons were in phase, one of them was always available for holding the nuclei together. And Sir Robert Robertson, who was the President of the Faraday Society, in his opening remarks at the session on organic chemistry, noted the 'awakening of the necessity' among the chemists to take account of the findings of the physicists and the effort of certain members of the community 'to bring home to chemists that electronic conceptions are to explain much and to foretell a great deal in the domain of chemical philosophy'.[3]

A few years later, towards the end of 1926, it appeared that many chemists were slowly becoming aware of the amazing explanatory power of the quantum mechanics, yet it was difficult for them to see how this newly developing explanatory framework would be assimilated into the culture of chemistry. Quite a few of them became rather apprehensive that such an assimilation may bring lasting, and not altogether welcome, changes to their culture. But, for some, it was a risk worth taking. N. V. Sidgwick, in his influential book *The Electronic Theory of Valency*, would have no inhibitions about letting the new quantum mechanics invade the realm of chemistry. He expressed an unreserved enthusiasm about the new quantum mechanics, and adopted Lewis' theory for the non-polar bond. Faced with the full development of the new mechanics by Heisenberg and Schrödinger, but not with an application of the theory to a chemical problem, in the very first lines of the preface to his book, Sidgwick attempted to clarify the methodological stumbling block that he sensed would be in the way of his fellow chemists. He talked of the courses open to the chemist in developing a theory of valency. One alternative, he thought, would be to use symbols with no definite physical connotation to express the reactivity of the atoms in a molecule, and to 'leave it to the subsequent progress of science to discover what realities these symbols represent'. But the chemist could also adopt the concepts of atomic physics and could try to explain the chemical facts in terms of these. Sidgwick chose the latter path, he was obliged to accept the physical conclusions in full, and was convinced that he 'must not assign to these entities properties which the physicists have found them not to possess'. As he emphasized, the chemist must not use the terminology of physics unless 'he is prepared to recognize its laws'. This was in 1927, just prior to the work of Walter Heitler and Fritz London which would remove any ambivalence from the feelings of the chemists about the use of the new quantum mechanics in chemistry.

London in Zürich

In 1926, London applied for a 'fellowship in science' from the International Education Board, which later became part of the Rockefeller Foundation. He was, at the time, Ewald's assistant at the Technical College in Stuttgart. In December 1926, he was granted the fellowship and he wrote to the Board that he intended to go to Zürich and to work with Schrödinger; this decision would bring a lasting change to his scientific agenda. He would no longer work on the problems related to quantum theory and he would start to work on problems requiring the application of quantum mechanics – first to chemistry and, then, to low temperature physics.

London arrived in Zürich in April 1927. Almost certainly, London did not meet Schrödinger upon his arrival, since the latter was on an extended visit to the USA, returning to Zürich on April 10. The rest of Schrödinger's stay at the University of Zürich would be for a very short period. He had been offered the Chair of Theoretical Physics at the University of Berlin. A committee appointed by the Philosophical Faculty of the Friedrich Wilhelm Universität had completed their search for Planck's successor. Their first choice was Sommerfeld who did not wish to leave Munich. The second choice was Schrödinger, who accepted the offer. Schrödinger moved to Berlin in August 1927 and, in October, London informed the Board that he would become Schrödinger's assistant in Berlin as of November 1. London was with Schrödinger in Zürich and Berlin, and they were together in Oxford after both left Germany in 1933. It was almost ten years of a relatively close, yet enigmatic, relationship between two people with diverging lifestyles, scientific interests, styles of work, philosophical preferences and differing views on social conventions. Yet there was a genuine mutual respect, and there are no indications of any misunderstandings between them. Their correspondence was meagre and was exclusively about scientific matters; all the indications are that their contacts were few and formal.

In Zürich – and in Berlin – London did not work with Schrödinger, since Schrödinger hardly ever collaborated with anyone else. But during his short stay in Zürich, London, together with Heitler, managed to solve one of the most intriguing problems of chemistry by using the wave mechanical methods of Schrödinger. They showed that the mysterious chemical binding of two neutral hydrogen atoms to form a hydrogen molecule could be understood only in terms of the principles of the new quantum mechanics.

Undoubtedly, the simultaneous presence of both Heitler and London in Zürich was one of those unplanned happy coincidences. Walter Heitler was born in Karlsruhe, in 1904, to a Jewish family, and his father was a professor of engineering. He became interested in physical chemistry while he attended lectures on the subject at the Technische Hochschule and in these lectures he came into contact with

Figure 4. Fritz London and Walter Heitler near Bristol, England, in 1934 or 1935. (Courtesy of Edith London.)

quantum theory. He had also acquired a strong background in mathematics. Wishing to work in theoretical physics, he first went to Berlin, but found the atmosphere inhospitable, since students were expected to be independent in choosing a problem and writing a thesis. Only after its completion would the holy men of Berlin examine it. After a year in Berlin, he went to Munich and completed his doctoral thesis, on concentrated solutions, with Karl Herzberg. The writing of his thesis coincided with the development of the new quantum mechanics, but, because of the kind of problems he was working on, he never had the opportunity to study the new developments in any systematic manner. After completing his thesis, Sommerfeld helped him to secure funding from the International Education Board, and he went to Copenhagen to work with J. Bjerrum on a problem about ions in solutions, but he was not particularly happy there. Determined to work in quantum mechanics, he convinced Bjerrum, the International Education Board and Schrödinger to allow him to spend the second half of his funded period in Zürich.

When they met in Zürich, Heilter and London decided to calculate the van der Waals' forces between two hydrogen atoms, considering the problem to be 'just a small "by the way" problem'. Nothing indicates that London and Heitler were given the problem of the hydrogen molecule by Schrödinger or that they had detailed discussions with him while they were proceeding with their calculations. The acknowledgements in their paper were quite diplomatic: they thanked Schrödinger for his hospitality and 'for the kind and supportive interest with which he has guided our work'. Linus Pauling, who was also in Zürich during the same time as Heitler and London, noted that neither he nor Heitler and London discussed their work with Schrödinger. However, Schrödinger knew what they were doing, since, in 1927, he told Mulliken, who was on a trip to Zürich, that there were two persons working in his Institute who had some results 'which he thought would interest me very much; he then introduced me to Heitler and London whose paper on the chemical bond in hydrogen was published not long after'.[4] It may be worth noting that Ewald remembered that the problem of the homopolar bond was in London's mind before going to Zürich, and Pauling recalled discussions with Heitler about bonding when they were both in Munich in 1926.

Binding forces

Heitler and London's initial aim was to calculate the interaction of the charges of two atoms. They were not particularly encouraged by their first results, since the attraction due to the 'Coulomb integral' was too small to account for the homopolar bond between two hydrogen atoms. But they were puzzled by the presence of the 'exchange integral', whose physical significance was not evident at all. Heisenberg's work on the quantum mechanical resonance phenomenon which had already been published was not of particular help to Heitler and London, since the exchange was part of the resonance of two electrons, both situated in the same atom, but where one was in the ground state and the other was excited. Heitler remembered that they were stuck and 'we did not know what it meant and did not know what to do with it'.[5]

> Then one day was a very disagreeable day in Zürich; [there was the] Fohn. It's a very hot south wind, and it takes people different ways. Some are very cross . . . and some people just fall asleep . . . I had slept till very late in the morning, found I couldn't do any work at all . . . went to sleep again in the afternoon. When I woke up at five o'clock I had clearly – I still remember it as if it were yesterday – the picture before me of the two wave functions of two hydrogen molecules joined together with a plus and minus and with the exchange in it. So I was very excited, and I got up and thought it out. As soon as I was clear that the exchange did play a role, I called London up; and he came as quickly as possible. Meanwhile I had already started developing a sort of perturbation theory. We worked together until rather late at night, and then by that time most of the paper was clear . . .. Well, I am not quite sure if we knew it in the same evening, but at least it was not later than the following day that we knew we had the formation of the hydrogen molecule in our hands. And we also knew that there was a second mode of interaction which meant repulsion between two hydrogen atoms – also new at the time – new to the chemists too. Well the rest was then rather quick work and very easy, except, of course, that we had to struggle with the proper formulation of the Pauli principle, which was not at that time available, and also the connection with spin . . .. There was a great deal of discussion about the Pauli principle and how it could be interpreted.[5]

The paper was sent for publication on June 30, 1927. From the different handwriting, it is clear that London wrote the introduction, discussion of results and conclusion, whereas Heitler did the calculations and wrote the discussions of the Pauli exclusion principle and molecular forces. Heitler and London started their calculations by considering the two hydrogen atoms coming slowly closer to each other. Electron 1 could either belong to atom *a* and electron 2 to atom *b* or vice versa. Because the electrons were identical, the total wave function of the system was the linear combination of the wave functions of the two cases:

$$\Psi = c_1\psi_a(1)\psi_b(2) + c_2\psi_a(2)\psi_b(1)$$

The problem now was to calculate the coefficients c_1 and c_2. This they did by minimizing the energy

$$E = \frac{\int \psi H \psi d\tau}{\int \psi^2 d\tau}$$

They found two values for the energy:

$$E_1 = 2E_0 + \frac{C + A}{1 + S_{12}}$$

$$E_2 = 2E_0 + \frac{C - A}{1 - S_{12}}$$

The integrals C (Coulomb integral) and A (exchange integral) had negative values, but A was larger than C. E_1 implied $c_1/c_2 = 1$ and E_2 implied $c_1/c_2 = -1$. Hence, the wave function of the system could now be written as:

$$\psi_{\mathrm{I}} = \psi_a(1)\psi_b(2) + \psi_a(2)\psi_b(1)$$

$$\psi_{\mathrm{II}} = \psi_a(1)\psi_b(2) - \psi_a(2)\psi_b(1)$$

Up to now, the spin of the electrons had not been taken into consideration. The symmetry properties required by the Pauli exclusion principle were satisfied only by ψ_{I}. This was the case when the electrons had antiparallel spins. But ψ_{I} corresponded to E_1. E_1 was less than $2E_0$, the sum of the energies of the two separate hydrogen atoms, and, hence, it signified attraction. ψ_{II}, which (when spin was taken into consideration) was a symmetrical combination, corresponded to E_2. But E_2 was greater than $2E_0$, and it implied repulsion. The mechanism responsible for the bonding between the two neutral hydrogen atoms was the pairing of the electrons, which became possible only when the relative orientations of the spins of the electrons were antiparallel. To form an electron pair, it did not suffice to have only energetically available electrons, but the electrons had to have the correct spin orientations. The homopolar bonding turned out to be a pure quantum effect, since its explanation depended wholly on the electron spin, which had no classical analogue. As Heitler and London noted in their paper, such a result could be described only very artificially in classical terms.[6] They found the bond energy was 72.3 kcals and the internuclear distance was 0.86 angstroms, compared with the experimental values of 109.4 and 0.74 respectively.

The numerical results for the interatomic distance and the binding potential derived by the original calculation of Heitler and London did allow for further more exact approximations. Their goal was not to calculate the most exact numerical values possible, but 'to gain insights into the physical conditions of the homopolar bond'.[7] There had been several attempts to derive improved results for the hydrogen molecule. Within a few months, Sigiura (1927) published the results of the first-order perturbation calculation. In Göttingen, Born showed Heitler the newly completed calculation by Sigiura, but he did not make the kindest of remarks: 'It appears that there are in the world some hard working dwarfs. But – honestly now – would *we* be able to manage it at all? . . . It seems to me ridiculous, especially if one sees what kind of perturbation calculation has been done . . .. It is either utterly stupid or a very clever situation.' He wrote to London that now there would not be any need to use the Institute of Computing at Berlin.[8] In a few years, however, Eisenschitz & London (1930) found that the perturbation method was not particularly satisfactory, since the calculation of the second-order perturbation calculation gave results which were in serious disagreement with the experimental values, when compared with the zeroth and first-order perturbation calculations. Wang (1928) used a more complicated expression for the effective charge on the nucleus in terms of a 'screening' parameter whose value was determined by giving a minimum to the value of the variational integral which generated Schrödinger's equation. This minimum value also gave the approximate energy between the two atoms. A further improvement was achieved by Rosen (1931) who did not consider a spherical 1s electron cloud, but included the polarizing effect of a second nucleus on the electron cloud of the first atom and, as a result, considered a cloud elongated in the direction of the second nucleus. He used the variational principle, since the second-order perturbation calculation of Eisenschitz and London gave worse results than the first-order calculation: 'one has the comforting assurance that one cannot go astray.' Weinbaum (1933) proceeded to make calculations by including the ionic terms where both electrons were on one of the atoms. It was found that the ionic states were of higher energy than the non-ionized states, and were, therefore, less probable. Most of these methods continued to have the same shortcomings as the original method: they implied that each atom preserved its individuality, since the atomic wave functions were used for the treatment of the molecule and they did not take into account the mutual repulsion of the two electrons. James & Coolidge (1933) constructed a molecular wave function by taking into consideration the mutual repulsion of the electrons and by not making the assumption that the atoms preserve their individuality. They determined the values of all the parameters by using the variational principle – deriving results which were in almost perfect agreement with the experimental measurements.[9]

Heitler and London soon realized that the proposed exchange mechanism obliged them to be confronted with a fundamentally new phenomenon. As Heitler remembered, they had to answer questions posed by experimental physicists and chemists, like 'What is really exchanged? Are the two electrons really exchanged? Is there any sense in asking what the frequency of exchange is?'[5]

> It became gradually clear to me that it has to be taken as a fundamentally new phenomenon that has no proper analogy in older physics. But I think the only honest answer today is that the exchange is something typical for quantum mechanics, and should not be interpreted – or should not try to interpret it – in terms of classical physics.[5]

In all their early writings, London and Heitler repeatedly stressed this 'non-visualizability' of the exchange mechanism. It is one aspect of their work which in the name of didactic expediency has been consistently misrepresented.

The Pauli exclusion principle

Though it appeared that the treatment of the homopolar bond of the hydrogen molecule was an 'extension' of the methods successfully used for the hydrogen molecular ion, there was a difference between the two cases that lead to quite radical implications. It was the role of the elusive Pauli exclusion principle. In the case of the hydrogen molecule ion, its solution was a successful application of the Schrödinger equation and the only forces determining the potential were electromagnetic. A similar approach to the problem of the hydrogen molecule leads to a mathematically well defined, but physically meaningless solution where the attractive forces could not be accounted for. There was a need for an additional constraint, so that the solution would become physically meaningful. At least part of the theoretical significance of the original work of Heitler and London was that this additional constraint was not in the form of any further assumptions about the forces involved. Invoking the Pauli exclusion principle as a further constraint led to a quite amazing metamorphosis of the physical content of the mathematical solutions. These solutions became physically meaningful and their interpretation in terms of the Pauli exclusion principle brought about the new possibilities provided by the electromagnetic interaction.

John Heilbron (1983), in his study of the origins of the Pauli exclusion principle, talked about one of the most unusual instruments in microphysics and said that Pauli's first enunciation in December 1924 had the form not of a dynamical principle but of the Ten Commandments. During the ceremony at the Institute for Advanced Studies, at Princeton, to honour Pauli's receiving the Nobel Prize in 1945, Hermann Weyl talked of Pauli's exclusion principle as something which revealed a 'general mysterious property of the electron'.[10] In his subsequent publications, London proceeded to devise a formulation of the Pauli exclusion principle for cases with more than two electrons, which was to become more convenient for his later work in group theory: the wave function can at most contain arguments symmetrical in pairs; those electron pairs on which the wave function depends symmetrically have antiparallel spin. He considered spin to be the constitutive characteristic of quantum chemistry. Since two electrons with antiparallel spin are not identical, the Pauli exclusion principle did not apply to them and, thus, one could legitimately choose the symmetrical solution. With the Pauli exclusion principle it became possible to comprehend 'valence' saturation, and, as it will be argued in the future work of both Heitler and London, spin would become one of the most significant indicators of valence behavior and, in the words of van Vleck, would forever be 'at the heart of chemistry'.[11]

The early years in Berlin

From Zürich, London was already in contact with Sommerfeld for a possible position in Munich. Sommerfeld did not have a position for an assistant, but suggested that London could become a *Privatdozent* in Munich. Sommerfeld felt that London would profit from the Institute's 'more concrete direction' as opposed to London's 'abstract approach'. Sommerfeld suggested that London should include in the Habilitation the work about the quantum theoretical version of Weyl's theory. If London had accepted such an arrangement, his lectures would have started in the summer term of 1928 and, hence, he would not have had to move to Munich till then. Sommerfeld also encouraged London to go to Schrödinger who, Sommerfeld hoped, would 'come to Berlin in the interest of Germany'.[12] By mid-September 1927, there was a specific offer by Schrödinger who asked London to become his assistant at the Institute for Theoretical Physics at the Friedrich Wilhelm Universität. Schrödinger had already written to the Ministry and had also promised to discuss with Planck and the faculty at Berlin the prospect for London's Habilitation, in case he decided not to go to Munich.[13] London was an assistant for less than a year. On May 7, 1928, London officially received the title of *Privatdozent* from the Faculty of Philosophy at the University of Berlin. The talk he was required to give was on the many-body problem. While in Berlin, London taught *Quantum mechanics and chemistry* during the spring of 1929; *Principles of quantum mechanics* during the fall of 1929; *Atomic physics and the chemical bond* during the spring of 1930; *Generalized Mechanics* during the fall of 1930 and *General Theory of Relativity* in the fall of 1931. The first three were basically the same course, and they were repeated until 1933. All the courses contained a historical introduction and he followed a fairly standard format.

London, of course, attended the weekly colloquia held on Wednesday afternoons. They were attended by the unique group of people who constituted the physics faculty at the University of Berlin: Einstein, Nernst, von Laue, Lise Meitner, Ladenburg, Paschen, and, of course, Planck and Schrödinger. The younger members like London, Kallmann,[14] Wigner, Möglich, and Szilard presented papers about the current developments. After a couple of hours, the meeting would adjourn to continue at a local tavern where a special room was reserved.[15]

Figure 5. Erwin Schrödinger and Fritz London in Berlin in 1928. (Courtesy of Edith London.)

Reactions to the Heitler–London paper

Right after the publication of the Heitler–London paper, it became quite obvious that it was the beginning of a new era in the study of chemical problems. But it also signified the formation of a subdiscipline – that of quantum chemistry. W. M. Fairbank was London's colleague at Duke University in 1953 and the coauthor with C. W. F. Everitt of the entry on Fritz London in the *Dictionary of Scientific Biographies*, and he recalled London telling him that Schrödinger did not expect that his equation would be able to describe the whole of chemistry as well. As London told A. B. Pippard, they were not expecting to find any such force since they had started working on the problem as a problem in van der Waals forces.[16] Born and Franck were very enthusiastic about the paper. Sommerfeld had a rather cool reaction, but he also became very enthusiastic once Heitler met him and explained certain points. The application of quantum mechanics led to the conclusion that two hydrogen atoms formed a molecule but this was not the case for two helium atoms. Such a 'distinction is characteristically chemical and its clarification marks the genesis of the science of subatomic theoretical chemistry' remarked Pauling.[17] A similar view, with a slightly different emphasis, was put forward by van Vleck:

> Is it too optimistic to hazard the opinion that this is perhaps the beginnings of a science of 'mathematical chemistry' in which chemical heats of reaction are calculated by quantum mechanics just as are the spectroscopic frequencies of the physicist? . . . The theoretical computer of molecular energy levels must have a technique comparable with that of a mathematical astronomer.[18]

Both Louis de Broglie (1974) and Max von Laue (1950) regarded the paper as a classic work. Eugene Wigner was 'impressed with the skill' of London's papers.[19] In their book on quantum mechanics for chemists, Linus Pauling and E. Bright Wilson hailed the paper as the 'greatest single contribution to the clarification of the chemist's conception of valence'[20] which had been made since Lewis's suggestion of the electron pair. In an address to the Chemical Section of the British Association for the Advancement of Science in 1931, Heisenberg considered that the Heitler–London theory of valence had 'the great advantage of leading exactly to the concept of valency which is used by the chemist'. Buckingham quoted McCrea who recalled his own attempts to solve the problem of the hydrogen molecule bond:

> This was the most important problem I considered these days, but I got nowhere. Then one day in 1927, I was able to tell Fowler that a paper by Walter Heitler and Fritz London apparently solved the problem in

> terms of a new concept: a quantum mechanical exchange force. He grasped the idea at once, and bade me to expound it in the next colloquium – which is how quantum chemistry came to Britain.[21]

In April 1926, Linus Pauling, supported by a Guggenheim Fellowship, arrived at Munich where he planned to work with Sommerfeld at the Institute of Theoretical Physics. He was twenty-five years old and had already received his doctorate from the California Institute of Technology, working with Roscoe Dickinson on the structure of molybdenite. When they met, Pauling did not follow Sommerfeld's advice to work on the spinning electron, since his main interest was in chemistry. When he was in Munich, he discussed the problem of the chemical bond with various people. He had been impressed by Edward Condon's treatment of the hydrogen molecule; his clever numerology and empirical methods were not at all unattractive to Pauling, especially since Condon 'got results as good as Heitler and London got later'.[22]

After Munich and before going to Copenhagen in late 1927, he spent three months in Zürich where he also met Heitler and London. Right after the appearance of the Heitler–London paper, Pauling published a short note to bring attention to an unforgivable omission: Lewis was mentioned nowhere in the paper and Pauling wanted to emphasize that the Heitler–London approach was 'in simple cases entirely equivalent to G. N. Lewis' successful theory of the shared electron pair, advanced in 1916 on the basis of purely chemical evidence',[23] acknowledging at the same time that the quantum mechanical explanation of valence was more powerful than the old picture. In this paper, Pauling mentioned for the first time that the changes in quantization might play a dominant role in the production of stable bonds in the chemical compounds. That was the first hint about the hybridization of orbitals. Perturbations to the quantized electronic levels might produce directed atomic orbitals whose overlapping would be better suited for the study of chemical bonds. Pauling, in that note, suggested the direction along which he would move to derive some new results and he explicitly stated his methodological commitments. When Pauling informed Lewis about his short note,[24] Lewis's response contains what is, perhaps, his most insightful statement about valence.

> I was very much interested in your paper as I had been in London's, and there is much in both papers with which I can agree ... I am sorry that in one regard my idea of valence has never been fully accepted. It was an essential part of my original theory that the two electrons in a bond completely lose their identity and cannot be traced back to the particular atom or atoms from which they have come; furthermore that

this pair of electrons is the only thing which we are justified in calling a bond. Failure to recognize this principle is responsible for much of the confusion prevailing in England on this subject.[25]

In 1928, there appeared two review articles which exerted a strong influence on the chemists. Both were published in *Chemical Reviews*, were written by Americans (Linus Pauling and John van Vleck), and had the explicit aim of 'educating' the chemists in the ways of the new mechanics. In Pauling's article, the Heitler–London treatment of the structure of the hydrogen molecule was considered to be 'most satisfactory' and it was repeatedly stated that spin and resonance would provide a satisfactory explanation of chemical valence. Pauling felt that the agreement with the qualitative conclusions of Lewis was one of the advantages of the work by Heitler and London.

Van Vleck's (1928) review of quantum mechanics concentrated on explaining the principles and the internal logic of the new theory. He was quite sympathetic to matrix mechanics. In this paper, van Vleck gave full credit to the work of Heitler and London whose results were 'already yielding one of the best and most promising theories of valency'.[26] The achievements of quantum mechanics in physics were summarized in ten points and the section about chemistry was appropriately titled: *What the quantum mechanics promises to do for the chemist*. Great emphasis was placed on the importance of spin for chemistry and it was shown that the Pauli exclusion principle could provide a remarkably coherent explanation of the periodic table. Some years later, its importance was stressed even more dramatically: 'The Pauli exclusion principle is the cornerstone of the entire science of chemistry.'[27] Nevertheless, if quantum mechanics was to be applicable in solving problems in chemistry, it was essential not just to explain the periodic table, but to understand why and how atoms combine. Van Vleck propounded his reductionist attitude and thought that the dynamics which were so successful in explaining atomic energy levels for the physicist should also be successful in calculating molecular energy levels for the chemist.

Polyelectronic molecules and the application of group theory to problems of chemical valence

The first indications that the work they started in their joint paper could be continued by using group theory are found in a letter to London by Heitler in late 1927. In September 1927, Heitler had become Born's assistant at Göttingen and wrote to London saying how very excited he was about the physics at Göttingen and especially about Born's course in quantum mechanics, where everything was presented in the matrix formulation and then one derived 'God knows how, Schrödinger's equation'.[28] Heitler felt that the use of group theory was the only way the many-body problem could be dealt with, and he outlined his program to London in two long letters.

His first aim was to clarify the meaning of the line chemists drew between two atoms. His basic assumption was that every bond line meant exchange of two electrons of opposite spin between two atoms. He examined the case with the nitrogen molecule and, by analogy with the hydrogen case, among all the possibilities, the term containing the outermost three electrons of each atom with spins in the same direction (i.e. ↑ ↑ ↑ and ↓ ↓ ↓) was picked out as signifying attraction. He felt that the general proof for something like this could not be given, except by group theory. His hunch was that it involved the theory of reducible representations, which as representations of the totality of permutations cannot be further reduced. 'Let us assume for the moment that the two atomic systems ↑ ↑ ↑ ↑ ↑ ... and ↓ ↓ ↓ ↓ ↓ ... are always attracted in a homopolar manner. We can, then, eat Chemistry with a spoon.'[29]

This overarching program to explain all of chemistry got Heitler into trouble more than once. Wigner used to tease him, since he was sceptical that the whole of chemistry could be derived in such a way. Wigner would ask Heitler to tell him what chemical compounds between nitrogen and hydrogen his theory could predict, and 'since he did not know any chemistry he couldn't tell me'.[30] Heitler confessed as much in his interview: 'The problem was to understand chemistry. This is perhaps a bit too much to ask, but it was to understand what the chemists mean when they say an atom has a valence of two or three or four.'[31] Both London and Heitler believed that all this must be now within the reach of quantum mechanics.

London agreed with Heitler that group theory might provide many clues for the generalization of the results derived by perturbation methods. The aim of such a program was to prove that, from all the possible combinations of spins between atoms, only one term provides the necessary attraction for molecule formation. It took London a while to familiarize himself with the new possibilities provided by group theory and he was not carried away by the spell of the new techniques – as Heitler was, in the company of Wigner and Weyl at Göttingen.[32] 'He thought it was too complicated and wanted to get on in his own more intuitive way.'

In Göttingen, Heitler started to study group theory intensively. Wigner's papers had already appeared and there was a realization that group theory could be used for classifying the energy values in a many-body problem.

In a significant paper, Heitler & Rumer (1931), were able to study the valence structures of polyatomic molecules and to find the closest possible analogue in quantum mechanics to the chemical formula which represented the molecule by fixed bonds between two adjoining atoms. They found that the emerging quantum mechanical picture was more general and that the bonds were not strictly localized. Nevertheless, the dominant structure was, in general, the one corresponding to the chemical formula. But there were other structures which were not only significant but also quite useful in understanding chemical reactions. Heitler thought that it was in this paper that he could really understand what the chemists meant by a chemical formula. There were a few new things for the chemists: one was that for each chemical formula there was a corresponding wave function, but the reverse was not true, since for each wave function there was, in general, a combination of several chemical formulae. A long time before the Heitler–Rumer paper, London was the first to show that the activation energies in the treatment of the three hydrogen atoms could be understood only in terms of quantum mechanics.

> Later Pauling called this a resonance between several structures, which is a name perhaps not quite in agreement with the use of the word resonance by physicists . . . a further point which was violently objected to by the chemists was that both London and I stated that the carbon atom with its 4 valences must be in an excited state . . . all this was later accepted by the chemists, but at that time I don't think the chemists did find this of much use for them.[33]

The theory of the irreducible representations of the permutation group provided the possibility of dealing with the problems of chemical valence mathematically in view of the difficulties with the many-body problems. And though this unavailability of reliable methods for tackling many-body problems haunted London all his life, many years later this difficulty was strangely liberating and helped him to articulate the concepts related to the macroscopic quantum phenomena as a means of superceding such difficulties.

Convinced that it was impossible to continue his work in chemical valence by more analytic methods, London eventually turned to group theory. But not everyone – even among those who thought of London's work very highly – was enthusiastic about the group theoretical techniques. Douglas Hartree was one such person.

> I am afraid that having studied physics, not mathematics, I find group theory very unfamiliar, and do not feel I understand properly what people are doing when they use it. (In England, 'Physics' usually means 'Experimental Physics'; until the last few years 'Theoretical Physics' has hardly been recognized like it is here [at the time Hartree was in

> Copenhagen], and, I understand, in your country. In Cambridge particularly the bias has been very much towards the experimental side, and most people now doing research in theoretical physics studied mathematics, not physics.) I have been waiting to see if the applications of group theory are going to remain of importance, or whether they will be superseded, before trying to learn some of the theory, as I do not want to find it is going to be of no value as soon as I begin to understand something about it! Is it really going to be necessary for the physicist and chemist of the future to know group theory? I am beginning to think it may be.[34]

Born had also responded in a surprisingly negative manner to the use of group theory. His objection was not because group theory was not easy to use, but as he said: 'in reality it is not in accord with the way things are'.[35]

London's first major paper on the application of the group theoretical methods to problems of chemical bonding appeared in 1928. London's group theoretical approach to chemical valence was formed along three directions. Firstly, anything that may give a rather strong correlation between qualitative assessments of a theoretical calculation and the 'known chemical facts' provided a rather strong backing for the methodological correctness of the chosen approach which expressed the observed regularities as rules. Secondly, since calculations were hopelessly complicated and in most cases impossible, the use of group theoretical methods was especially convenient when one was dealing with the valence numbers of polyelectronic atoms, since the outcome was expressed either as zero or in natural numbers. Thirdly, the overall result was that the interpretation of the chemical facts was compatible with the conceptual framework of quantum mechanics.

By the middle of 1928, London drew up a program to tackle what he thought was the most fundamental problem of atomic theory, he said: 'the mysterious order of clear lawfulness, which is the basis for the immense factual knowledge of chemistry and which has been expressed symbolically in the language of chemical formulas'.[36] It was a three-pronged program. Firstly, he intended to deal with the problem of the mutual force interactions between the atoms. Secondly, he wished to examine whether it was possible to decipher the meaning of the rules that the chemists had found in semiempirical ways and to place those on a sound theoretical basis. Thirdly, he would attempt to determine the limits of these rules and, if possible, to initiate a quantitative treatment of them.

But he was not at all certain that the principles considered so far in atomic theory could, in fact, be used for the realization of such a program. This was because the characteristic interaction of the chemical forces deviated completely from other familiar forces: these forces appeared to be activated and they suddenly vanished after the exhaustion of the available valences. Group theory did provide a way out, and the interpretation of the mathematical results turned out to be formally equivalent to the chemical model, i.e. it produced the same valence numbers and it satisfied the same formal combination rules as those expressed in the symbolic representation of the structural formulas of chemistry. In particular the fact that the valences were 'saturated' proved in this context to be an expression of the restriction of the Pauli exclusion principle.

London's 'spin theory of valence' dealt mainly with those cases where each electron in a pair comes from a different atom. He examined the conditions whereby electrons from different atoms can pair with each other so that the resultant spin of the pair was zero. An electron already paired with another electron in the same atom was not considered to be in this schema of pair formation for bonding. Two electrons in the same atom were said to be paired if they had opposite spins and all their other quantum numbers were the same. But an electron that was already paired could become available for bond formation with an electron from another atom, if it could be unpaired without the expenditure of too much energy. London claimed that an electron can be unpaired provided that the total quantum number (n) of that electron does not change. Such an unpairing was considered by London to be an intermediate step in the formation of a compound.[37]

A rather interesting result of group theory was the formulation of a theory of activation. This was a way to explain the exchange reactions in chemistry $A + BC \rightarrow AB + C$. The activation energy which was required to separate BC and then to have B change its attachment was shown to come from the saturation of the bond and the ensuing repulsion of a foreign atom. This process was continuous and the changeover of the bond was also continuous. There was, thus, an intermediate state where the bond was not localized and this was the transition state. Such a state was characteristic of quantum mechanics and, like so many other aspects of London's work, could not be understood properly within the classical framework.

Chemists as physicists?

Among the meetings where questions related to chemical bonding and valence were exhaustively discussed, two in particular were quite suggestive of the changes occurring among the chemists. The first was the 'Symposium on atomic structure and valence', organized by the

American Chemical Society in 1928 in St Louis. The second was organized by the Faraday Society in 1929 in Bristol and its theme was 'Molecular spectra and molecular structure'.

In his opening remarks at the 1928 meeting of the American Chemical Society, G. L. Clark noted some of the difficulties associated with atomic physics, but ascribed them to the failure of the chemists to test 'their well-founded conceptions with the facts of physical experimentation, and that far too few physicists inquired critically into the facts of chemical combination'. He thought that physicists and chemists were 'firmly entrenched, each in his own domain, a certain long-range firing of static cubical atoms against infinitesimal solar atoms has ensued, with few casualties and few peace conferences'.[38]

Clark was not alone in attempting to specify the newly acquired consciousness about this strange relationship between the physicists and the chemists. Worth Rodebush, one of the first to receive a doctorate in 1917 from the newly established Department of Chemistry at Berkeley under the chairmanship of Lewis, went a step further than Clark. He asserted that the divergent paths of physicists and chemists were being drawn together after the advent of quantum theory and especially after Bohr's original papers. But in this process the physicist seemed to have yielded more ground and to have learned more from the chemist than vice versa. Rodebush gracefully remarked that it was to the credit of the physicist that he could now calculate the energy of formation of the hydrogen molecule by using the Schrödinger equation. But the outstanding tasks for a theory of valence were to predict the existence, and absence, of various compounds and the nature of valence which can be expressed by a series of small whole numbers leading to the law of multiple proportions. The 'brilliant theories' of Lewis accounted for the features of valence 'in a remarkably satisfactory manner, at least from the chemist's point of view'.[39] London's treatment of valence by group theory was considered to be an important piece of work, even though it did not provide answers to all the queries of the chemists; for example, why there were differences in degree of stability between chemical compounds.

Perhaps the most cogent manifestation of what would become the characteristic approach of the American chemists was Harry Fry's contribution to this symposium. He attempted to articulate what he called the 'pragmatic' outlook. He started by posing a single question: what would be the kind of modifications to the structural formulas so as to conform with the current concepts of electronic valency? Such a question, he suggested, should by no means lead to a confusion of the fundamental purpose of a structural formula which is to present the number, the kind and the arrangement of atoms in a molecule as well as to correlate the manifold chemical reactions displayed by the molecule.

> The opinion is now growing that the structural formula of the organic chemist is not the canvas on which the cubist artist should impose his drawings which he alone can interpret . . . On the grounds that practical results are the sole test of truth, such a simple system of electronic valence notation may be termed 'pragmatic'.[40]

'Chemical pragmatism' resisted the attempts to embody in the structural formulas what Fry considered to be metaphysical hypotheses: questions related to the constitution of the atom and the disposition of its valence electrons. It was the actual chemical behavior of molecules that was the primary concern of the pragmatic chemist, rather than the imposition of an electronic system of notation on these formulas which was further complicated by the metaphysical speculations involving the unsolved problems about the constitution of the atom. Fry had to admit the obvious fact that, as chemists would know more about the constitution of the atom, they would be able to explain more fully the chemical properties. He warned, though, that premises lying outside the territory of sensation experience are bound to lead to contradictory conclusions, quoting Kant and becoming, surely, the only chemist to use Kant's ideas in order to convince other chemists at a conference about an issue in chemistry!

London's first contacts in Berlin

In early 1928, Schrödinger asked London to write an article based on a talk he had given in Haber's seminar at the Kaiser Wilhelm Institute for Physical Chemistry and Electrochemistry, which was on the quantum mechanical interpretation of valence. The idea had come from Eucken, who at the time was the editor of *Fortschritte der Chemie, Physik und physikalische Chemie*. Eucken suggested to London that his article should examine the problem of chemical valence from the point of view of quantum mechanics. He told London that the present situation had prompted the chemists to have a great interest in quantum mechanics and its implications. For chemists like Eucken, the adaptation to the new theory was exceedingly difficult and he thought that such an article by London could help them to overcome these difficulties. He felt that it was important to deal analytically with the subject, since, he stated: 'a lot of chemists want to go deeper than only learn the results . . . In about half a year the question of quantum mechanics for chemists will mature'.[41] London never wrote the article. Perhaps he did not share Eucken's views about how much quantum mechanics the chemists should know. Three years later, London continued to be ambivalent about this issue. In a letter to Schrödinger discussing his teaching commitments, he did not think that quantum mechanics was necessary for the chemists' understanding of the chemical processes. On the contrary, he thought that a course in quantum mechanics for chemists might frighten them.[42]

Sidgwick, whose book *The Electronic Theory of Valency* appeared a short while before the publication of the Heitler–London paper, sent London a copy of his book and asked his opinion on a number of questions. He wanted to see how consistent his formulations were in view of the new developments. Among his ideas was the proposal of a single electron bond, but he had a feeling that London would not agree with such a mechanism. While diplomatically praising the book, London insisted on his own approach.[43] In June, London was one of the five speakers at a small meeting organized by Debye, at Leipzig University, with the overall theme of quantum mechanics and chemistry; his talk was titled *Quantum theory and the chemical bond*. The other speakers were Kossel, Hinshelwood, Fermi and Dirac.[44] He stayed at a house together with Kossel and Dirac. On December 5, 1928, it was Sommerfeld's sixtieth birthday. Debye took the initiative to celebrate it with a volume containing papers from all Sommerfeld's students, so that it would 'become possible to document how fruitful his teaching was'.[45]

Ehrenfest invited London to visit Leiden in November 1928 and to participate in the discussions of their group in Leiden. 'Maybe the most beautiful thing in our small circle is the ease with which we make the most stupid mistakes without ever concealing our stupidity and lack of knowledge ... We hit each other and give some blows to our guests with force and love.' Ehrenfest's group was mostly interested in clarifying the foundational principles upon which recent work in quantum mechanics was based. He would have liked very much to have had long talks with London about the many new things of which he said: 'either I do not understand at all or understand them very badly. And since, unfortunately, I understand very slowly, many hours are needed for these conversations.'[46]

London continued to receive encouraging reactions to his work in quantum chemistry. Wigner, who was in Göttingen, was delighted with a paper sent by London concerned with the group theoretical treatment of the chemical bond and thought it was a big advance.[47] Sommerfeld wrote to London from the USA. He was going to visit Berkeley, where he was planning to meet Lewis with whom, he hoped, to discuss London's theory of the homopolar bond.[48] London was invited by the Faraday Society to be present and to contribute to the discussion for a conference to be held at the University of Bristol in

September 1929 on *Molecular spectra and molecular structure*.[49] He did not attend the conference, where Lewis was also present and where his joint paper with Heitler was discussed extensively. A particularly interesting aspect of this meeting of the Faraday Society was the systematic articulation of the molecular-orbital approach as a way of providing a quantitative dimension to the possibilities made explicit by the group theoretical considerations for valence by London (and to a lesser extent by Heitler). This is not to imply that these considerations were the sole reason for the formulation of the molecular-orbital approach, nor that all its adherents had the same starting point. But it is something to be emphasized, since the widely held view which regards this approach as the antipodes of the valence-bond method, though it is methodologically justified, is historically untenable. Hund's (1929) contribution was an attempt to alleviate a weakness of the group theoretical approach where chemical binding could not be understood in terms of energetics, and only the saturation of the valences was explainable in terms of spin. He suggested mechanisms for molecule formation to account for the characteristic difference between hydrogen and HeH. In the same meeting, Lennard-Jones (1929) proposed a set of rules for the assignment of electrons to molecules, which was consistent with the implications of the group theoretical considerations of both London and Heitler.

Though London was not working on philosophy any more he did accept an invitation by Hans Reichenbach to give a talk at the *Gesselschaft für empirische Philosophie*, founded in Berlin, which, together with the Vienna Circle, was to become one of the centers in the development of logical empiricism.[50] On May 6, 1930, he gave a talk titled *The philosophical problems of quantum mechanics*. London's talk – a month after Ostwald's talk on biology – was the first given at the Society on such a topic since its founding in 1928.[51] Reichenbach asked London whether he was willing to publish his lecture in the new journal Reichenbach was planning to start with Carnap and Schlick.[52] London politely declined, being unwilling to be included among those whose writings, in the first issue, would have been regarded as part of the programmatic statement of this new trend in philosophy. In the first issue of *Erkenntnis*, there was an article by Reichenbach on the philosophical issues in modern physics.

Marriage

Fritz London and Edith Caspary met in February 1928 at the Mardi Gras Ball – the Rhinelander Ball – which took place at the Berlin Art Academy. While each was dancing with a different partner, the two

Figure 6. Fritz London and Edith (née Caspary) in 1930, about a year after their marriage. (Courtesy of Edith London.)

couples met and 'we looked at each other and we both dropped our partners and have been together ever since'.[53] But 'dancing was not the highest art that Fritz was involved in' and they sat down, spending the whole evening talking to each other. Before meeting Edith, London was only once seriously involved with a girl living in their house in Bonn. After Franz's death, London's mother, Louise, decided to rent a couple of rooms in their house. Students – among them female students – often occupied the rooms. London's mother was very unhappy about his relationship with one of these students, and did her utmost to end it. When Fritz invited Edith to meet his mother for the first time, he said to Louise London in no uncertain terms that she should not repeat her earlier behavior. The 'Gothic Lady', as one of Fritz's childhood friends called Louise London, obliged.

Edith was born in Berlin. Her mother, Ullendorf, was the youngest of a family of five children. Her father was an umbrella manufacturer and was first the coowner and, then the sole owner of 'Phillip and Company'. The factory employed fifty workers, and had representatives in many other European countries. Both parents were Jewish: Edith's mother's family was very conservative, in contrast to that of her father. Her father's oldest brother was one of the ten representatives of the Berlin Jewish Community. Jewish holidays were celebrated according to the traditions of her mother's family. Edith had a sister who was nine years younger than her, and there was a boy who died when he was eleven months old. Edith started attending the only Gymnasium for girls in Berlin, but because of health problems she followed the doctors' advice to continue her studies in the Lyceum. She graduated from the Lyceum in 1922 at the beginning of the period of the economic crisis in Germany and her father asked her to help him with his business. During the day she worked in the office, and in the evenings she took lessons for reciting poetry – 'together with art, one of my two dreams'.[53] Edith's father had made arrangements with the firm's Italian representative Signore Springorum so that Edith could spend a year in Milan learning the Italian language.

During their first date, Edith was told by London that he had been baptized when he was seven years old. 'I was literally speechless. He was the first person whom I had met that was baptized. He said "Why do you have such a tragic look?" I said "I feel very sorry for you that you were deprived of something very beautiful".'[54] It was not uncommon for many middle class Jews to baptize their children and it was more common among people associated with universities. Franz London had to wait for many years before he was offered a professorship. He thought that, by baptizing the children, they would not suffer any discriminations in their professional lives because of their faith. For many Jews, in the words of Heine, baptism became the ticket 'of admission to European civilization' and by the turn of the century many Jews considered themselves to be *Deutsche Staatsburger juridischen Glaubens* (i.e. German citizen of Jewish faith). The tolerance of Edith's liberal environment and, especially, of her Rabbi were greatly appreciated by both Fritz and Edith. Having been baptized did not really mean anything to London. His only written views on the subject are from a letter to Rosenthal, some years later after the Nazis came to power.

> After the law about civil servants, I have no hope to continue working. Of course, I do not belong to the Jewish religious community because my father had baptized me a Christian when I was a child. The issue of

> returning to Judaism, for which I long internally and to which I consider myself to be faithful – my wife is a practising Jew – has been emphasized to me for some time. Nevertheless, I was indecisive about making the first step, because I did not want to give the impression that I simply turn where the wind blows, since it appears to me more important to have a religion in my heart even if in this way I am exposed to the danger of giving the impression that I am sitting between two chairs![55]

Fritz and Edith were married on March 12, 1929, with a civil ceremony. They went to southern Italy for their honeymoon, first to the island of Ischia, and then they took the boat and sailed to Trieste. The boat had many American athletes of Greek descent, and it made an extra stop at Piraeus, which was and still is the main port of Athens. They had only six hours and rushed off the boat for a trip to the Acropolis. They were the first to get off amid cheers by a crowd which had gathered to welcome the unexpected stopover of the Greek-American athletes. There was a small hitch to an otherwise deeply satisfying visit: Fritz was disappointed that people could not understand what he was saying in ancient Greek!

London never felt at home in Berlin and never missed a chance to express his dislike of the city. In contrast, his wife Edith loved everything about it. London insisted that no place in Germany could surpass the elegance of Bonn, an elegance which came about by the French influence. When he was in the Gymnasium, London wrote an essay titled *Der Rhine, Deutschlands Strom, nicht Deutschlands Grenze* (i.e. The Rhine, Germany's river not Germany's borderline). The essay extolling the virtues of French influence in Bonn, did not fall on sympathetic eyes. The teacher did not consider that the piece was a very patriotic diatribe. This got London into trouble and his father was asked to go to see the teacher. London was always very happy with the French influence, and he had confided to Edith after their honeymoon that he did not feel at home in Germany. Fritz was very often teased about the positive things he had found in Edith, a veritable Berliner. He would respond that 'Edith is not a Berliner!'

Job offers

During the early years in Berlin, London received two job offers. In the spring of 1929, Born explored the possibility of having London as his assistant to replace Leon Rosenfeld.[56] Born had already talked to Schrödinger, who had no objection. The offer was very attractive. Born told London that he was able to secure some further funds from people in industry, and, hence, his salary as an assistant would be

Figure 7. Fritz London in Berlin in 1930. (Courtesy of Edith London.)

supplemented. In fact, he asked London to tell him the amount he needed to come to Göttingen. He mentioned that his collaboration with Jordan to coauthor the second volume of his book on quantum mechanics was becoming more and more problematic, and asked if London would like to work on the book together with Born. London was quite unwilling to leave Berlin for Göttingen. 'You know' he wrote to Schrödinger 'that I do not have a great emotional inclination towards the scientific life at Göttingen.' He asked Schrödinger whether he could stay as his assistant after his contract ended in November 1929 – unless Schrödinger thought that a position of an assistant should not be continuously occupied by the same person. 'But if I do not have any prospect of continuing my very dear activities with you, I would not want to let go of such an attractive offer by Professor Born.'[57] In the meantime, London received a letter from Born telling him that a difficulty arose concerning his offer to London. When Born mentioned this offer to Heitler, he was told that there was an 'endless bickering' between the two. Born was unwilling to proceed with any arrangements unless the problem between Heitler and London was resolved to the mutual satisfaction of both.[58] London responded immediately that any such problem should not be prohibitive. London had chided Heitler because he did not actively seek London's cooperation in the papers involving the group theoretical approach to chemical bonding. London was annoyed that the situation did not change during the eighteen months since the misunderstanding first occurred and that Heitler chose to bring the issue up with Born in this particular circumstance.[59] By the end of April, London decided not to go to Göttingen but to stay in Berlin. The more formal work at Göttingen did not attract him, and in Berlin he had all the freedom to work on any problems which interested him with no intervention by Schrödinger. Later, in 1935, in Oxford, Fritz was discussing this offer with his wife, and he could not specify why he had not accepted it: 'He was a little unconvinced about the spirit of Göttingen, that he may not fit quite in that spirit.'[60]

In 1929, Alpheus Smith, who was Chairman of the Department of Physics at Ohio State University, inquired whether London would be interested in a position there. They were planning for a major appointment in theoretical physics, mainly for graduate teaching, research work and close contact with experimental work and they were offering a salary of $4000–5000 per year.[61] London responded by saying that he would like to work for just one year in the USA, since at this point of his carreer he could not stay longer. He was informed that the Department was also discussing the matter with L. H. Thomas, W. Gordon and A. Lande, and were looking for someone who would be willing to consider staying permanently after the one year trial period. London wrote to them that it would not be possible for him to leave Berlin even for a year.[62]

Intermolecular forces

As London knew all along, the application of perturbation methods to chemical bonding was a hopeless undertaking and the use of group theoretical methods could not be profitably used beyond the systematization of what was already known experimentally. In 1930, he turned his attention to the study of the attractive forces between the molecules. London untangled two of the serious difficulties associated with the results of Debye and Keesom by his quantum mechanical reading of the underlying mechanisms. The dispersion forces which are proportional to r^{-7} still bear his name, and his papers on this subject provided the first proper quantum mechanical treatment of the long range attractive forces, often referred to as van der Waals forces.

Discussion about these forces started at the end of the eighteenth century during the attempts to solve the problem of capillary action. Laplace was the first to formulate a theory to explain capillary action as a phenomenon due to the surface tension produced by the short range forces acting along the lines joining the centers of the molecules. The systematic treatment of these forces was first given by van der Waals in his doctorate; he was led to the corrections of Boyle's law by the study of the capillarity problem. That was in 1873 and even though it was the mystery of the short range repulsive forces which haunted van der Waals, even in his dreams, as he admitted in his Nobel speech, it was the long range attractive forces which, understandably, became the focus of attention of the community.

The phenomenological treatment of both these forces by Lennard-Jones was quite decisive for most of the subsequent developments. He proposed a formula for the total force which gave surprisingly good results, and, hence, suggested the kind of dependence on the intermolecular distance:

$$F = Ar^{-n} - Br^{-m}$$

The first term represented the repulsive forces and the second the attractive forces, and this was guaranteed by demanding that $n > m$. The two theoretical derivations which set the stage for London's work examined the effects of molecular dipole moments. Two molecules with permanent dipole moments develop an attractive force between them because of their average orientations due to thermal motions – which can be shown not to give a zero result. This was Keesom's view and he had found a formula which did have the correct distance dependence, but, also, depended on the inverse of the temperature. Debye proceeded in a slightly different direction: he calculated the force between two molecules when one or both have a polarizability without necessarily having permanent dipole moments. Thus, he calculated a force caused by induction. Both calculations assumed the presence of a permanent dipole moment in at least one of the interacting pair of molecules. It was, however, known that not all molecules have such moments. Keesom tried to remedy the situation in involved calculations where he assumed quadrupole moments that many molecules possess, but the unwarranted high temperature behavior cast a doubt on the reliability of the calculations.

In contrast to valence forces, the van der Waals forces cannot be saturated. In other words, they are additive, in that the force between a pair of molecules is relatively unaffected when a third molecule is brought into their vicinity. Neither Keesom's alignment forces nor Debye's induction forces could express this property, even though Debye's induction forces remained attractive at high temperatures. This became the starting point of London's criticism in 1930.

Characteristically, his criticism of what appeared to be a technical point originated from his understanding of the unifying character of the van der Waals forces which were the common cause of various phenomena. The identity of forces in both the liquid and the gaseous state, the phenomena of capillarity and adsorption, and the sublimation heat of molecular lattices were some of the phenomena which could be understood in terms of molecular forces. London started from the continuous motion of the electrons, even in the ground states of atoms. Because of the uncertainty principle, such a motion creates rapidly fluctuating dipole moments. Molecules with no permanent dipole can, thus, acquire an 'instantaneous' dipole which can induce an 'instantaneous' dipole in another molecule. Such a moment of one atom can induce a moment in another atom, and so on. London showed that such an induction force, which is not conditioned by the existence of a permanent dipole moment but by the quantum mechanical behavior of the electrons, led to long range attractive forces which were also additive, even in the case of the generalized calculations. In his calculations, it was the second-order perturbation term which gave the r^{-6} energy dependence, and London called them dispersion forces because perturbation terms were expressed in terms of the same oscillator strengths as those which appeared in the equations for the dispersion of light. They were independent of temperature and, therefore, could be meaningfully used for understanding the behavior of substances at low temperatures. In the paper written with Eisenschitz, they were the first to point out the possibility of resonance between two like molecules when a quantum of energy could be emitted by one and absorbed by the other. This, again, was a purely quantum phenomenon and it became significant in the collisions between like molecules.

One of the persons who had worked extensively on various problems of molecular forces was Henry Margenau, who later became one of the outstanding philosophers of science. After receiving his doctorate from Yale, Margenau was awarded a Sterling fellowship, and part of the time he spent in Europe was at the Institute for Theoretical Physics at the University of Berlin. 'My work in Berlin was dominated almost entirely by the genius of Fritz London, who had just published a first paper on the quantum theory of van der Waals forces. Realizing that the best way to learn quantum mechanics was to apply it, I extended London's work.'[63]

The book which could not be written

Arnold Berliner, the influential editor of *Naturwissenschaften*, had recommended to Springer Publishers that they should commission London to write a book about the 'significance of quantum mechanics for chemistry'. Springer Publishers agreed and their first contact with London was on October 30, 1929. From the very start there was a rather serious disagreement between London and Springer. London asked Springer to refuse to publish any manuscript other than his. Springer refused to agree to such a proposal, saying that it would be worse if such a book were to be published by another publisher. They did, though, come to an understanding that if another author went to Springer with a similar proposal, Springer would inform London about the intended contents of such a book.[64] The contract was signed at the end of December 1929, and London was given one year to prepare the book. In mid-December, however, D. Whyte informed London[65] that R. H. Fowler, who was one of the editors of Cambridge University Press, would be interested in commissioning London to write a book 'on the foundations of chemistry in quantum mechanics, along the lines of much of your recent work'. A small, but characteristic point: to Whyte's suggestion of a book about the 'foundations of chemistry in quantum mechanics', London's answer was a book about 'quantum mechanics and chemistry'. London wrote to Fowler telling him about the binding conditions with Springer and proposed the translation of the book, but such an agreement was possible only if the original book and its translation were published simultaneously. Springer did not agree with a simultaneous translation, and Fowler could not see a later translation being published in the series, even though he thought it was a book 'that ought to be translated'.[66]

About a year after London signed the contract with Springer, he sent them a letter saying that it was not possible to complete the *Quantum Theory and Chemistry* by the agreed date. There were new results by other researchers and he wanted to understand the new developments more fully. He asked for an extention of the deadline. 'Things are very unusual in this uneasy time when the subjects are developing faster than the books.'[67] Springer agreed to the request and advised him to give himself a deadline.[68]

There was a long period of silence till March 1933. Springer sent

London a letter saying that the deadline set by the contract was long overdue and that, since the publisher did not expect London to finish the book, they should both agree that the contract was no longer valid.[69] London immediately wrote to Berliner telling him that he had already finished three-quarters of the book he now called *Atomic and Molecular Forces*.[70] He felt that the slow pace of the book should not be considered to be a disadvantage, since it would have been quite premature to have written such a book in 1929. He mentioned Fowler's offer, and, before doing anything further, he wanted to know whether Springer would reconsider its decision, since he preferred to remain with them. London asked Berliner to intervene and to advise him accordingly. After discussions with Berliner, Springer agreed to continue the contract.[71] These exchanges were just before the Nazi decrees obliged London to resign from his position.

The manuscript followed the fate of the author. When London moved to Oxford in 1933, there was an offer from Clarendon Press. They urged London to reach a decision, since it was a 'field which is active and where we may have other proposals'.[72] London agreed to have the book published by Clarendon Press, and started translating it, expecting to finish the job in about six months. He felt confident that Springer would not create any difficulties and would release him.[73] By mid-1935, London realized that it would be a futile task to complete the book. 'But now I am afraid time has passed away ... and so I have decided to abandon it definitely ... this decision has not been a very easy one for me as a lot of troublesome work is invested in the matter.'[74]

In the nearly complete manuscript which is in the Fritz London Archives at Duke University, London developed the formalism of quantum mechanics and put a great emphasis on the issues concerning symmetry, discussing in length the Pauli exclusion principle and presenting analytically the valence-bond method. The last part included discussions about the van der Waals forces as well as the alternative approach of molecular orbitals. In the appendices, London presented various aspects of group theory and the methods of Slater and Weyl.

But this was not the only opportunity to write a book on issues related to quantum chemistry. In 1931, E. Rabinowitsch, in his capacity as an editor of a series called *New Problems of Physics and Chemistry* to be published by Hirzel, asked London to write a book. The books in the series were not to be popular books, but were intended for scientists who were more or less familiar with the subject and could have an overview of an area in a short monograph. The first book of the series was planned to be *The Structure of the Nucleus* by Gamow. There would follow the *Courses in Foundational Principles of Physics* by Born and *Introduction to Geochemistry* by Goldschmidt. Rabinowitsch asked London to write something about valence forces, van der Waals forces or anything else which might interest him. But if London was not willing to write a book on these topics, he could still contribute to the series.

> I discussed in Berlin with Einstein as to who would be the most appropriate person for a report on the present state of the 'General theory of relativity and its relations with cosmology'. He referred me to you, because you recently gave a lecture on this subject in Berlin. I will be very glad if such a proposal will interest you.[75]

London liked the general idea of the series, but referred to his contract with Springer, saying that he would like the book to be published in the series, if Springer changed its mind, because of the delay. The suggestion by Einstein acquires an added significance, since the books in the series required authors who were well acquainted with the recent advances and had made original contributions to the subject. Einstein must have been impressed by London's presentation and must have had in mind London's earlier work on Weyl's unified theory. 'Einstein's recommendation to write the book on relativity flatters me very much, but does not make me the most appropriate person, since I have not worked at all on this area.'[76] He was pressured into writing a shortened version of the book he was preparing for Springer for the series, but nothing came out of all this.[77]

Leningrad and Rome

At the end of March 1931, M. Polanyi, who was at the Kaiser Wilhelm Institute in Berlin, asked London whether he would accept an invitation to spend a few weeks in Leningrad. The invitation was by Simon Roginskiy on behalf of A. Joffe who was, then, the Director of the Leningrad Physicotechnical Institute. London was very enthusiastic about such a prospect and asked if he could be accompanied by his wife.[78] Polanyi went to Leningrad in April and told Joffe about London's willingness to visit them. Upon his return, he informed London of what he would be expected to do there: to give a basic course in quantum mechanics with emphasis on atomic and molecular forces; and to participate in a weekly colloquium about problems of chemical kinetics, primarily related to the recent work at the Institute. The Institute would pay, in German marks, the amount withheld from the University of Berlin during his absence, the travel costs for him and his wife and all expenses while they were in the Soviet Union.[79] On June 1, 1931, London received the official invitation to deliver a series of lectures in quantum mechanics, from September 10 to October 10.

London's willingness to visit the Soviet Union was not independent of his interest about what was going on there. Though non-committal about the Party politics, the whole 'experiment' and the promise of a society with no poor and uneducated people appealed to his humanist ideals. The Londons decided to sail from Stettin, since London felt that one ought to go to a harbour city like Leningrad by boat and not by train. The beauty of the city and the warm welcome by his colleagues there compensated for the rather uncomfortable sea trip. There are no notes from his lectures there, but in a letter to Semenoff prior to his departure he wrote that a large part of the lectures would be a systematic exposition of the fundamental aspects of quantum mechanics, since, without these, one could not 'hope to have a clear understanding of the meaning and the field of application of the special propositions that interest the chemists'. He said 'To be honest, in this way it will be possible to realise a plan I have, to include these questions in a book.'[80]

The lectures were held at the Institute between 7 pm and 9 pm, and, thus, Edith and Fritz had a lot of time to walk around during the day. In one of these walks there was an unpleasant incident. They were on top of St Isaac's Cathedral, with a panoramic view of the city, and London was taking pictures, including the harbour with the Soviet Fleet there. A sign in Russian warned that photography was forbidden. A plain clothes policeman approached them and asked them to follow him. London managed to take the film out of the camera, but they spent hours in the police headquarters being interrogated. When questioned, London denied he had the film on him – 'it was the only time I was furious at Fritz'. Leaving the building they noticed a shocking sight. There was a net around one of the middle floors so that people who had been arrested could not commit suicide by jumping down. They found that their hotel room had been thoroughly searched. Their hosts got very worried and decided to accompany the Londons on all their future walks.

Economically, it was a tough period for the Soviet Union, but London found that his colleagues had high spirits and were full of plans for the future. London was quite impressed with what he saw and was touched by older people who would hear them speak in German, and then tell them that now they felt like they were in paradise compared to the way they were treated before the Revolution. But London was fully aware of the differences with the West and the difficulties ahead. 'Some day it will come to a clash' he confided to Edith. They left the Soviet Union via Odessa after London had given a talk in Moscow as well. The end of the visit was marked by an incident which greatly upset London. While in Leningrad, London was offered a professorship there. Though pressured to decide while they were in the Soviet Union, London wanted to think the matter over and give his answer from Germany. But, after they left, there was news in the German papers that London had been offered a job in Leningrad and had accepted. London was not planning to accept the job anyway, but he considered this to have been orchestrated by the Soviets to put pressure on him.

The boat they took for the return trip from Odessa made a stop at Thessaloniki. The second largest city of Greece has some of the best-preserved samples of Byzantine architecture and, at the time, the city still preserved a lot if its memories from centuries of Ottoman occupation, having itself been liberated nearly a century after southern Greece had gained independence in 1821. At the time, Thessaloniki had a very large and active Jewish population who had first settled there after the expulsions from Spain. While the Londons were visiting the Jewish cemetery, Edith realized that they had to rush to catch the boat. As they were going towards the harbour, she noticed that London was not walking as fast as she had seen him on other similar occasions. As a result they missed the boat. London was unperturbed, and appeared to be enjoying the whole thing, making all the arrangements to find another vessel to take them to Athens, where they eventually joined their original ship. On the way south to Athens, none of the islands went by without being correctly named by London.

In June 1931, London applied for another grant from the Rockefeller Foundation to stay in Rome with Fermi's group during the winter of 1931–1932,[81] and was granted a fellowship for six months starting from November.[82] The stay in Rome was a memorable period. They travelled extensively. Edith attended art classes at the British Academy and she was able to acquire a special pass and go to the Forum every day, drawing sketches. 'Fermi did not invite me to his home yet, nearly every evening he studies the new papers together with his students, where even the most repulsive papers are translated into melodious Italian',[83] London wrote to Schrödinger. He worked almost exclusively on his book, but could not finish it. Slater and Pauling had published their significant contributions, obliging London to do some hard thinking about the state of the theory of the chemical bond and molecular forces. While London was in Rome, Victor Weisskopf taught his courses in Berlin.[84] The prospect of being offered a professorship did not seem to be particularly imminent, and soon after returning to Berlin they decided to buy a house and to settle down. Their house was near Grunewald where many other members of faculty from the university lived.

Difficulties with group theory

In 1931, Born planned to write an article dealing with a series of issues related to the chemical bond and its quantum mechanical treatments. The article was to appear in *Ergebnisse der exakten Naturwissenschaften* and was quite critical of some aspects of the group theoretical approach of Heitler and London. Born, in effect, thought that various terms neglected in these papers were significant and that their contributions to the activation temperature were non-trivial. London agreed that both he and Heitler had made unwarranted assumptions in their group theoretical approach. But he felt that such a criticism overlooked the substance of their approach, especially since it was an approach which attempted for the first time to make inroads into a field that was exceedingly complicated and was not amenable to quantitative calculations. He reminded Born that his aim in these papers was to work out an idealized situation. Clearly, London had different theoretical priorities in mind from those advocated by Born. 'The possibility to visualize (if we can talk this way) the framework of a model which depends only on potential forces, seems to me to be of some interest.'[85]

Born did not appear to be convinced by London's answers. He was rather uncompromising on one point. Born felt that London should not claim the priority of the idea that the potential threshold in general meant activation. Born had found similar results in his own work on potential centers, and he felt that London's originality was in showing that such processes resulted in a straightforward manner from London's theory of the covalent bond. He promised London to present 'your beautiful thoughts without my objections and, thus, to publicize them. But you should trust me, so that you should tolerate a little criticism'[86] and that 'the new generation to which you belong is writing in such a manner that one needs useless effort to be able to understand the work'.[87] He promised to defend London's positions – which, in fact, he did in his article.

In London's papers on group theory, there were the first indications of a process that he would, eventually, be able to complete in the 1940s with his work in superfluidity. He started to flirt again with the ideas he first expressed in his philosophy thesis. At a time when it was panegyrically reasserted that physics was the unquestionable paradigm for a 'good and proper' scientific discipline, London started to articulate an agenda based on the non-reductionism of quantum chemistry to physics. It was an agenda with hardly any followers, especially among the community of physicists who were under the spell of promises delivered by Dirac's reductionist program, expressed as a theoretically correct, but practically meaningless dictum:

> The underlying physical laws necessary for the mathematical theory of a large part of physics and the whole of chemistry are . . . completely known, and the difficulty is only the exact application of these laws.[88]

The Heitler–London approach to the homopolar bond was neither the only approach nor the most practical one for molecules with many electrons. Their group theoretical proposals were mathematically quite involved and not conducive to producing quantitative results. The methods which could produce such quantitative results were developed by two Americans – Linus Pauling and Robert Mulliken. As these methods were being developed, Heitler and London perceived them as antagonistic to their own and progressively realized that the chemical community was less and less willing to adopt their approach to the chemical bond. Drawing up a program to examine the nature of the chemical bond presupposed a particular attitude on how to construct a theory in chemistry, on how much one borrowed from physics and what the methodological status of empirical observations for theory building was. Concerning the work of Heitler, London, Pauling and Mulliken, two different research traditions were formed. Heitler and London insisted on an approach which, even though it was not as reductionist as Dirac's pronouncement of 1929, followed this path of orthodoxy. Pauling and Mulliken had a strong preference for semiempirical methods whose only criterion for acceptability was their practical success. To suppose that the question of a stronger command over the mathematical details was the sole differentiating criterion between the two traditions is quite misleading. The difference could be understood only by examining the two different styles for doing quantum chemistry. It is a matter of explicating the internal, theoretical and methodological coherence of each proposed schema, and realizing that they constitute two diverging programs. In 1935, London and Heitler reestablished contact, and started writing frantically to each other, trying to see what could be saved of their common approach from the onslaught of the new methods. Before discussing their correspondence, let me briefly discuss Pauling's and Mulliken's approaches.[89]

Linus Pauling's resonance structures

Almost everything in the series of Pauling's papers, starting in 1931 and titled *The Nature of the Chemical Bond*, are included in his book of the same title. There are, however, some details of significance. In the opening paragraph of the first paper in the series, Pauling stated his assessment of the situation concerning the work which had already been done on the chemical bond as well as the method he would follow.

> During the last four years the problem of the nature of the chemical bond has been attacked by *theoretical physicists*, especially Heitler and London, by the application of quantum mechanics. This work has led to an approximate theoretical calculation of the energy of formation and of other properties of simple molecules . . . and has also provided a *formal justification* of the rules set up in 1916 by G. N. Lewis for his electron bond. In [this] paper it will be shown that many more results of *chemical significance* can be obtained from the quantum mechanical equations, permitting the formulation of an *extensive and powerful set of rules* for the electron-pair bond *supplementing* those of Lewis.[90] (Emphasis added.)

Texts of this sort are, in a way, pace setting texts; they are rhetorical texts contributing to the formation of the chemists' culture, to the way chemists view others and themselves. According to Pauling, it was the theoretical physicists who applied quantum mechanics to a chemical problem, but, at the same time, Pauling considered his own work as an extension of their program. Pauling insisted that his applications would provide 'many more' results which could be obtained in the form of rules supplementing other rules – by Lewis, in fact, who had formulated them much earlier than the advent of wave mechanics! But since Lewis's cardinal rule – that of electron pairing – had been given formal justification, one could formulate new rules supplementing Lewis's rules! Interestingly, concerning the question of the relationship of the various alternatives to the Lewis schema, Heitler and London thought that their work replaced that of Lewis, whereas Pauling and Mulliken considerd that their work supplemented it.

Pauling proposed six rules for the electron-pair bond. Not all these rules were derived from first principles, but were mostly inferred from rigorous treatments of the hydrogen molecule, the helium atom and the lithium atom. Pauling exploited maximally the quantum mechanical phenomenon of resonance, and was eventually in a position to formulate a comprehensive theory of chemical bonding. The success of the theory of resonance in structural chemistry consisted in finding the actual structures of various molecules as a result of resonance among other more basic structures. In the same manner that the Heitler–London approach provided a quantum mechanical explanation of the Lewis electron-pair bond mechanism, the quantum mechanical theory of resonance provided a more sound theoretical basis for the ideas of tautomerism, mesomerism and the theory of intermediate state.[91]

But the ontological status of these more basic structures was rather problematic. In 1944, George Willard Wheland, who was a student of Pauling's and one of the strongest propagandists of the theory of resonance, published his book *The Theory of Resonance and its Applications to Organic Chemistry*. Appropriately, the book was dedicated to Pauling. Wheland's view was that:

> Resonance is a man made concept in a more fundamental sense than most other physical theories. It does not correspond to any intrinsic property of the molecule itself, but instead it is only a mathematical device, deliberately invented by the physicist or chemist for his own convenience.[92]

At the time, Pauling did not seem to disagree with Wheland's assessment. However, when Wheland's book was revised in 1955, a lively correspondence ensued between the two scientists about the actual nature of resonance theory. Wheland added to his original statement that:

> In anthropomorphic terms, I might say that the molecule does not know about the resonance in the same sense in which it knows about its weight, energy, shape and other properties that have what I would call real physical significance.[93]

Pauling disagreed and wrote: 'I feel that in your book you have done an injustice to resonance theory by overemphasizing its man-made character.'[94] Their correspondence continued and neither appeared to be convinced by the other. What Pauling emphasized was not how arbitrary the concept of resonance was, but its immense usefulness and convenience which 'make the disadvantage of the element of arbitrariness of little significance'.[95] According to Pauling, this became the constitutive criterion for theory building in chemistry. It was the way, as he had noted, to particularize Bridgman's operationalism in chemistry. In fact, as he noted in his interview, Pauling felt more at ease with the Schrödinger approach than with matrix mechanics and did not worry about questions of interpretation in quantum mechanics. 'I tend not be be interested in the more abstruse aspects of quantum mechanics. I take a sort of Bridgmanian attitude toward them.'[96]

In his analysis of resonance, Pauling expressed, in the most explicit manner, his views about theory building in chemistry. He asserted that the theory of resonance was a chemical theory, and, in this respect, it had very little in common with the valence-bond method of making approximate quantum mechanical calculations of molecular wave functions and properties. Such a theory was 'obtained largely by induction from the results of chemical experiments'.[96] The development of the theory of molecular structure and the nature of the chemical bond, Pauling asserted in his Nobel speech in 1954: 'is in considerable part empirical – based upon the facts of chemistry – but with the interpretation of these facts greatly influenced by quantum mechanical principles and concepts'.[97]

Both the discussions with Wheland and a vicious attack against his theory by chemists in the Soviet Union[98] prompted Pauling to include a discussion of the nature of chemical theory in the third edition of his book in 1960. The theory of resonance was not simply a theory embodying exact quantum mechanical calculations. Its great extension has been 'almost entirely empirical, with only the valuable and effective guidance of fundamental quantum mechanical principles'. Pauling emphasized that the theory of resonance in chemistry was an essentially qualitative theory, which: 'like the classical structure theory, depends for its successful application largely upon a chemical feeling that is developed through practice'.[99] Pauling himself has repeatedly stressed the rather empirical character of his theory of resonance.

> My work on the nature of the chemical bond and its application to the structure of molecules and crystals has been largely empirical, but for the most part guided by quantum mechanical principles. I might even contend that there are four ways of discussing the nature of the chemical bond: the Hund–Mulliken way, the Heitler–London way, the Slater–Pauling way, and the Pauling semi-empirical way.[100]

Robert Mulliken's molecular obitals

Though the method of molecular orbitals was first introduced by Hund, it was Mulliken who provided both the most thorough treatment of the different kinds of molecules as well as the theoretical and methodological justifications for legitimizing the molecular-orbital approach. Mulliken was born in 1897 and received his doctorate from the University of Chicago in 1921, working with D. W. Harkins on isotope separation, especially of mercury. He had worked at the University of Chicago and at Harvard as a National Research Fellow, and by 1926 he was an assistant professor at New York University. When he did his foundational work on the method of molecular orbitals he had moved to the University of Chicago and had spent some months travelling in Europe, before his extended stay there in 1930 as a Guggenheim Fellow.

After his work on band spectra and the assignment of quantum numbers to electrons in molecules, Mulliken was getting ready for 'an attack on the Heitler and London theory of valence' as he wrote to Birge in 1931, since he was becoming more and more convinced that one could 'understand chemical binding decidedly better and more intimately, by a consideration of molecular electron configurations than by Heitler and London's method'.[101] Mulliken proceeded to formulate his approach to the problem of valence in a series of papers published in 1932. The theory was, in a way, the outcome of a program whose aim was to describe and to understand molecules in terms of (one-electron) orbital wave functions of distinctly molecular character. The attempt was to articulate the autonomous nature of molecules through a process that depended on the extensive data concerning band spectra and on analogies with atoms. In fact, his theory became an alternative mode to the treatment of the problem of valence by Heitler, London, Pauling and Slater. Holding the view that the concept of valence itself is one which should not be held too sacred, Mulliken assumed that there were not only bonding and non-bonding electrons, but also 'anti-bonding electrons, i.e., electrons which actively oppose the union of the atoms'.[102] For his was the molecular point of view where the emphasis was on the existence of the molecule as a distinct and autonomous entity and not as a union of atoms held together by valence bonds. Therefore, from such a molecular point of view the understanding of the mechanism of uniting atoms became of secondary importance.

Unshared electrons were described in terms of atomic orbitals, and the notion of molecular orbitals was introduced to describe shared electrons. Electrons were divided into three categories according to their roles in the binding process: shared electrons (at least for diatomic molecules) were either bonding or antibonding electrons; unshared electrons were the non-bonding electrons. The latter occured in diatomic molecules only when accompanied by a larger number of bonding electrons. In Mulliken's method, molecular orbitals were conceived as 'entities quite independently of atomic orbital'.[103]

Mulliken urged the distinction between Heitler and London's valence theory and their 'valuable perturbation-method for calculating energies' of molecule formation. He thought that the theories of Heitler and London, Pauling and Slater (HLPS) might be called *electron-pairing* theories, whereas Lewis's theory was an *electron-pair* theory.[104] Time and again, Mulliken emphasized that the concept of the bonding molecular orbitals was more general, more flexible and certainly more 'natural' than the Heitler–London electron-pair bonding – even though the latter may turn out to be more convenient for quantitative results for a number of problems.

The assignment of the various quantum numbers to the molecular orbitals led to an alternative explanation of homopolar valence that did not depend on resonance, but rather on the redefinition of the notion of promotion to be used for the one-nucleus viewpoint of the nuclei in the molecule. The new concept Mulliken introduced was that of 'premotion'. Then, bonding electrons were regarded, in effect, as unpremoted or sometimes as slightly premoted electrons whereas antibonding electrons were considered, from an energetic point of view, as premoted electrons. Therefore, chemical combination of the homopolar type was the result of the shrinkage and consequent energy-decrease of atomic orbitals in the fields of the neighboring nuclei, when such orbitals were shared with little or no premotion. In Mulliken's schema, the exchange integrals of Heitler and London corresponded to the electron density of the molecular orbitals: compared with the densities that would have resulted by the overlapping of the electron densities of the orbitals of isolated atoms, bonding orbitals had a higher electron density and antibonding orbitals had a lower density in the regions between the nuclei. Mulliken's theory was the antipodes of that of Heitler and London. He insisted that the occurrence of the electron pairs in the molecules had 'no fundamental connection with the existence of chemical binding'. The Pauli exclusion principle could adequately explain the fact that each type of molecular orbit can be occupied by just two electrons.

These remarks by Mulliken about the Heitler–London approach were less than graciously received by London, who urged Heitler to look up the assessment of their work by Mulliken and implored him to decide for himself: 'whether we are neglecting something or not, when we leave unanswered these kinds of distortions. And they are not at all isolated cases'.[105] Mulliken had recognized that the Heitler–London approach produced results in agreement with the experimentally measured values for the ground states of simple molecules, but he warned: 'experimental data show that it is unsafe to generalize too far from calculations made for a limited number of cases'. Mulliken's main objection was methodological: the approach of Heitler and London required long calculations in order to make the quantitative predictions, but, as he wrote, 'qualitative predictions can usually be made much more easily by a consideration of electron configurations of atoms and molecules'.[106] It was these kinds of pronouncements that deeply angered London. He did not mind there being a theory superior to his own approach, but one had to play the game according to the rules and should not devise new rules along the way. So much the worse when these rules were nothing but rationalizations of experimental data! Some years later he would be furious when he thought that Lev Landau was doing the same thing in the field of superfluidity.

Earlier, in 1928, Mulliken had made some attempts to give due credit to the work of Heitler and London. He considered that their joint work, and the subsequent papers using group theory, together with Hund's papers, promised: 'a suitable theoretical foundation for an understanding of the problems of valence and of the structure and stability of molecules'.[107] In a paper published a year later, Mulliken (1929) postulated that London's group theoretical approach was a translation of Lewis's theory into quantum mechanical language. But such credit slowly waned. In 1933, he did not refer at all to the Heitler–London paper, but rather to the theory of Slater and Pauling which, together with the molecular-orbital approach, he considered to illuminate Lewis's theory from more or less complementary directions. Mulliken did become progressively more reserved about the usefulness of the Heitler–London paper. In his Nobel speech of 1966, he referred to the paper as merely initiating an alternative approach to the molecular-orbital method. He did not even recognize that it provided the quantum mechanical explanation of the Lewis schema, since he stated: 'electrons in the chemical molecular orbitals represent the closest possible quantum mechanical counterpart to Lewis' beautiful pre-quantum valence theory'.[108] London's work in group theory, however, was mentioned nowhere in Mulliken's 1931 article in *Chemical Reviews* where Mulliken expressed in a detailed manner his objections to the Heitler–London method and theory.

In a paper titled *On the method of molecular orbitals*, published in 1935, Mulliken expressed his views on what he considered to be the most characteristic and differentiating aspects of his theory. The Heitler–London method, he said: 'follows the ideology of chemistry'[109] and treated each molecule as being composed of atoms. Mulliken thought that the most notable success of the conceptual scheme proposed by Heitler–London was its ability to explain empirical rules of valence and to proceed to calculations of energies of formation. He was adamant that his method of molecular orbitals departed from 'chemical ideology . . . and treats each molecule, so far as possible, as a unit'. This apparent difference in terminology highlighted the theoretical issues involved in the study of molecular physics.

> It is the writer's belief that, of the various possible methods, the present one may be the best adapted to the construction of an exploratory conceptual scheme within whose framework may be fitted both chemical data and data on electron levels from electron spectra.[110]

Mulliken had realized that one of the reasons for the poor quantitative agreement using the molecular-orbital approach was because of the inability of this theory to include the details of the interactions between the electrons. But even though their inclusion would make a theoretical calculation from first principles an impossible job, he wrote: 'their qualitative inclusion has always formed a vital part of the method of molecular orbitals used as a *conceptual scheme* for the interpretation of empirical data on electronic states of molecules'.[111]

Such considerations, in fact, led to the qualitative explanation of the paramagnetism of oxygen – one of the main weaknesses of the valence-bond approach.

The appearance of Kronig's book in 1935 did nothing to alleviate London's feelings that he and Heitler should take a strong stand against what he considered to be distortions of their theory. Kronig's book, written almost exclusively from a chemist's viewpoint, was indeed quite harsh towards Heitler and London, and welcomed the approach of Slater and Pauling. In assessing Lewis's theory, it claimed that: 'from the standpoint of the chemist it comprised the most complete formulation of the facts concerning valence in a great variety of compounds'. Such a positive assessment of Lewis's theory was rather unusual in 1935. Kronig mentioned several shortcomings of the Heitler–London approach. Firstly, it could not deal successfully with atoms which were not in their ground state. Secondly, it was not possible to explain the numerous compounds between oxygen, sulphur and the halogens. Thirdly, ascribing the quadrivalence of the carbon atom in many of its compounds to the existence of a low ^{5}S-term was considered to be wholly hypothetical. This approach could not account for the compounds where unpaired electrons contributed to the binding. The calculations were only in first approximation and this made many of the results doubtful, since near the normal states of the increasing atoms there were other states. Though it was recognised that the Slater–Pauling approach gave the same results as that of Heitler and London for atoms that had only s-electrons outside the closed shells, its unquestionable advantage was the interpretation of the directed nature of valence bonds in the case of the p-electrons. For the Pauling approach Kronig stated that it was, in fact, a perturbation method involving atoms which were initially an infinite distance from each other. But Kronig's criticism concerning the applicability of Pauling's theory was not as severe as for the Heitler–London theory, since all the low-lying atomic states were taken equally into account.[112]

Trying to save what could not be saved

Apart from the letters they exchanged in late 1927 about the possibilities offered by group theory to the problems related to the chemical bond, there was no correspondence between Heitler and London until 1935. Recent publications had shown a marked preference for the methods proposed by Pauling and Mulliken over their own. In 1935, Heitler and London started writing to each other trying to save their own schema. Their letters were prompted by the developments of Pauling's resonance theory and Mulliken's molecular-orbital approach and are quite revealing. Their correspondence shows the attitude of each about the possible development of the approach laid down in their common paper – the tension between them as well as the search for the means to consolidate their theory at a time when the Americans appeared to be taking over the field of quantum chemistry. Their correspondence also reflects the different styles of the intellectual heritage of each. Faithful to the Göttingen spirit, Heitler was more mathematical, whereas London continued in the Berlin tradition of theoretical physics, and his inclination was to examine intuitive proposals. Both had left Germany after the rise of the Nazis and had settled in England. Heitler was in Bristol; London was in Oxford. Both were working on different problems, far removed from the problem of chemical bonding. Heitler was working on the theory of radiation[113] and London had just published the paper with Heinz London on superconductivity.

In their initial letters, Heitler and London discussed the possibility of writing an article in *Nature* to present their old results and to include some new aspects which had not been emphasized properly in their earlier papers. These were the activation of spin valence and the possibility of a bond that would not depend on spin saturation. 'That is what I meant in a past note – vaguely and wrong – with the term orbital valence.'[114]

London suggested that they should write a short monograph, possibly with Eisenschitz. Heitler did not appear to be so keen on this idea and he proposed that Eisenschitz should write a review article which, nevertheless, might be difficult to publish, especially after van Vleck's review. Some years before, Fowler had asked Heitler to write a book on chemical valence for the Oxford series and Heitler had declined. Now would be an opportunity to reconsider this offer for Eisenschitz.

London asked Eisenschitz to undertake the project, especially since the latter had the same clear view of the whole field as London and Heitler, and London thought that an article in *Nature* would not attract enough attention. He did not agree that all three should coauthor a book, since this would imply that: 'we agree on this general dance of books . . . The whole thing would be considered as a manoeuver for defence, and everything will be spoiled'.[115] Eisenschitz did not appear to have a definite idea about the structure of such a paper, and he wanted to see whether it would be possible to describe the different approximations applied up to now, as special cases of one formal, but exact, solution. He promised to send an outline.[116] In this outline, Eisenschitz proposed an interesting starting point. Since a large part of the chemical events could be comprehended with the formalism of valence, which described the atoms and their chemical behavior in terms of integers, first, one should examine whether this formalism could be derived from the integer numbers of quantum theory or whether, in order to understand it, it was necessary to proceed to a detailed investigation of the atomic model. He was strongly in favor of the group theoretical approach, thinking that such an approach was more advantageous than the other viewpoints, especially since he wrote: 'the conceptually blurred perturbation calculation is made rather superfluous by the application of the symmetry properties of the exact wave function'. The formalism of valence should not be confused with a taxonomy of fixed energy bonds, and the article should include something about the relative stability of a bond (i.e. activation energies). Eisenschitz concluded his proposals by noting that: 'the empirical study of the theory is, fundamentally, a study of the valence formalism'.[117]

Heitler's attitude concerning the approach by Slater and Pauling was that they were correct about the principles they adopted and he was quite sympathetic about the direction of their researches, even though a series of results did not follow exclusively from their theory. He thought a polemic against them was quite unjustified. 'I simply find that the importance of this theory has been monstrously overrated in America.'[118] Heitler and London felt that among the missed opportunities was their lack of insistence about the oxygen molecule: 'It is only due to our negligence that now comes v. Vleck (after the publication of the matter!) and writes that O_2 is a triumph of Mulliken–Hund, because in our theory is "less elementary".'[119] London insisted that the 'essence of a discovery is to know what one is doing'.[120]

Doubts were expressed for the first time about the character of the attractive forces. It was conceivable that these forces may not be due only to spin. There were other attractive forces with the same order of magnitude and those did not follow from their original theory of spin valence. It was wrong to believe that these forces could originate only from the directional degeneracy of the ground state. It might be the case that these forces resulted in the formation of a molecule only if there were also spin valences. At this point, nothing much could be said about the claim that these forces did not have the characteristics of valence. They admitted that they did not know much about what happened when more than two atoms were near each other, since the mechanism of these forces differed from the mechanism of spin valences. The chemists, they thought, were not so fussy about it and named as valence forces whatever forces formed molecules. 'This is exactly our job. To say that there are also other forces of molecule formation except our old ones and *which* phenomena of chemical valence depend on those, and, especially, that our old scheme can be extended.' Heitler's feeling was that there had been no attack against them by the Americans, except for the case of the oxygen molecule whose paramagnetism they could not explain. The nucleus of their theory was the spin valence and he insisted that their theory was the only one that explained the mechanism of repulsion in a qualitatively exact manner. 'You could perhaps include the above discussion under the title: *Delineation of completeness* ... In any case, we should stress that the extension could be realized on the basis of our theory and, substantially, it includes whatever one could wish (this last thing only as a footnote for us).'[121]

London's answer was not exactly an eulogy to the chemical profession.

> The word 'valence' means for the chemist *something more than simply forces of molecular formation*. For him it means a substitute for these forces whose aim is to free him from the necessity to proceed, in complicated cases, by calculations deep into the model. It is clear that this remains wishful thinking. Also the fact that it has certain heuristic successes. We can, also, show the quantum mechanical framework of this success ... the chemist is made out of hard wood and he needs to have rules even if they are incomprehensible.[122]

They progressively realized that part of the problem was their isolation, and this realization bred even more frustration. The fact that they had not even been attacked was not an indication of the acceptance of their theory. They felt that their theory may have even been forgotten or that it could 'be combatted much more effectively by the conscious failure to appreciate and avoid mentioning it'.[122]

Heitler did not agree with London that their theory was 'fought by the most unfair and secretive means'.[123] 'It may be true for some people in America. Not all people, however, are rascals (e.g. I would not believe it of van Vleck), but only silly and lazy. And we should accept that our theory was quite complicated. I would gladly like to look at the books of Sidgwick and Pauling. I cannot get them here.'[124] Heitler did not agree with London's suggestion to write a book, especially since he had just finished writing his *Quantum Theory of Radiation*, but his main worry was that they may not have many new results to include.[125]

Heitler visited Oxford at the beginning of December 1935. Both were now fully aware that the Americans were starting to dominate the field. As soon as Heitler was back in Bristol, he read a paper by Wheland. In the paper it was noted that the Heitler–London–Pauling–Slater method was developed originally by Slater as a generalization of Heitler and London's treatment of hydrogen, and had since been simplified by Pauling. These thoughts exhausted the last vestiges of tolerance displayed by the more objective of the two: Heitler was vitriolic in his response to London:

> I propose in the future to talk only about the Slater–Pauling theory of the chemical bond, since, in the last analysis [it explained only] the H_2 – well now what can this be compared with the feats of the Americans ... I am afraid that the reading of the papers that we have voluntarily undertaken shall be the purgatory of our souls. If you cannot restrain me, I think I will write a very clear letter to this Pauling (he should give a better upbringing to his students) ... It would be really good to write something which will discuss those things they are stealing from us in America. Do not think I am exasperated because (in the case of Pauling) it involves my paper with Rumer, but because of our common cause. Your achievements disappear equally in the lies ... For van Vleck I notice that his papers are more dignified than his report and [he] does not thank Slater and Pauling for free.[126]

They decided to find an excuse to write to Pauling, admitting that the work of Pauling and Slater did, in fact, go beyond the version of their original theory. London took it upon himself to read their papers carefully. 'We should find many points where it will be *evident* that the passages were written in bad faith ... The best thing would be to have as an excuse a substantial question or a criticism to Pauling's papers. [Slater's] shameless behaviour starts from 1931.' Slater had claimed that the theory of Heitler and Rumer was valid only when the bond energy was small with respect to multiplet dissociation and, therefore, it had no physical meaning. This was not correct, and it was because Slater confused multiplet dissociation with the separation of terms via the Coulomb interaction. Heitler, then, made a specific proposal to London:

> The local chemists, in hordes, torment me with that wretched B_2H_6. It is a typical case where there should be special reasons for bonding. Their examination would be useful because: (1) the opportunity is given to underline your view that it is possible to have special forces, but what is *generally valid* is the formation of pairs; (2) it would impress many chemists; (3) it would let the wind out of the sails of certain ill intentioned or silly people.[127]

They planned to meet in London, UK, in mid-February. They were both back to study chemistry. Heitler felt that Slater and Pauling were: 'so proud about something which is not so bad, but which, under no circumstances, is so distinguished. It gives a *general formula* for the bond, that corresponds to the pair bonding and the repulsion of the valence lines'. Their approximation was as rough as Heitler's semiclassical theory and it was superior only because it included the directional properties. One, however, totally lost the activation energy and the non-additivity of the bond energy.

> It is needless to say that it is fully based on our ideas . . . We should not, though, fall into the error and regard this work as bad or insignificant (as *these* people do ours). It is a branching from our work, from about the point where we strictly suppose that the atoms are in only one state . . . Generally, I believe that we made a mistake [in not giving] *more concrete* applications of the theory, it was a mistake to leave it to the chemists (who are nearer to this kind of work) . . . I find, though, that our direction is not being neglected (apart from the details) in Europe.[128]

There are not many places where we can read the opinions of either of them concerning the molecular-orbital approach. Heitler thought that their basic objection to 'Hund's people' – who both agreed were not the most unpleasant of their enemies – was not related so much with the actual results derived from this method. Sufficient patience with the calculations and a lot of semiempirical considerations gave correct results, in fact. 'Nevertheless, no one could name this a general theory – much less a valence theory – since all the *general and substantive* points are forever lost.'[129] After Born had himself expressed the intention to write an article on the valence-bond method, Heitler and London were seriously thinking of asking him to write it, since, they said 'it is very difficult for us to correct the situation with the necessary emphasis'. London informed Born of all the developments and wrote to Heitler giving details of those whom he informed and with whom he talked, so that Heitler should be sufficiently informed about everything, since he was the 'sole representative of our enterprise in England'[130] – London, in the meantime, having moved to Paris.

An article by Lennard-Jones gave the opportunity for some further clarifications of London's position. In all his publications, Lennard-Jones preferred the 'one-electron-orbital-bonds' and presented the version by Heitler and London 'as not so beautiful and as inadequate'. London asked Born's advice on how to get out of the deadlock they found themselves in. London felt that the big publicity of the molecular-orbital theory, which, due to its simplicity harmed the reliability of their theory. He thought it was a mistake to have been silent on questions of principle concerning the molecular-orbital approach. 'Both of us thought it was superfluous, because we both "transcended" this same phase of Lennard-Jones–Mullikan in the beginning of our observations in 1927, and we were very proud when we realized that we get the exchange degeneracy because of the similarity of the *electrons*.' They thought that the molecular-orbital approach was inherently contradictory and, London wrote:

> Maybe, for this reason we did not take it seriously. Recently, I talked very often with Heitler about this lost ground and repeatedly we tried to find a way to make up for it. We continuously fail . . . We have, undoubtedly, made a mistake by not taking seriously our competitors . . .. The situation has become clear since 1932–33, since when we should have thought about finding new issues and not making enemies with our polemics.[131]

Born did not wish to publish anything about the problems of valence, since he had not followed the developments closely. But he thought that it was absolutely necessary that London and Heitler should take a position and publish something that would be accessible to chemists. He even promised that he would encourage his publisher to have a new series and that they would be the first to write something about the chemical bond.[132]

At long last, Heitler and London realized that 'in the last analysis the pressure to do what is necessary falls on us. What is needed to keep the more dangerous of our colleagues, those, in other words, who work with our method, from falsifying history (Eyring, Pauling, etc.) in their place, is a good standard book. Would you not want to write it?'[133] Oxford University Press suggested that Heitler should write another book, especially after the success of the book on radiation, and so he toyed with the idea of writing one with London about quantum mechanics and chemistry.

London's move to Paris and the incomparably more pleasant conditions there in comparison to Oxford; Heitler's success with his book and his work in quantum electrodynamics; London's success with the theory of superconductivity – somehow, one cannot help feeling that both of them could now afford to be gracious. Suddenly, as if by magic, there was no more talk about these issues – maybe all the reading and the discussions did become the 'purgatory of our souls' as Heitler had suggested in one of his first letters.

Quantum chemistry developed an autonomous language with respect to physics and, above all, its eventual acceptance by the chemists involved a series of issues concerning the way chemists (should) practice the new subdiscipline. Coulson has talked of the period through the Second World War as a period when the chemists were concerned about 'escaping from the thought forms of the physicist'.[134] Usually, the two different approaches to the basic problem of quantum chemistry – that of valence – are compared with each other, and then the relative merits and disadvantages of each method are assessed. What appeared to be disputes over methods were, in fact, discussions concerning the collective decision of the chemical community about methodological priorities and ontological commitments. In many instances the scientific papers had a strong rhetoric propagandizing various changes in the chemists' culture. During the 1930s, the discussions and disputes among chemists were, to a large extent, about the new legitimizing procedures and consensual activities to be incorporated into the chemists' culture.

In two of his interviews, Thomas Kuhn discussed the issue of differences in style concerning the different approaches to chemical bonding. He asked Mulliken whether there was a series of schools – a Heitler–London school, a molecular-orbital school etc. – with some geographical localization and whether in certain places one approach was being used and in other places another approach was being used 'so that people were somewhat past each other'. Mulliken was noncommital in his answer: 'I do not know. The way I was thinking was not in such terms as to notice things quite in that framework. I would say there were some people who were stronger for one thing than for another, but whether they were more abundant in one particular place I do not know.'[135] Wigner, on the other hand, stated that he never felt this opposition, since it was very clear to him from the very beginning that these approaches had different objectives. For example, the molecular-orbital method does not speak about the bond, but rather it has molecular orbitals which extend over the whole molecule. 'This is too far away from the very useful and very fruitful chemical concepts.'[136]

Friedrich Hund, a German, is considered to be among the founders of the molecular-orbital approach. In 1928, Hund had completed a paper where he discussed various points concerning the molecular orbitals. Just before sending his paper for publication, he was sent a preprint by Mulliken who had essentially done the same calculations. But, as he wrote to London, Hund decided to go ahead and to publish his paper, since: 'Mulliken's paper is rather American, e.g. he proceeds by groping in an uncertain manner, where one can decide theoretically the cases for which a particular claim is valid'.[137]

These differences were rather eloquently expressed by van Vleck in a review article he wrote with Sherman in 1935. Anyone who was looking for straightforward calculations from the basic postulates of quantum mechanics was bound to be disappointed. How, then, van Vleck and Sherman asked, can it be said that we have a quantum theory of valence? In order to give a satisfactory answer, they proposed that one 'must adopt the mental attitude and procedure of the optimist rather than the pessimist'. The latter demands rigorous calculations from first principles and does not allow questionable approximations or appeals to empirically known facts. The optimist, according to van Vleck and Sherman, is content with approximate solutions of the wave equation and 'he appeals freely to experiment to determine constants, the direct calculation of which would be too difficult'. The optimist's attitude is that the approximate calculations give one an excellent 'steer' and a very good idea of how things go, permitting the systematization and understanding of what would otherwise be a maze of experimental data codified by purely empirical valence rules.[138] It is not clear whether van Vleck and Sherman used the words 'optimist' and 'pessimist' for pedagogic purposes and to imply their own preferences, despite the fact that they promised to 'adopt a middle ground between the two extreme points of view'. One can, of course, speculate whether the optimists referred to the enthusiastic Americans and the 'pessimists' to the reserved Germans.

The beginnings and the establishment of quantum chemistry involved a series of issues which transcend the question of the application of quantum mechanics to chemical problems. The outstanding issue to be settled in the community turned out to be the character of theory for chemistry and, therefore, a reappraisal of the praxis of the chemists. To the Germans, American pragmatism appeared to be flippant. To the Americans, the German mania to do everything from first principles appeared to be an unnecessary torture of the chemists' psyche. There was a confluence of many different styles of research and, at the same time, there was an uneasy feeling that not all differences could be reconciled. Whether such differences may be suggestive of differences that will allow one to talk of different national styles, is an intriguing question. And it is a question to which there cannot be a definite answer, yet, it is possible to exclude a definitely negative answer.

Heitler and London wrote to each other again in 1951. In a letter sent by Heitler in 1951 to thank London for sending him the first volume of his book *Superfluids*, he talked of life in Zürich nearly twenty-five years after the time they were both there. 'It may interest you that these days Café Globus has been demolished . . . I have not noticed whether they have put up a plate with "The chemical bond was born here".'[139]

London met with Heitler in Zürich in 1953 on his way to Leiden to receive the Lorentz Medal. Heitler was, at that time, a Fellow of the Royal Irish Academy and the Royal Society of London. He bitterly remarked: 'at least something good has happened to one of us'.[140] London thought of Heitler some days before his sudden death in 1954. He thought of writing a critical review of the state of theory concerning the chemical bond. He did not write to him, but said to his wife that he 'must get in touch with Heitler'. In 1955, a year after London's death, Heitler was invited to deliver the opening paper at a

Conference on Quantum Chemistry. His presentation of the ideas about the valence bond in view of the recent developments was enthusiastically received by the audience and, as he noted in his letter to Edith London: 'perhaps your husband was right when he said what is right will find its way quite on its own'.

Heitler – unlike London – returned to the subject of the chemical bond in review articles he wrote in 1955, in 1967 and in his book (Heitler, 1945). In all these writings he appeared to be a strong advocate of the approach he initiated with London and of the subsequent developments in their separate work. He did not miss an opportunity to stress that the four valences of carbon and the paramagnetism of the oxygen molecule were given acceptable explanations by London and himself in terms of the non-crossing of the potential curves:[141] they had shown that because of non-crossing, the repulsive potential of the ground state increases the attraction due to the excited state, by pushing further down the curve for the attractive potential of the excited state.

Why is it that Heitler and London never managed to coauthor an article or a book together? Obviously their different scientific interests after 1933, their wanderings and professional insecurity did play a role in this. Hard feelings and misunderstandings about the way each one of them started publishing in group theory were overcome by 1933, so in 1935–1936 they were not a factor in undermining a possible collaboration. The main reason should be sought from their different views about the role of physics in such an approach. Heitler's strong reductionism was evident in all his writings. The last paragraph of his book *Elementary Wave Mechanics with Applications to Quantum Chemistry* reads: 'It can be said that wave mechanics is the tool for a complete understanding, on a physical basis, of all the fundamental facts of chemistry'.[142] This view was also expressed by Heitler in his last published paper where he stressed that the purpose of the early investigations in chemistry was to reduce the chemical phenomena to the laws of atomic physics and that after the advent of quantum mechanics 'the two sciences of physics and chemistry were amalgamated'.[143] He insisted that the fundamental problem of chemistry was a problem in physics.[144] Heitler was interested in the physical justification of the formulas used by the chemists which he believed was provided by the valence structures implied by these formulas.[145]

But such an approach was particularly unappealing to London and not even its success could convince him. Despite the fact that their joint paper was a classic example of a reductionist approach to which Heitler remained faithful to the end of his career, in London's subsequent work in quantum chemistry there was a confluence of timidly articulated trends expressing London's search for an alternative, non-reductionist approach. One senses his trying to find the fringes of a net he had so successfully woven in his school essays and, especially, in his philosophy thesis. It is ironic, that this first important paper of his was such a pronounced deviation from his grand schema to view theories as wholes! But, by the end of 1936, when the correspondence with Heitler ended, London was well on his way to make amends and to give a big boost to his neglected program: having rediscovered the fringes, he was beginning to weave a new net to accommodate the notion of a macroscopic quantum phenomenon.

Notes to Chapter 2

1. In fact, London referred only to Kossel's 1916 paper and never mentioned his talk to the Bunsen Society in 1920 about the chemical bond. In this talk, Kossel had not mentioned Lewis' or Langmuir's work. He had argued it was implausible that the homopolar bond could be due to anything except electrostatic forces, stating that it was in fact a 'purely physical problem'. That the community of physicists and chemists in postwar Germany was quite unaware of the work of Lewis and Langmuir is also corroborated by the absence of any such reference in the historical introduction to Kossel's talk by Fritz Arndt (Kohler, 1975, 445–6). Interestingly, in his later papers, London referred only to the 1916 papers of Lewis, and not to his book of 1923, where Lewis had made some modifications with respect to his earlier ideas, mainly as a result of the Bohr papers of 1921.
2. Thomson (1924), 450.
3. Ibid., 483.
4. Mulliken (1965), S7.
5. Interview with Walter Heitler by John Heilbron (March 18, 1963. AHQP).
6. One of the main drawbacks of such an approach is the non-existence of the contribution of the ionic terms. Though they do not lead to any appreciable differences in the case of the hydrogen molecule, their *a priori* absence in similar kinds of calculations has been systematically examined by van Vleck & Sherman (1935).
7. Heitler & London (1927), 456.
8. Heitler to London (September 1927, exact date unknown).
9. See James' recollection of his collaboration with Coolidge and their excitement when they achieved better and better results with every new term (in Sopka, 1988, 278–80).
10. Weyl (1946), 216.
11. Van Vleck (1970), 240.
12. Sommerfeld to London (August 8, 1927; September 2, 1927).
13. Schrödinger to London (September 16, 1927).
14. Kallmann was not in the Berlin faculty but at the Kaiser Wilhelm Institute for Physical Chemistry and Electrochemistry under the Directorship of Fritz Haber.
15. Moore (1989), 243.
16. Private communication by Professor Sir Brian Pippard to K. G.
17. Pauling (1928*a*), 174.
18. Van Vleck (1928), 506.
19. Interview with E. Wigner by T. S. Kuhn (December 4, 1963. AHQP).
20. Pauling & Wilson (1935), 340.
21. Buckingham (1987), 113.
22. Interview with Linus Pauling by John Heilbron (March 27, 1964, AHQP). An alternative approach to the valence bond method was first attempted by Edward Condon (1927) a few months before the appearance of the paper by Heitler and London.
23. Pauling (1928*b*), 359. Heitler and London had only mentioned Kossel's work. In fact, Sommerfeld, in his book *Atomic Structure*, had also mentioned only Kossel's work. This had prompted Langmuir to write that Lewis's 'is the only theory which is in accord with chemical facts'. Langmuir to Sommerfeld (March 16, 1920).
24. Linus Pauling to G. N. Lewis (March 7, 1928): 'It pleases me very much that in the new atomic model the salient features of the Lewis atom have been reproduced as much as those of the Bohr atom.'
25. G. N. Lewis to Linus Pauling (May 1, 1928).
26. Van Vleck (1928), 500.
27. Van Vleck & Sherman (1935), 173.

28. Heitler to London (September 1927, exact date not specified).
29. Heitler to London (September 1927, exact date unknown).
30. Interview with E. Wigner by T. S. Kuhn (December 4, 1963, 14. AHQP).
31. Interview with Heitler (see note 5).
32. Heitler to London (December 7, 1927).
33. Interview with Heitler (see note 5).
34. Hartree to London (September 16, 1928).
35. Born to London (October 8, 1930).
36. London (1928*d*), 61.
37. The following results were derived by using group theory:

    1. Two hydrogen atoms can combine to form a hydrogen molecule.
    2. It is not possible to have a helium molecule from two helium atoms in their ground states, but it is possible to have one if one of the atoms is excited.
    3. Inert gases cannot exhibit valences.
    4. The valences of halides are 1, 3, 5, 7. Fluorine is an exception and its valence may have the value 1.
    5. The valences of sulfur, selenium and tellurium may take the values 0, 2, 4, 6. Oxygen may have only 0, 2.
    6. The valences of phosphorus, arsenic, antimony and bismuth may take the values of 1, 3, 5. Nitrogen may have only 1, 3.
    7. The valences of carbon, silicon and germanium may take the values of 0, 2, 4.

38. Clark (1928), 362.
39. Ibid., 363.
40. Fry (1928), 558–9.
41. Eucken to London (March 1, 1928); also see London to Eucken (February 23, 1928).
42. London to Schrödinger (April 1, 1931). In the spring semester of 1932, he taught an introductory course in quantum mechanics and helped Schrödinger's wife, Anna, in her attempts to organize student welfare.
43. Sidgwick to London (May 12, 1928); London to Sidgwick (May 29, 1928).
44. Debye to London (May 4, 1928; June 9, 1928).
45. Debye to London (April 23, 1928).
46. Ehrenfest to London (September 30, 1928).
47. Wigner to London (June 29, 1928).
48. Sommerfeld to London (February 20, 1929).
49. Secretary of the Faraday Society to London (April 24, 1929).
50. See Feigl (1969).
51. Lise Meitner had talked in 1929 on *The elementary constituents of matter*. See *Erkenntnis*, 1, 72–3, 1930.
52. Reichenbach to London (April 25, 1930).
53. Interview with Edith London by Steve Heims (February–May, 1988, at Durham, North Carolina, USA).
54. Ibid.
55. London to A. Rosenthal (May 23, 1933).
56. Born to London (March 25, 1929).
57. London to Schrödinger (April 9, 1929).
58. Born to London (April 12, 1929).
59. London to Born (April 18, 1929).
60. Interview with Edith London (see note 53).
61. Smith to London (May 23, 1929).
62. Smith to London (June 11, 1929); London to Smith (July 12, 1929). Eventually, there were two appointments and Thomas and Lande were offered the positions.
63. Margenau (1978), xxii. Margenau (1939) and Margenau & Kestner (1969), are still among the very best accounts of the subject, and include a historical background to the subject. Molecular forces are exhaustively studied in Hirschfelder, Curtiss & Byron Bird (1954).
64. London to Springer (November 20, 1929); Springer to London (November 23, 1929).
65. D. Whyte to London (December 14, 1929).
66. R. H. Fowler to London (January 14, 1930).
67. London to Springer (November 24, 1930).
68. Springer to London (November 25, 1930).
69. Springer to London (March 7, 1933). The letter by Springer was written one day after the elections of March 6, when the Nazis gained the majority in Parliament.
70. London to Berliner (March 9, 1933).
71. Springer to London (March 15, 1933). This was the last letter exchanged with Springer.
72. Clarendon Press (K. Sisam) to London (July 16, 1934; January 16, 1935).
73. London to Clarendon (January 22, 1935).
74. London to Clarendon (June 30, 1935).
75. Rabinowitsch to London (November 23, 1931).
76. London to Rabinowitsch (November 26, 1931). London suggested that Lanczos should write the book on general relativity, something that Rabinowitsch did not like too much, since he did not

consider that Lanczos' style was fit for the series.
77. Rabinowitsch to London (December 11, 1931).
78. Polanyi to London (March 25, 1931); London to Polanyi (March 31, 1931).
79. Polanyi to London (April 25, 1931).
80. London to Semenoff (June 19, 1931).
81. London to Rockefeller Foundation (June 11, 1931).
82. Rockefeller Foundation to Schrödinger (August 5, 1931); Rockefeller Foundation to London (August 26, 1931). The stipend was $200 per month plus travel expenses.
83. London to Schrödinger (November 14, 1931).
84. Weisskopf to London (December 12, 1931). In a letter to London, Weisskopf sent greetings to Placzek who was in Rome, and then there was a cryptic passage, he wrote: 'I want to curse Placzek for his indicreetness concerning my urge related to you and Mussolini. Since you have just come back from the Soviet Union, you may be able to explain this aggressive urge of mine materialistically'. Unfortunately, Professor Victor Weisskopf could not remember the incident.
85. London to Born (August 28, 1931).
86. Born to London (September 3, 1931).
87. Born to London (September 12, 1931).
88. Dirac (1929), 714.
89. For a detailed discussion of the work in quantum chemistry by Linus Pauling and Robert Mulliken, see Ana Simoes (1993) and Gavroglu & Simoes (1994).
90. Pauling (1931*a*), 1367.
91. These theories considered that it was possible for the actual state of a molecule to be not identical with that represented by any single classical valence-bond structure, but to be intermediate between those represented by two or more valence-bond structures. The quantum mechanical resonance approach led to an understanding of the conditions under which a molecule can be expected to exist in an intermediate stage or mesomeric state as well as accounting for the greater stability of those molecules that are the result of resonance.
92. Wheland (1944), 31.
93. Wheland to Pauling (January 20, 1956).
94. Pauling to Wheland (Jaunary 26, 1956). Wheland's papers are at the University of Chicago Library, USA.
95. Pauling (1960), 95.
96. Pauling (1939), 219.
97. Pauling (1954), 92.
98. Kursanov, Gonikberg, Dubinin *et al.* (1952); Tatevskii & Shakhparanov (1952); Moyer Hunsberger (1954).
99. Pauling (1960), 220.
100. Private communication by Pauling to K. G. (February 12, 1991).
101. Mulliken to Birge (March 26, 1931).
102. Mulliken (1931), 349.
103. Mulliken (1932*b*), 51.
104. Ibid., 54–5.
105. London to Heitler (November 1935, exact date unknown).
106. Mulliken (1933), 498.
107. Mulliken (1928*a*), 189.
108. Mulliken (1967), 17.
109. Mulliken (1935), 376.
110. Ibid., 377.
111. Ibid., 378.
112. Kronig (1935), 198–202.
113. In 1933, after the Nazis came to power, Heitler went to Bristol University, UK. There was a standing arrangement between Göttingen and Bristol for one of Born's assistants to spend a year at Lennard-Jones' laboratory. Martin Stobbe had spent the previous year at Bristol, and when he left, since he was not a Jew, he 'exchanged' places with Heitler. Stobbe was, however very strongly opposed to the Nazi Government, and later resigned. Mott had just become the Professor of Theoretical Physics at Bristol University, and was delighted with Heitler's going there. Mott wrote to Tyndall (April 8, 1933) saying 'I would be absolutely delighted if he came for the winter, at any case that would make us the strongest school in theoretical physics in England' (as quoted in Keith & Hoch (1986)). At the time, in Bristol, Powell's group was conducting his experiments with cosmic rays. Heitler started working on quantum electrodynamics and, in 1936, he published his well received book *Quantum Theory of Radiation* which was to become one of the standard books on the subject. By 1935, he had become a member of the staff of the Physics Department at Bristol University.
114. Heitler to London (November 4, 1935).
115. London to Eisenschitz (November 1935, exact date not specified).
116. Eisenschitz to London (before November 13, 1935).
117. Eisenschitz to London (November 13, 1935).
118. Heitler to London (November 12, 1935).
119. Heitler to London (November 7, 1935).
120. London to Heitler (November 6, 1935).
121. Heitler to London (November 12, 1935).

122. London to Heitler (October or November, 1935, exact date not specified).
123. London to Heitler (November 17, 1935).
124. Heitler to London (November 22, 1935).
125. Heitler to London (November 28, 1935).
126. Heitler to London (beginning of December 1935).
127. Heitler to London (December 13, 1935).
128. Heitler to London (February 6, 1936).
129. Heitler to London (October 7, 1936).
130. London to Heitler (November 1936, exact date unknown).
131. London to Born (October 1, 1936).
132. Born to London (October 10, 1936).
133. Heitler to London (October 7, 1936).
134. Coulson (1970), 259, 287. Coulson was a mathematician by training and the author of one of the standard books on valence theory. He was the first student of Lennard-Jones at Cambridge University, UK, when the latter moved from Bristol to Cambridge in 1928 to the first (and for many years only) chair of theoretical chemistry in Britain.
135. Interview with R. S. Mulliken by T. S. Kuhn (February 1, 1964, 17).
136. Interview with E. Wigner by T. S. Kuhn (December 4, 1963, AHQP).
137. Hund to London (July 13, 1928).
138. Van Vleck & Sherman (1935), 168–9.
139. Heitler to London (February 18, 1951). Heitler continued his work in quantum electrodynamics. In late 1939, an unpublished memorandum by Heitler and Fröhlich discussed the possibility of a chain reaction with uranium. In 1940, Heitler was interned for about three months together with some other colleagues from Bristol University: H. Fröhlich, Heinz London, K. Hoselitz, P. Gross and his brother Hans. In 1941, he moved to Dublin, Eire, having been offered the post of assistant Director at the Dublin Institute of Advanced Studies. Schrödinger was the Director. He worked mainly on meson theory and taught a course on wave mechanics for chemists. In 1949, he moved to Zürich, where he was offered the directorship of the Institute for Theoretical Physics at the University. He continued to work on meson theory and various aspects of cosmic ray physics as well as on questions related to the relatively recent techniques of renormalization in quantum field theory. Starting in the early 1960s, Heitler diverted his attention to what is commonly referred to as philosophical writings. These writings, where he examined the relation of science to the human condition and religion, were rather sophisticated statements about a series of moral issues and religious questions. His emphasis was on humanism and the ways to discover God. His book *Die Natur und das Goettliche* won the Literary Prize of the Foundation for Western Thinking in 1979. When Heitler died in 1981 in Zürich, Sir Neville Mott noted in his obituary in the *London Times* that he sought to 'find a synthesis between Christian belief and natural science'.
140. Interview with Edith London (see note 53).
141. Heitler & Poschl (1934).
142. Heitler (1945), 190.
143. Heitler (1967), 14, 35.
144. Ibid., 13.
145. Heitler (1955), 156.

3

Oxford and superconductivity

Like the great majority of Jews living in Germany, Fritz and Edith London did not read the signs of the ominous events that culminated with the formation of the Nazi Government in 1933 and the decrees it issued during its first six months in power. London was obliged to resign and soon afterwards Lindemann offered him a research fellowship at Oxford University, UK.

Lindemann was very keen to set up a group to study low temperature physics. He had brought Simon, Mendelssohn and Kurti to Oxford, and, in 1934, London's brother Heinz joined them. A couple of months after Heinz's arrival, the two brothers worked out the electrodynamics of the superconductors and offered a theoretical schema for the explanation of superconductivity – twenty-two years after the phenomenon was first discovered. In 1933, Meissner and Ochsenfeld discovered that, in contrast to all expectations, superconductors were diamagnetic. In view of this result, the Londons considered the expulsion of the magnetic field, rather than the infinite conductivity, to be the fundamental characteristic of superconductors and proceeded to formulate their equations. Hence, the vantage point for superconductivity shifted and what was considered, for over twenty years, to be a phenomenon of infinite conductivity came to be regarded, primarily, as a case of diamagnetism at very low temperatures. Their theory was not a microscopic theory but explained the phenomenon in terms of the dynamics of the electrons. Even though the scientific community reacted favorably to their theory, Max von Laue had many objections to it. Von Laue, the only one among the well known German physicists who stayed in Germany and strongly opposed the Nazis, knew London from Berlin. The objections, which form part of a long correspondence between the two, were not free of personal antagonisms and had an ugly ending twenty years later.

In 1935, during a discussion about low temperature phenomena at the

Royal Society of London, London formulated his most radical concept: the superconducting electrons acquired a rigid wave function with a wavelength the size of the superconductor and, hence, because of the uncertainty principle, superconductivity could be understood in terms of order in momentum rather than phase space. According to London, superconductivity became a macroscopic quantum phenomenon. During his stay in Oxford, London started thinking about the structure of helium at a temperature of absolute zero. Helium remains liquid all the way down to absolute zero, and it becomes solid only under pressure – to overcome its relatively large zero-point energy.

In 1936, the funding of the research fellowship at Oxford stopped and it was nearly impossible to find a proper academic job in England. London accepted an offer for a research position at the Institut Henri Poincaré in Paris.

The rise of the Nazis

In early 1929, less than two years after London's arrival in Berlin, the signs of a serious deterioration in the German economy, with its ensuing social repercussions, became all too evident. Germany was not alone, of course, since the crisis in Wall Street on October 25 was a rather strong indication that the USA and many European countries would be faced with immense problems, whose severity was unknown at the time. And no one could imagine their specific form in Germany. The anaemic policies of the Government could not even slow the snowballing unemployment. A bloody clash between the unemployed and the police during the May Day demonstrations of 1929 in Berlin left many dead and wounded, and scores of demonstrators were arrested. The National Socialist German Workers' Party, the Nazis, gained 8% of the seats in the Berlin City Council during the elections of November 1929, and in less than a year they acquired over 18% of the votes nationally, becoming the second party after the Socialists in the general elections of September 1930. During that fatal year, the street fights, especially between the communists and the Nazis, had become part of daily life. The ineffectiveness, and to many people's mind the unwillingness, of the authorities to intervene and to stop violence, and the fear of communism by the republican parties, gave the Nazis a free rein. One of the results of street fighting was that the Nazis emerged with a well organized and disciplined private army of tens of thousands. Slowly and shrewdly, they articulated an ideology which, on the one hand, appealed to the embittered unemployed and, on the other, bolstered nationalistic feelings. Its bottom line was a catastrophic synthesis of two typically European phobias and prejudices each with a history of its own in Germany: age long antisemitism merged with contemporary anticommunism.

The universities were naturally not immune from such a crisis. Violent demonstrations were frequent and many students demanded a limitation to the number of Jewish students enrolling in the Universities. Neither the demonstrations nor the economic crisis had, in the first stages, a direct effect on the faculty members who were living in their isolated suburban homes.

The national elections of November 1932 legitimized the Nazis' aspirations. Though their strength declined from what it had been in the July elections, Hitler was asked to form a Government by Hindenburg and, on January 30, 1933, he became Chancellor. Both the prime minister and the generals, as well as some of the leading industrialists, had been seeking such a solution for some time. Feelings among many people that the responsibilities of government would dampen some of the excesses of the Nazis were quickly dispensed with. On February 4, 1933, the new Chancellor, again with the approval of Hindenburg, invoked the Enabling Clause of the Constitution and a law was passed controlling all public assemblies, parades and publications. The storm troopers almost certainly set on fire the Reichstag Building on February 27, and the communists were blamed for it – but were acquitted by the courts some time later. Using the Reichstag fire as a pretext, a law was issued at the end of February which suspended those articles of the Weimar Constitution which guaranteed freedom of speech, press, assembly, privacy of the various forms of communication, and protection against house searchings. A second law gave authority for the arrest of the communists and the crushing of the left wing press. On February 28, the Government spokesman declared: 'there is no Communist Press anymore. It is doubtful whether there would be any socialist press again either'. After the elections of March 6, where the communists were not allowed to take part, the Nazis gained the majority. In some districts their strength increased dramatically. Having dispensed with the communists, antisemitism became truly the order of the day.

On the whole, the excesses of the Nazis were met complacently. Many of the developments since the end of the Great War could not be regarded as making the intervening years an exemplary period. Any kind of opposition to the new regime could easily be taken to mean a yearning for the immediate past. The politics of the Nazis carried a message in which many found a refuge. Strong government was an answer to the insecurities of the middle classes, whereas the working class saw their aspirations reflected in the demagogic rhetoric. A number of aspects of university life were frequently mentioned in the press as manifestations of deeply worrisome trends developing among the cultured elite: anarchy among the students; highly paid professors who were totally isolated from what was going on around them; the breeding of all kinds of radicalism for which the Jews were not innocent and who had proportionately more positions than they 'deserved'. These totally marginal issues of university life were brought to the fore as being the essential characteristics of the world of higher learning.

There were, of course, reactions from outside. In a speech at the dinner given in his honour in London by the Friends of Palestine in Parliament, Chaim Weizmann expressed his shock that a 'great people like the Germans relapsed into barbarism in their attitude towards a small and law abiding minority of the citizens of their State'. The *Times*, in their leader of March 4, 1933, expressed fear that Germany was marching towards the establishment of a dictatorship and warned that the Nazis' violence would not be restricted to the communists alone.

Einstein, who had declared on March 9, 1933, that he would not return to Germany because of the policies of the government, inquired about the steps he had to take in order to renounce his German citizenship. Einstein resigned from his post in the Prussian Academy, since he could not remain in an Academy which was neglecting its duty to defend freedom of speech. He declared that 'the German public is so cowed by the reigning terror that they have lost all power to resistance'.[1]

On March 10, the Interior Minister, Frick, announced that the communists would be taught to become useful members of society and would be given this opportunity in the concentration camps! About 1500 political prisoners were already in just one concentration camp near Stuttgart and they worked on construction in the camp. By the end of March, various laws drastically limited the number of Jews in the medical and legal professions. There were also some promises that the Reich and the State Governments would take steps to put an end to the outrages and acts of vengeance against the Jews. The Central Union of German Citizens of the Jewish Faith, being the largest national Jewish organization with 535 000 members, issued a carefully worded and rather mild statement welcoming such measures.

On March 24, the Parliament, using its new Enabling Act invested Hitler with extraordinary powers. The regime could govern without a constitution for four years. 'Herr Hitler's old paradox has come true', wrote the *Daily Telegraph* (London) in its leader the next day, 'Parliamentary democracy has voted its own death warrant. The two parties who agree in hating it, have been confirmed in their joint tenure of power by the heaviest national poll on record'. The *Echo de Paris* noted gravely that the Nazi experiment would develop in the same way as the Italian situation, but with a Hitler who was taking charge of a Germany where trains would leave and arrive on schedule.

On March 28, it was officially announced that on April 1 there would be a nationwide boycott of Jewish shops and Jewish professionals. Nazi pickets blocked the doorways of all Jewish stores; Jewish lawyers could receive no clients; Jewish doctors could attend no patients. The Berlin Committee of Action had given orders that photographers and film operators would photograph persons who still wanted to buy in Jewish shops and their portraits would be published in Nazi newspapers and shown in the cinemas. On the day of the boycott, one of London's best childhood friends died. Wolfgang Herbert was the conductor of the Berlin Radio Orchestra, and he died after stomach surgery. Edith and Fritz London had to traverse the whole city for the funeral, and Edith remembered that: 'it was one of the saddest times of our life'.

The decree of April 4 stipulated the kinds of punishments for 'treasonable activity' and acts of terrorism. The property of communists and, subsequently, of all State enemies was designated for confiscation by the decrees of May 26 and July 14, 1933. On July 7, the Social Democratic Party was outlawed and by the decree of July 14, the National Socialist Workers' Party was declared to be the only legal party in Germany. On the same date a law was passed whereby it became possible to revoke the citizenship of 'undesirables' who had become citizens under the Weimar Republic. This was aimed primarily at the Jewish immigrants from Eastern Europe.

On April 6, in a speech to the German doctors' professional association, Hitler talked about the speedy elimination of the excess of Jewish intellectuals from the cultural and intellectual life of Germany. The next day the *Law for restoration of the career civil service* was passed, and a series of memoranda were sent to the universities with the explicit aim of dismissing Jewish faculty members. The application of the racial law to all personnel of the universities was formally announced on May 6. Though its supplementary regulations were not completed until August 1935, it now became possible to dismiss civil servants on grounds of inadequate training, political unreliability (especially membership of the Communist Party) and for the purpose of simplifying the administration. The first supplement (dated April 11, 1933) to the law specified that an official was 'non-Aryan' if one or more of the grandparents were of Jewish blood or religion. 'Non-Aryans' could also be dismissed, except for cases where the particular person was in office before August 1, 1914, had fought in the Great War or had lost a father or a son in the War.[2] The right of the Jewish lawyers to plead in the courts was to be granted to a quota of Jews, calculated as the proportion of the Jewish community to the total population. These laws were followed by two other significant developments. On April 12, the new Prussian student constitution was officially recognized and, in a few days, it was decided to extend it to the whole of Germany. One of its main objectives was the restructuring of the entire curriculum with respect to the criteria of the new spirit. Despite the decree of April 18, 1933, stipulating that the text-books would not be changed for the immediate semester, a series of measures was passed, partly revising the old instruction in special fields, partly introducing new subjects altogether, and partly stressing general lines of emphasis to be pursued in all fields of study. By the end of 1934, the Minister of Education was quite explicit about the aims of the curriculum: 'It is the task of the German universities to put scientific research into the closest possible relationship with the national political needs of out people.'[3]

On May 10, under the initiative of students from the University of Berlin who had founded the Committee for Fighting the Un-German Spirit, 20 000 books and periodicals collected from public and private libraries were burned in the square outside the Berlin Opera House. Part of the story in the London *Morning Post* ran as follows:

> When the bonfire was burning furiously, a student standing on the platform above it, picked up in turn and threw symbolically into the flames: first the works of Karl Marx, as an enemy of social solidarity; then of Heinrich Mann and Erich Kastner, as enemies of morality; of the pacifist Forster, as an enemy of patriotism; of Freud, as an advocate of excessive subservience to the instinct; and, then, among others Remarque, on the grounds that he was disrespectful to the German soldier.[4]

Having instigated all these extreme measures, the Government proceeded to portray a mellower face to quell the reactions from abroad. A declaration by Hitler's Government on May 16 was the occasion. It presented a passionate, yet very balanced, account of the deadlocks created by the Versailles Treaty and saw MacDonald's proposals as providing an opportunity for a peaceful settlement in Europe. The speech did have the desired effect. In England the next day, the *Times* found the speech was moderate and representative of the views of official Germany; the *Daily Telegraph* thought that one speech could not wholly restore lost confidence; the *Morning Post* found the speech a pleasant relief. In the USA and France, the mild tone of Hitler's speech brought a feeling of relief. It was a short lived lull. In less than a week, Vice Chancellor Franz von Papen delivered an impassioned speech against pacifism. The USA proceeded to launch its economic boycott of Germany.

The changes at the University

The great majority of the Jews working in the universities were unprepared, neither individually nor collectively, to cope with the events of April 1933. Most of their colleagues were sympathetic, yet non committal, towards their plight, despite those rare occasions where some of the leading figures were somewhat more assertive. Members of faculty were, after all, civil servants. As argued by Alan Beyerchen in his study of scientists under the Nazis, almost everyone would adhere to the tradition in the Prussian civil service which expected an official to defend himself against accusations and rumors, not by words but by means of the propriety of his actions.

> The relative ease of emigration was all too apparent to those leaders of the German physics community unaffected by the Nazi ordinances. Planck, von Laue, Sommerfeld, Heisenberg and others signed petitions, counseled those who were included in the provisions of decrees and sought as best as they could to hold their community together. The watchword was that those who could should stay.[5]

There are no indications that London foresaw any of the developments. In March, he wrote an enthusiastic letter to his mother after his last lecture of the winter semester which he had given under the new flag. He had talked about his own work and felt good when he saw 'the great interest among the students'.[6] Even after the decree of April 7, there was still no change of plans for the courses to be offered the next semester. The first signs of the despairing London are in a letter to Ewald a week later.

> I would like to have your opinion about the known problem. My two superiors are trying to intervene in the best possible manner at every level and they are trying to console me. But the hope of keeping me here appears to me very faint, and it is highly questionable if whatever arrangement would be of a long term settlement. The tendency to minimize the damage incurred would be shared by all visionary patriots – as it is with my more active superiors . . . , I would like to 'get out' of the temporary solutions and to be included in a program where I can, again, work. Do you think I should wait till something comes from outside Germany or should I chase it myself? Please tell me your opinion. The correct thing would be for an interested side to take the situation in its own hands and there should be a *general* settlement.[7]

On April 21, London received a letter from von Laue to complete a questionnaire. Von Laue was embarrassed to have to ask London to do this in view of the stipulations of the new law, and repeated his advice that London should not resign, but should take a leave of absence. 'You should be convinced that everything that is humanly possible will be done so that you can continue your scientific activity, even though something like this appears impossible right now . . ., on a personal level everything remains as before'.[8] London notified the Dean on April 27 that he would be teaching a course called *Corpuscular theory of light*. The next day, though, he received a notice from the Administrative Director of the University, addressed to all honorary professors, professors and *Privatdozents*. The Minister of Science, Arts and National Education, was 'advising' all those who were subject to the stipulations of the law about civil servants: 'not to practice their *venia legendi*, otherwise it will be necessary to proceed to measures concerning the danger for the public order and security of the University, something that would be detrimental to the prestige of the University teachers and to the Institution'. Also on April 28, all categories of teachers received a document clarifying the rules for the election of their representatives to the University Senate and for electing the Rector. The notice ended with the necessary stipulation: 'According to the meaning of the directives that have been given for the new election, the gentlemen who are subject to the decrees of the new law for civil servants concerning the non-Aryans are asked not to participate in the meeting and the elections.'

On May 5, London sent a letter to the Dean requesting leave of absence for the semester and asked him to replace the notice about the courses with the following: 'The series of lectures on the *Corpuscular theory of light* that was announced in the course directory has been cancelled. The reviews of the new papers in theoretical physics with Prof. Schrödinger, and Drs Kallmann, Möglich, Szilard will commence on May 9 . . . Concerning the reviews of the new papers in the physics of the atomic nucleus see notice by Prof. L. Meitner.' The next day, May 6, it was announced that the racial law would apply to all instructors in the universities.

After a week, London visited Göttingen and wrote his impressions to Franz Simon:

> I was at Göttingen for a few days . . . I do not think that E. is planning to return to Berlin. He has been informed about the [situation] exactly from v. L. and others . . . I do not know if one should wait for long. People like H., with whom I talked again at Dahlem, have totally discouraged me. What do you think? I confess that I cannot stand the role of the bride waiting for the groom. At Göttingen I found the people totally peaceful, as if they were on the moon. Born has gone. F. is a real oasis. He is a good man, who finishes whatever he starts and is always convined about what he is doing. Unfortunately, he has placed himself in a position from which, it appears to me, there is no retreat.[9]

Before returning to Berlin, London visited Paris and met Langevin there. [10]

Schrödinger and von Laue were London's immediate superiors. Schrödinger's opposition to the Nazis was quite evident, although without having any practical manifestations. An exception was the day of the boycott of the Jewish businesses. He could not control himself when he saw storm troopers looting Wertheim's, one of the largest department stores in Berlin. He was attacked by the troopers and rescued by *Privatdozent* Friedrich Möglich, who was there and wore the Nazi paraphernalia.[11] Life in Berlin was progressively becomiong stifling and soon he decided that he should leave.[12]

Max von Laue was the notable exception among the senior people at Berlin University because of his vocal and unswerving opposition to the Nazis. He first showed his public disapproval by summoning an extraordinary general meeting of the Prussian Academy, whose secretary he was, in order to withdraw a statement denouncing Einstein issued by the Academy. Before the summer of 1933, he did not encourage colleagues to resign. During the summer he slowly realized that it would be a long wait. He decided to stay in Germany, confident that the Nazis would not be in power for long. As he noted in his *Autobiographical Notes*, von Laue wanted to be present to witness the collapse of the Third Reich, hoping for a cultural reconstruction upon the ruins of the Reich. Von Laue's speech as the presiding officer at the annual Physicists' Conference in September was a scathing attack against the Nazis, comparing them to the Inquisition in Galileo's time. He was reprimanded by the Prussian Ministry of Education for his speech and for the two obituaries he published after Haber's death in January 1934. He opposed Stark's attacks on Einstein's theory of relativity and Stark's candidacy to the Prussian Academy. In 1942, Berliner, who was his friend and the founder and editor of *Naturwissenschaften*, committed suicide when the Nazis burst into his apartment to seize him. Von Laue was among the very few who attended the funeral at the Jewish cemetery. In fact, when Ewald visited Einstein in 1935 and asked him whether he cared to send any messages to people in Germay, Einstein replied 'greet Laue for me'. When Ewald inquired whether he wished to send messages to anyone else, Einstein just repeated 'greet Laue for me'.[13] Among London's papers, there is an undated letter addressed to 'Herr Professor'.[14] It is a letter where he leaves no misunderstandings about his sentiments and the reasons for his resignation:

> In a period when activities are forbidden to people like Born and many others, it is impossible to behave as if nothing happened, as if the whole matter did not concern me. The fact that up to now I was not explicitly forbidden to teach as a *Privatdozent* is a simple coincidence, and I would not like to let it slip. As long as the dominant attitude is that the Jews propagate a spirit that has to be fought and is detrimental to the

German spririt, our pride dictates that we should withdraw in a manner that is dignified for the public opinion, and we should not fight for the rights we still have. For many, as you know, it is impossible to continue their duties.

I am not Franck, and do not feel motivated to provide the Dean with such a detailed explanation of my reasons. But I am obliged to be open with you, who have trusted me throughout this period with such a great friendship. I am truly sorry that I have to do something that does not have your approval from the beginning.

PS. Could you, please, give this letter to Prof. von Laue, to whom it is also addressed?

About a month later, London heard from Lindemann, who made an official offer on behalf of Oxford University. In August 1933, just before leaving for England, London visited a few of his colleagues to say good-bye. Planck was adamant that London was doing the wrong thing.

Going to Oxford

London's short visit to Holland and Denmark in May 1933, was, of course, a trip to investigate the job possibilities there. Ehrenfest had asked London to go from Göttingen to Berlin via Leiden.[15] Rosenfeld, who at the time was very active at the Club de la Fondation Universitaire, informed London that he had indications that there might be a job offer in England and that everything possible was being done by people at Oxford.[16] London wrote to the Academic Assistance Council (AAC) and told them that they could get references for him from von Laue, Schrödinger, Ewald, Born, Franck, Weyl, Schur, Toplitz and Hausdorff. He also informed them that Slater and Pauling in the USA were doing similar research to himself and that they knew his work.[17] When London was informed by Heinz London, who was at the University of Breslau, that Oxford University was considering making him an offer, he wrote immediately to Mendelssohn, who was already in Oxford, urging him to tell the administration that he was definitely interested.[18] Mendelssohn got in touch with Lindemann, and the latter verified the rumors: he told Mendelssohn that Oxford was in a position to make a definite offer and that he would very much like to meet London during his planned trip to Germany.[19]

When he was asked by Lindemann, Sommerfeld recommended London and Hans Bethe as the two outstanding theoretical physicists for the post in Oxford. Bethe had already been offered a position at Manchester University whose Department of Physics was incomparably better than Oxford's. Einstein, who had visited Oxford three times before the events in Germany (the last time as Herbert Spencer Lecturer on the methods of theoretical physics) convinced Lindemann that it would be useful to induce a couple of German theoretical physicists to seek refuge in Oxford for two or three years. He, also, had in mind Fritz London and Hans Bethe.[20]

London's work fitted perfectly with Lindemann's opinion about what Oxford needed: 'a man who can work out a problem and get an answer'. Lindemann's intense dislike of 'high brow' theoretical physics[21] was well known and often it was an embarassment to his colleagues. London's practical style of doing theoretical physics, preferring the relatively simple mathematics and continuously seeking physical counterparts to his mathematical formulations, came close to what Lindemann considered to be proper theoretical physics. On August 10, 1933, London was officially offered a fellowship at Oxford University from the Imperial Chemical Industries (ICI). He wrote: 'I am very glad to have this opportunity of thanking you for your generosity in enabling me so promptly to live in England and so to continue my research work that was interrupted in Germany'.[22] Edith and Fritz moved to Oxford in September 1933 and stayed at Hill Top Road.

Lindemann, Simon and Heinz London

Three people were of particular significance during London's Oxford years. Lindemann had done his best to bring London to Oxford and did not do anything to hold him there after his contract with ICI was terminated in 1936. Simon's experimental work in low temperature physics was decisive in London's reorientation of his own researches. Heinz London went to Oxford in 1934 and, soon afterwards, the two brothers published the first successful theory of superconductivity.

Frederick Alexander Lindemann (Viscount Cherwell, 1886–1957) had received his doctorate in Berlin with Nernst in 1910. He was elected to be Dr Lee's Professor of Experimental Philosophy at Oxford in 1919. He remained as the head of the Clarendon Laboratory till 1956. In the new situation created after the rise of the Nazis, he saw a great opportunity for Clarendon. McGowan, a friend of Lindemann's, who had succeeded Mond as chairman of ICI, agreed to Lindemann's plan that employing the best scientific brains expelled from Germany would be to Britain's advantage. The objectives of such a plan were stated to be: the direct benefits accruing to ICI through advances made in scientific research; the improvement in the training of University students who would afterwards become members of staff or teachers in British universities; and the infusion of new ideas and new methods of research.[23] By May 1933, ICI had decided to finance Jewish scientists in England for two years with a possibility of renewal. In mid-June there were already thirteen grant recipients. Lindemann's vision of a laboratory whose work would not be interrupted by considerations of industrial applications was, also, shared by Vickers of the shipbuilding firm and Nuffield of Morris Motors, who later contributed substantial funds as well.

Lindemann's greatest success was to bring Simon to Oxford. Simon went on to build one of the very best centers for research in low temperature physics. When, in 1922, Lindemann acquired an apparatus for liquid air, Oxford was the only English university without such a machine. In 1930, he inquired whether the Clarendon Laboratory could buy a helium liquefier designed by Simon. He preferred him to Keesom, who had come to Oxford in 1926 to discuss the same question, since he knew Simon from his days in Berlin and also because Simon had measured specific heats at low temperatures. Simon agreed to help, and Mendelssohn was sent to Oxford to install the liquefier by Christmas of 1932. Mendelssohn had no position in Germany and was offered a two-year contract at Oxford with funds from ICI, getting the position before Simon's arrival. By January 1933, the first liquid helium in Britain was produced in Oxford – Lindemann duly publicized the event in *The Times*[24] and *Nature*[25]. It was an important first for Oxford. Kapitza, working hard in the better equipped and better financed Mond Laboratory at Cambridge, was able to liquefy helium in 1934. The only other places which could produce liquid helium and, thus, were able to do work in low temperature physics were Leiden, Berlin and Toronto.

In 1912, after having completed the classical stream in high school, Simon (1893–1956) entered the University of Munich to study physics, mathematics and chemistry. Then, he went to Göttingen, but in 1913 he was called for his military service. He completed his doctorate on specific heats at liquid hydrogen temperatures in 1921 with Nernst in Berlin, who was then the Director of the Physikalisch Chemisches Institut. His first appointment was assistant to Nernst. In 1931, he was offered the Chair of Physical Chemistry at the Technical University at Breslau to succeed Eucken, who had been appointed to a Chair in Göttingen. Immediately afterwards, he became actively involved in low temperature physics – especially superconductivity. While in Berlin, Simon tactfully avoided working on this subject, since the more senior Meissner was doing research in superconductivity. Deeply convinced that the Nazi ascension to power would be something which Germany would have to bear for a long time, Simon was one of the very few scientists who did not entertain any false hopes about his future in Germany, even though he was exempt from the first decrees, since he had fought in the Great War.

Simon visited England in May 1933. In June, he resigned his post at Breslau and accepted the offer from Lindemann to go to Oxford under the ICI scheme. When he arrived in England in August 1933, Simon arranged with the administration of the University of Breslau to exchange his office furniture for two helium liquefiers which he took with him to Oxford. In order to be able to start a group in low temperature research, he asked that two of his assistants should be offered positions. One was Nicholas Kurti, who arrived with Simon. The other was Heinz London, who was going to join them as soon as he finished his doctorate in Germany.[26]

Simon's presence and his plans to set up a low temperature group were undoubtedly the main factors which must have influenced London gradually to leave quantum chemistry and to start work in low temperature physics. A mutual understanding developed that London would be the theoretical advisor to Simon's group. In fact, this group was the only place in Oxford where London could go to discuss physics – also discussing various problems over lengthy telephone conversations. The arrival of Heinz London strengthened his ties with the group.

Fritz London's decision to work in low temperature physics is quite suggestive of his overall attitude to his agenda and is also different from the decisions of most of the physicists of his generation. He never worked on problems of nuclear physics – though he toyed, for a while, with the idea while he was in Paris – and field theory. The fact that quantum chemistry and nuclear physics were quickly becoming very competitive should not be ignored in trying to understand the factors which influenced London's choice. He disliked the new competitive spirit and the kind of publications which were not thought out thoroughly and which were quickly becoming the dominant mode of communication in the community. He had, in a way, completed his agenda in atomic-molecular forces. Simon's presence set the stage for the kinds of problems to be studied, and his brother Heinz had arrived with some very interesting ideas about superconductivity.

After leaving Germany, the person with whom Fritz had the strongest scientific rapport was his brother Heinz. Together, they formulated their theory of superconductivity and corresponded extensively for almost twenty years from the time Fritz left for Paris in 1936 till his death in 1954. Heinz London was born on November 7, 1907. Fritz felt that, because he was the oldest male of the family, he should take over the guidance of Heinz, which he did throughout his lifetime. Heinz was a lonesome child. He was baptised also, but he never made anything out of it, being an atheist. He abstained from games with other children, since he felt that he was not fit enough to deal with the harsh demands of his mates. At school, he excelled in the sciences and showed no particlar zeal for the courses in the humanities. From 1926, when he started his higher studies, he attended the University of Bonn, the Technische Hochschule in Berlin and the University of Munich. Upon the insistence of Fritz, he did an extra year of chemistry, and Heinz always thought that this was a worthwhile investment. On one occasion, Fritz was told by a professor of Heinz that Heinz should learn to say less and be less cocky in discussions – something which did not particularly embarrass Heinz.

In 1931, he started his graduate studies, initially working with Gerlach, but was very unhappy with him. Then Fritz talked to Gerlach about Heinz and told him that Simon had agreed to take Heinz to Breslau. Heinz's strong attachment to the thermodynamic approach and his insistence on always examining the possibilities provided by thermodynamics, was, no doubt, a result of Simon's influence. 'For the second law, I will burn at the stake' was a favorite saying of his. Heinz investigated the variation of the resistance with high frequency currents. The original suggestion for such a problem had come from Schottky at Siemens, who wanted to know whether there would be heat produced. Heinz could not get anywhere because of poor cryogenic facilities and, as Mendelssohn wrote: 'let us admit it now, after more than thirty years, by a certain lack of manual dexterity'.[27] He could not detect anything in his first attempts, and it was some years before Heinz produced high enough frequencies to get the first conclusive results. Interestingly though, while he was conducting these experiments which were not sensitive enough for detecting the frequency dependence of the resistance in a superconductor, he developed some theoretical ideas concerning the electrodynamics of superconductors. To explain the possible appearance of resistance at high frequencies, he proposed two ideas. The first was that the inertia of the electrons forced the superconducting current to flow in a small and finite 'penetration depth'. The second idea was the presence of normal electrons in a superconductor. This was the first and most primitive formulation of the 'two-fluid' framework which was to play such a decisive role in understanding the low temperature phenomena. Assuming a two-fluid model with superconductive and normal electrons, he predicted that at high frequencies the former would be unable to shield the latter completely, and that Joule heat would be developed.

Concerning this point there have been rather partisan accounts about who first developed the idea of penetration depth and the two-fluid model. Neither idea first appeared in print under Heinz's name. The equation of the acceleration of the electrons and the ensuing mechanism of penetration depth was proposed by Becker, Heller & Sauter (1933). In 1925, a similar proposal was made by Mrs de Haas, Lorentz's daughter, in answer to a query by de Haas about how a magnetic field could exert an influence on a superconductor, since the latter was screened by surface currents. The notion of the coexisting two fluids was first published by Gorter and Casimir in 1934. Never keen on publishing anything he had not thoroughly examined, Heinz expressed these ideas systematically in his thesis in 1934. Nonetheless, it was within this particular framework of the ideas expressed in Heinz's thesis, that Heinz and Fritz were able to assess the fundamental character of the Meissner effect, which led to the London equations and, eventually, to the novel interpretation, by Fritz, of superconductivity as a macroscopic quantum phenomenon.

Heinz did not go to Oxford in 1933 with Simon and his group. He followed Simon's advice to stay behind to complete his doctorate and when he received it late in 1933 he must have been one of the last Jews to be awarded a degree by a German university.[28] Soon afterwards, he joined the low temperature group at Clarendon. During the three years when both brothers were at Oxford, Heinz stayed with Fritz and Edith at their house in Hill Top Road. Apart from the joint work with Fritz in superconductivity, while in Oxford, Heinz showed experimentally that the penetration of a charge into a superconducting metal was not appreciably different from the case of a normal metal and he dealt theoretically with the equilibrium between the superconducting and normal phases. The notion of surface energy in superconductors was also due to Heinz.

Though Heinz liked to consider himself an experimentalist, he was neither a pure theoretician nor an experimentalist. His substantial contributions were in both directions and, maybe, it was his inventions which, more than anything else, bore the mark of his particular strand of work. Fritz thought that he would have become a mathematician if his father had not been one, and that Heinz would have become a theoretical physicist if Fritz had not been one. Fritz believed that Heinz was a very good theoretician and a very good experimentalist, though he felt that, unlike himself, Heinz was not very skilful with his hands. Simon was very pragmatic about this and had said that Heinz was fine – he did not need to become a watchmaker or a surgeon, but they would always tease Heinz whenever something broke in the lab.

Electricity in the very cold

The first systematic studies of the dependence of electrical resistance on temperature had been undertaken by Cailletet, Bouty and

Wroblewski in 1885. The researches of Cailletet and Bouty led them to the assertion that it would not be unreasonable to expect 'a zero value for the resistance for a temperature higher than −273 °C'.[29] The next set of exhaustive measurements of the electrical resistance of various metals were performed by Dewar & Fleming (1892, 1893). In 1896, they completed a study of the resistance of mercury at liquid air temperatures, and their results indicated that the resistance of mercury could vanish at 0 K.[30] But when the measurements were done at liquid hydrogen temperatures, it was found that, after reaching a minimum, the resistance started increasing, and they thought that: 'the parabolic connection between temperature and resistance is not longer tenable at very low temperatures'.[31] It was a disappointing result and it led Dewar to adopt Lord Kelvin's proposal that the resistance initially decreases because the electrons encounter fewer obstacles due to dampening by thermal motion, but when the temperature is further reduced, the electrons themselves freeze onto their atoms and, hence, due to this peculiar 'electron condensation' the resistance starts to increase and becomes infinite at absolute zero!

In 1911, after having liquefied helium in 1908, Kamerlingh Onnes at Leiden University, measured the resistances of platinum and pure mercury at helium temperatures. He found that at 3 K the value of the resistance of pure mercury became 0.0001 times that of solid mercury at 0 °C. Later that year the phenomenon was reaffirmed at 4.19 K. By 1913, it was realized that impurities did not play any role in hindering the disappearance of the ordinary resistance, and the phenomenon was for the first time called the 'superconductivity of mercury'.[32] In 1914, Kamerlingh Onnes discovered that an external magnetic field could disturb superconductivity by generating resistance in lead and tin. And, while studying the Hall effect, it was found that superconductivity was also destroyed when current above a certain threshold value passed through the superconductor.

By the beginning of the century, the form of the law correlating electrical resistance to temperatures (even at the liquid hydrogen range) was unknown, despite the successful theory of electrical (and thermal) properties of metals proposed by Riecke and Drude.[33] They treated the electric current in a metal as a drift of an electron gas under the influence of an electric field. Lorentz's theory of electrical conduction had, as a starting point, the statistical theory of Maxwell and Boltzmann, and he investigated the dynamics of the collision processes. Nevertheless, his theory could not account for the rapid fall in resistance at extremely low temperatures. The low temperature regions became a fundamental difficulty for the electron theory of electrical conduction.

In 1924, Lorentz drew attention to a remark originally made by

Maxwell concerning perfect electrical conductors: if a conductor has no resistance there will be no electric field inside it even when there is a current flowing. The physical meaning of this result was that any change in the external magnetic field induced currents on the surface of the metal, and the magnetic field of these currents inside the metal compensated for the change of external field, thus keeping the field 'frozen-in' the metal. This physical assumption was regarded as being so self-evident that there was no systematic experimental study of the predicted phenomenon. Kamerlingh Onnes & Tuyn's (1926) experiments with a lead sphere seemed to confirm this idea, but the sphere was hollow, and if one used hollow spheres or a sphere with some imperfections there would indeed be a frozen-in field, although smaller than that predicted by the simple theory of frozen-in fields.

It was Bloch who, in 1928, proposed a satisfactory electron theory of conduction on the basis of wave mechanics. The electrons in a metal were considered to be uncoupled, though the field in which any one electron moved was found by an averaging process over the other electrons. If the metal was at absolute zero, its lattice determined a periodic potential field for the electronic motions, and the electrical resistance by the immobile lattice was zero. An electron could move freely through a perfect crystal, and a finite free path could be due only to the imperfections in the lattice. In general, the imperfections were caused predominantly by the thermal motion of the atoms and were strongly temperature-dependent, increasing with increasing temperature. Impurities, however, also scattered the electrons, but in this case the free path would not vary appreciably with temperature. The resistance therefore consisted of the impurity resistance and the resistance due to the thermal motion of the atoms. According to Bloch's analysis of the motion of an electron in a perfect lattice, all the electrons in a metal could be considered to be free, but it did not necessarily follow that they were all conduction electrons. This theory accounted for metals, semiconductors and insulators, but not for superconductors.

But even though, in 1928, there was a successful theory of electrical conductivity, superconductivity, regarded as a phenomenon of infinite conductivity, was still not understood. Bloch himself tried unsuccessfully to solve the problem during 1928–1929. His interpretation was suggested through analogy with ferromagnetism. He showed that the most stable state of a conductor, in the absence of an external magnetic field, was a state with no currents. Since superconductivity was a stable state displaying persistent currents without external fields, it was difficult to see how a theory for superconductivity could be constructed. 'This brought me to the facetious statement that all theories of superconductivity can be disproved, later quoted in the more radical form of "Bloch's theorem"; superconductivity is impossible.'[34]

In 1930, the thermal conductivity of a metal in a superconducting state was found to differ from that of the same metal at the same temperature, in which superconductivity had been destroyed by a magnetic field.[35] The disappearance of resistance was no longer an isolated phenomenon, but was found to be accompanied by other changes, and these experimental results gave hints that a thermodynamic treatment of the transition to the superconductive state might be feasible.

In 1932, Keesom (with van der Ende) found a jump in the specific heat of tin at its critical temperature. This prompted Ehrenfest, in the last year of his life, to introduce the notion of phase transition of second order. Rutgers suggested its application to superconductivity. Gorter proceeded to calculate the difference in the Gibbs function of a superconductive sample in zero magnetic field and of the same sample in the normal state. Making this simple calculation as a straightforward application of 'Keesom's undergraduate course in thermodynamics', Gorter (1933*a*) obtained the field at which disturbance of superconductivity was thermodynamically possible, but did not discuss whether the mysterious irreversibility of the transition would shift it to still higher fields. The results were published in a short note in the rather obscure *Archives du Musée Teyler*. Rutgers declined when asked by Gorter to coauthor the paper, thinking that such a thermodynamic treatment could not be justified because of the implied irreversibility of the phenomenon of superconductivity. Also, it was during the same year that Landau attempted to show that the resulting superconductive state could have lower free energy than the state of random motion. Assuming the saturation current density to be uniform, Landau showed that it was possible to find a balanced system of local currents which would be electrodynamically stable.

The end of old certainties

At the beginning of November 1933, in a short letter in *Naturwissenschaften*, Meissner & Ochsenfeld (1933) presented strong evidence that, contrary to every expectation and belief of the past twenty years, a superconductor expelled the magnetic field. Superconductors were found to be diamagnetic. The letter noted several experimental arrangements, involving either a pair of solid cylinders made of tin or lead or a cylindrical lead tube. In each case, the sample, which was in a constant magnetic field, was cooled below its transition point. When the transition point was reached, a sharp increase of flux was registered. Meissner and Ochsenfeld concluded that the magnetic flux in the specimen did not remain constant, but the lines of force were driven out of the superconductor, thereby increasing the flux in its neighborhood.[36]

The assumption that there was a perfect shielding of the superconductors by their persistent currents ceased to be valid. In the experiment by Meissner and Ochsenfeld, it appeared that the magnetic field was pushed out after the transition to the superconducting state, and the magnetic flux became zero. The phenomenon of transition to the superconducting state turned out to be a reversible phenomenon: it did not matter whether the transition to the superconducting state had been realized in the presence of an external magnetic field or in the absence of such a field. The superconductor could no longer be considered to have a permanent memory of the magnetic field which was present when the transition was established and, thus, the transition could now be considered to be independent of its past history.

The history of the experiment itself, though, was not as incontestable as the results themselves. Some details are necessary in order to assess the later dispute and tension between London and von Laue.

In 1931, de Haas and Voogd had performed experiments in order to examine the reappearance of resistance in a monocrystalline tin wire with the magnetic field both parallel and perpendicular to the wire. From earlier experiments it was found that the threshold value was independent of the direction of the magnetic field. De Haas & Voogd (1931) found that the sharp transition occurred only when the field was parallel to the wire. When the field was perpendicular to the wire, resistance started to reappear when the field strength was only $1/2\,H_T$. Then, it increased gradually and the normal value was attained at the same field strength in both the parallel and perpendicular cases. Von Laue (1932) drew attention to the perfect shielding property of a superconducting body. In 1932, von Laue received the German Physical Society's Max Planck Medal, and in his acceptance speech, he gave an explanation of the difference between the effect of a longitudinal and of a transverse magnetic field. If one had a superconducting cylinder and switched on a transverse field, then there would be no induction inside the cylinder because of the eddy currents. The lines of force are compressed near the cylinder and the maximum tangential field would be twice the external field. So, von Laue reasoned, the field near the cylinder would reach the critical value when the external field was only half the critical value. Then it would start to penetrate and there would be a kind of transition. No such compression of lines of force could be found for longitudinal fields. De Haas was intrigued by such an explanation and decided to repeat the measurements. He reasoned that to see whether von Laue was right he would perform a different kind of experiment. If he kept the field constant and changed the temperature, then there should not be any field expelled, the lines of force would go straight through the wire and it should make no difference whether the field was transverse or longitudinal. These measurements were taken in February 1933 and they were reported at the *Physiker Tagung* at Leipzig,[37] a colloquium organized by Debye. De Haas found that one got essentially the same sort of behavior in a transverse field, independent of whether one kept the field constant and changed the temperature or whether one started cooling the sample in a zero field and then switched on the field. De Haas did not draw the conclusion that the magnetic field distribution was the same in both cases, since the belief in the frozen-in field was too strong and, as Casimir said, 'de Haas advocated another solution: von Laue's theory is not good'.[38]

But de Haas, who (together with Keesom) was a co-director of the Physical Laboratory of the University of Leiden, always contested that he had prepared the way for the discovery of the Meissner–Ochsenfeld effect. In 1937, London sent his booklet *Une Conception Nouvelle de la Supraconductibilité* to de Haas, who was annoyed by London's failure to discuss his role in the discovery of the diamagnetic character of superconductors. According to the Leiden people the correct order of events was as follows. The experiments of de Haas and Voogd on the monocrystalline wires in the transverse and longitudinal field had given von Laue the impetus for his observations on the field distribution. Although de Haas did not conclude that superconductors were diamagnetic, his experiments paved the way for Meissner's discovery, since he emphasized that the field distribution around the superconductors would have to be examined closely. De Haas would be satisfied with a reference to the effect that the new interest for the distribution of fields was based, at least partially, upon the measurements of the resistance of the copper-crystal wires. Casimir, who conveyed de Haas' views to London, claimed that Meissner started his experiments after (and as a result of) von Laue's observations. Both de Haas and Casimir accepted that Meissner was the first to show directly in an experiment, and in a manner that was unobjectionable, that the versions that were circulating about frozen-in fields were incorrect, and that de Haas' experiments in 1933 were the first to verify Meissner's discovery. Casimir wrote: 'Naturally de Haas accepts that he missed the essential discovery. But he thinks, rightly so, that the papers that have played an important role, have been totally forgotten now'.[39] London doubted Casimir's assertion that de Haas' experiments of 1933 were the first to verify Meissner's discovery. Initially, the published results of the measurements were related to totally different kinds of experiments. In other words these were experiments performed during the formation of the magnetic field, where the Meissner effect could not be observed.[40]

The thermodynamic treatment

Meissner's results were very exciting indeed. 'I myself am working on the phenomenological theory of superconductivity. Meissner's new results agree very well with my views and I will soon publish another note on the subject' wrote an enthusiastic Gorter to van Vleck.[41] Gorter immediately sent a note to *Nature*, suggesting $B = 0$ to be a general characteristic of superconductivity. This meant that the condition $B = 0$, assumed in the previous thermodynamic treatment, was not a restrictive hypothesis. In other words, after the Meissner–Ochsenfeld result, a superconductor could be regarded as being a perfect conductor as well as a perfect diamagnet. Hence, the phenomenon turned out to be reversible and thermodynamics could be justifiably used. In fact, Fritz and Heinz London initially moved along such a direction. Fritz London later told Gorter that, after reading Meissner's and Ochsenfeld's note, first they proceeded to do the thermodynamic analysis, but they were disappointed to find out that identical results had already been published in what 'he was right to consider a rather obscure journal'.[42]

Gorter & Casimir (1934) proposed a two-fluid model in which the particle and current densities were expressed as the sum of normal and superfluid (superconducting) components. The normal component, subject to the usual dissipation, came from electrons excited out of the condensate. Gorter and Casimir mentioned the four different 'investigations in the last year, which gave rather unexpected results which have considerably increased our knowledge of phenomena'.[43] First, there was the observation that there was a discontinuous change in the specific heat of tin in the transition temperature. Secondly, there was de Haas' discovery that the influence of a transversal magnetic field on the resistance of a wire of elliptic cross section, depended on the orientation of the cross section in the field. Thirdly, there was the Meissner–Ochsenfeld effect. Finally, there were recent results by de Haas and Mrs Casimir, who had studied the local distribution of the magnetic field inside a monocrystalline cylinder of tin at various external fields and temperatures. De Haas had shown them the results before they were published. The results showed that the magnetic field had the tendency to disappear entirely when the external field or the temperature were low enough. The remaining internal magnetic field was inhomogeneous and disappeared in a transverse field at first in the exterior parts of the cylinder. Taking into consideration all these developments, Gorter and Casimir developed a thermodynamic treatment of phase change, assuming $B = 0$. 'Though certainly the assumption of $B = 0$ cannot be considered as rigorously proved by all these measurements, this assumption offers undoubtedly the most simple and elegant way of explaining qualitatively the phenomenon observed, with which it is never in contradiction.'[44]

The Londons' theory of superconductivity

During his first months at Oxford University, London was not working on anything specific. In fact, when Hartree invited London to give a lecture at Manchester University, London unenthusiastically agreed to go, warning Hartree that 'in the last few years I have not succeeded in doing very much worth speaking about'.[45] However, he did not present his planned talk about superconductivity, because of unpleasantness surrounding objections raised by Peierls and Bethe concerning his way of determining the number of superconducting electrons. He wrote to Hartree as soon as he was back in Oxford that he had learnt very much during those days of rather continuous discussions and 'though the criticism of my German colleagues was sometimes too hard, it was at least instructive. Of course, scarcely was I in the train, when I meditated on the objection by which I was "murdered" by Bethe and Peierls, and, near Barrington, rising from the dead I saw I was quite right'.[46] Bethe and Peierls wrote a letter of apology to London which they signed the 'pseudo-murderers'. They had told Hartree and Bragg that they were wrong and that they had 'murdered' him unnecessarily. 'We take everything back, your suggestion is absolutely correct, we are idiots and the laws of nature have been explained without any contradictions.'[47] London answered back immediately: 'Murderers, I have not died at all!'[48] London told them that he thought they were joking rather than criticizing his proposal. He held them responsible for his unsuccessful debut, since he had not delivered his talk because of the objections Bethe and Peierls had raised. Hartree also wrote to London to say how glad he was that London was able to straighten things out and that Wigner, Bethe and Peierls, with whom he discussed the problem of determining the number of superconducting electrons, thought that London's answers were quite right.[49] That was early in the summer of 1934. By the fall, Fritz and Heinz London had their theory.

On many occasions, Fritz and Heinz worked together for hours. Much to Heinz's dislike, Fritz would continuously push Heinz to write down his thoughts in detail. Usually, Heinz would stay in the laboratory till very late, so that he could have an opportunity to talk with Fritz during the day, but, at the end of August 1934, Fritz and Heinz were also involved in long conversations during the nights as well. In a few days they were able to formulate their theory. Edith London's reminiscences are particularly vivid:

> Our house was a two storey house. I was in the kitchen cooking and suddenly the upstairs door was opened by Fritz. 'Edith, Edith come, we have it. Come up, we have it'. And maybe the wind closed the door. I do not know what had happened upstairs. I left everything, ran up and, then, the door was opened in my face. On my forehead I had a bruise for a week. Fritz said 'The equations are established. We have the solution. We can explain it'.

The Londons assumed that the diamagnetism must be taken to be an intrinsic property of an ideal superconductor, and not merely a consequence of perfect conductivity. They proposed that superconductivity demanded an entirely new relation in which the current was connected not with the electric, but with the magnetic, field. The breakthrough came when they realized that the original acceleration equation proposed by Heinz in his doctorate, and which involved a relation between time derivatives of the current and the magnetic field, could be integrated without having to add a constant of integration. Such an assumption would lead to the electrodynamics of a superconductor which were consistent with both the zero resistance and the Meissner effect. By the end of September 1934, Fritz and Heinz London had formulated the phenomenological theory of the electrodynamics of a superconductor. Their paper was sent to the Royal Society on October 22, 1934, by Lindemann–the referees needed less than a month to accept the paper with no major revision–and it was published in the Proceedings of the Royal Society in 1935.

The 'obvious' thing to do with the Meissner–Ochsenfeld result was to try to fit it into Maxwell's electrodynamics, but, with the permeability changing to zero, the equations became indeterminate. The first such attempt to supplement Maxwell's equations was made by Becker, Heller & Sauter (1933). They argued that in a superconductor, or rather in a body without any resistance, one cannot have any change of magnetic field, and they pointed out that, because of the inertia of the electrons, an applied electric field would accelerate them steadily. But the Londons objected to such an approach, feeling that the equations proposed by Becker, Heller and Sauter implied more than 'is verified by experiment'.

> Presupposing an acceleration without any friction implies a premature theory, the development of which has presented a hopelessly insoluble

problem to mathematical physicists. . ., [Our] new description seems to provide an entirely new point of view for a theoretical explanation.[50]

For a conductor with no resistance, Lorentz had shown that there would be no electric field inside it even when there was a current flowing. Maxwell's second equation of the electrodynamic field took the form:

$$\frac{\mathrm{d}\boldsymbol{H}}{\mathrm{dt}} = -\, c \text{ curl } \boldsymbol{E} = 0$$

and after integration:

$$\boldsymbol{H} = \boldsymbol{H}_0 \qquad \text{III-1}$$

where $\boldsymbol{H}$ was the field in the specimen when the latter lost its resistance. If there are n electrons per cm^3 of mass m, charge e and velocity v, the current density $J = nev$, and

$$\mathrm{E} = \frac{4\pi\lambda^2}{c^2}\frac{\partial J}{\partial t} \qquad \text{III-2}$$

where λ is a constant. Taking the curl on both sides of equation (III-1) and using Faraday's law

$$\frac{4\pi\lambda^2}{c^2} \text{ curl } \boldsymbol{J} = -\frac{\partial \boldsymbol{H}}{\partial t} \qquad \text{III-3}$$

Substituting in Maxwell's equation curl $\boldsymbol{H} = (4\pi/c)\mathrm{J}$,

$$\lambda^2\nabla^2\frac{\partial \boldsymbol{H}}{\partial t} = \frac{\partial \boldsymbol{H}}{\partial t} \qquad \text{III-4}$$

Integrating with respect to time, equation (III-4) becomes

$$\lambda^2\nabla^2(\boldsymbol{H} - \boldsymbol{H}_0) = \boldsymbol{H} - \boldsymbol{H}_0 \qquad \text{III-5}$$

where $\boldsymbol{H}_0$ is an arbitrary field – the field which happened to be inside the body when it last lost its resistance. Therefore, the general solution of equation III-4 meant that, practically, the original field persisted in the superconductor for ever.

As is often the case, the success of phenomenological schemata usually depends on finding a clever way of incorporating constraints which have been either experimentally found or theoretically proposed, into equations that are otherwise too general. Fritz and Heinz noted that equation III-1 had rather interesting implications.

From the magnetic properties of a perfect conductor the simpler result $\partial\boldsymbol{H}/\partial t = 0$ (equation III-1) was obtained instead of equation III-4. The novelty of equation III-4 was in showing that the value $\partial\boldsymbol{H}/\partial t = 0$ (or $\boldsymbol{H} = \boldsymbol{H}_0$) was also to be found only at a depth inside the metal greater than λ. Indeed, the solutions of this equation decreased exponentially as one receded from the surface, where they were fitted into the values of the external field. There was no point in developing this form of the theory any further, for equation III-3 merely led to the equation $\boldsymbol{H} = \boldsymbol{H}_0$ with the modification that the magnetic field penetrated the body to a small but finite depth. The Londons proposed that the connection between magnetic field $\boldsymbol{H}$ and current density J_S for the pure superconductive case may be given by the equation

$$\frac{4\pi\lambda^2}{c^2} \operatorname{curl} \boldsymbol{J} = -\boldsymbol{H} \qquad \text{III-6}$$

Equation III-6 can be obtained by time integration from equation III-3 if it is assumed that the constant of integration is zero ($\boldsymbol{H}_0 = 0$) and it was considered as a completion of Becker, Heller and Sauter's formalism by fixing the integration constant of the magnetic field according to the Meissner effect.

Equation (III-6) led to:

$$\lambda^2 \nabla^2 \boldsymbol{H} = \boldsymbol{H} \qquad \text{III-7}$$

For large specimens, the characteristic feature of the solutions to this equation is that they decay exponentially into the interior of the specimen. At a distance λ from the surface, the field is practically zero. Meissner's experimental result is represented by equation III-6 with one restriction, namely that the magnetic flux decreases, not abruptly on the surface, but continuously in a very small interval below the surface. Equations III-2 and III-6 described the zero resistance and the Meissner effect respectively. Equation III-6 says more than equation III-3, so far as it includes the Meissner effect. Proceeding from equation III-6 to III-3 by differentiating with respect to time, it is not possible to deduce equation III-2. Nevertheless, the following weaker statement is obtained from equation III-3:

$$\operatorname{curl}\left(\frac{4\pi\lambda^2}{c^2}\boldsymbol{J} - \boldsymbol{E}\right) = 0 \qquad \text{III-8}$$

which shows only that

$$\frac{4\pi\lambda^2}{c^2}\boldsymbol{J} - \boldsymbol{E} = \operatorname{grad}\mu \text{ or } \frac{4\pi\lambda^2}{c^2}\left(\boldsymbol{J} - \operatorname{grad}\frac{c^2\mu}{4\pi\lambda^2}\right) = \boldsymbol{E}$$

where μ is a scalar. On the other hand, equation III-2 leads not to equation III-6 but only to its time derivative, equation III-3. Thus, the propositions equation III-2 and equation III-6 'possess, so to speak, the same degree of generality'.[51]

It is not, then, unreasonable to take equation III-6 to be more fundamental than equation III-2, and this was an indication that a supercurrent could be regarded as a kind of diamagnetic current. In examining the relation between the behavior of a ring and the Meissner effect, Fritz London showed that equation III-6 could be expressed in such a way as to provide some clues to what was required of a fundamental theory of superconductivity. He suggested that the entire superconductor behaves as a 'single big diamagnetic atom'. He then went on to argue that if the ground state eigenfunction is *rigid* and, thus, not modified very much by an applied magnetic field, the current density would be proportional to the vector potential and would give the equation which describes the Meissner effect.

Fritz and Heinz London supposed that the electrons were coupled by some form of interaction: 'Then the lowest state of the electron may be separated by a finite distance from the excited ones'.[52] This may have been the earliest suggestion of an energy gap.

Some years later, in 1937, London confided to Casimir that the work of Gorter and Casimir had exerted considerable influence on the Londons' work: 'and if one had to give historical evidence, I would feel obliged to refer explicitly to those works that were *really* decisive for us'.[53] London continued the letter by expressing some thoughts about the origins of his work in superconductivity which are not to be found in any other document left by him.

> The paper by Gorter, and that of Gorter and youself, made a strong impression on me at that time, and incited me to engage myself with superconductivity. It is true that the Meissner effect could have been predicted from Gorter's ideas. The fact that the development of things did not take place this way and needed an experimental push, always appeared to me a sign that the acceptance of the *reversibility* was not at all self-evident, because of the fact that all the experiments displayed hysteresis and other non-reversibilities. It was at that time only a *hypothesis* that was constructed in the dark . . . and it was not proper to interpret in *exactly* that manner the objective non-reversibility and to pinpoint in *that* particular manner the assumption of reversibility. This is why the verification of this magnetothermic phenomenon seems to be so important for me. Because it means something more than a mere verification of thermodynamics. It also means, as far as I know, the verification that what we *assumed* to be reversible is *indeed* reversible.

At the beginning of October 1934, London was invited by Mott to talk at Bristol University. London wanted to postpone the seminar in Bristol for some weeks, since, he said, he would be ashamed to go with nothing to report.

> I believe in the past weeks to have made a real progress with the beloved supraconductivity, and I should like to discuss it with you when I come to Bristol. At the moment the matter is so very strange and displays so many new aspects every day, that I think it would be better to wait some weeks until I make clarifications. Now I am too occupied by the matter and too enthusiastic.[54]

Obviously, he wanted to be sure that he and Heinz would finish the manuscript and that it had been accepted for publication before giving the seminar. Mott replied: 'It is very exciting that you have something new about supraconductivity. We had Gorter here recently, and he told us about these funny thermodynamic things. We shall look forward very much to hearing about your theory'.[55] The talk was given on November 19, 1934.

Gorter and London made plans to meet in England in the fall of 1934. They could not meet, though, both regretting it very much, especially as London had 'so many questions to discuss' about superconductivity and Gorter 'had thought so much about it'.[56] But it may have been better that way. 'One often reaches a goal faster by doing some good thinking, rather than by premature conversations.'[56] London wanted Gorter to be the first to hear about the Londons' theory, before discussing it in public in Bristol. He wrote to Gorter that he and Heinz had made some progress and even though, as yet, they had not found an explanation for superconductivity, they had been able, as London wrote somewhat cryptically, to overcome a serious hurdle with the electrodynamics of the superconductor. Gorter could not exactly guess which aspects of the problem the Londons had solved. 'The fact that you do not promise to have a full theory, means that Bloch's dictum that all theories of superconductivity are contradictory, does not apply in your case.'[57] He arranged for London to give a seminar at Leiden; and London wrote:

> We are not sure at all that Bloch's dictum does not concern us. Still this cannot be an excuse for you not to subject us to 'full' criticism. I hope you do not expect too much, because the only thing we did was to think about what can substitute Ohm's law for a superconductor, and a lot of unexpected things result from this.[58]

In the letter with the details of London's visit to Leiden,[59] Gorter could not overcome the temptation to ask whether they had substituted Ohm's law for the equation:

$$\frac{\mathrm{d}J}{\mathrm{d}t} = \frac{e^2 \boldsymbol{E}}{m}$$

By this time the paper by Fritz and Heinz had already been accepted for publication, and Fritz sent the manuscript to Gorter at the beginning of December 1934. Gorter eventually became quite enthusiastic about the treatment of superconductivity by the Londons, considering the whole thing to be a 'very elegant treatment [which] has not yet found any direct quantitative confirmation or refutation'.[60] Soon after London's visit to Leiden, Gorter visited Germany and explained the Londons' theory to Meissner and Sommerfeld, who became very interested in it.

Initial reactions by von Laue

From the very beginning, the Londons were very keen to inform von Laue about the developments. Von Laue had visited Oxford at the end of May 1934, and he had stayed with the Schrödingers. He had met Fritz London, but they had not discussed anything about superconductivity. Von Laue's first reactions were encouraging: 'I read your work last night till 1.30 with increasing interest. It is a view which is in total contrast to your work up to now, and for this reason alone it has tremendous interest by itself'.[61] In 1935, after the talk at Leiden University, London talked about the theory in Paris. He wrote to von Laue that the audience, especially Brillouin, was at first very distrustful of the theory, but he felt that he had convinced them.[62]

When von Laue's detailed answer arrived, the two brothers had gone to see *The Importance of Being Earnest* where a young actor, Laurence Olivier, had received impressive reviews. Von Laue's letter disappointed them. He was sceptical about the thermodynamic implications, especially when considering the transition from the superconducting state to the normal state. He felt that their theory might have implied that an Ohm resistance appeared slowly during the transition, and, thus, created Joule heat. If von Laue's objections were correct, then the transition was non-reversible.[63] Furthermore, von Laue could not understand the proof that heat production was always positive. On the contrary, he could show that there were cases where heat production could be made positive or negative – something which would be fatal for the theory, since it would mean that heat could be transformed to electrical energy without any other change.[64] Von Laue informed London that his work would be discussed at their weekly colloquium at Berlin University.[65]

During March 1935, Edith visited her parents in Berlin for a few days. Fritz had informed von Laue about the visit and said that she would contact him. The meeting was disappointing, since von Laue was rather formal and cool, and it was a short and uneasy encounter. After some small talk, he told Edith that he did not believe the Londons' theory of superconductivity was correct. Both Edith and

Fritz London were greatly upset by such a response, not least of all because of the special respect they had for von Laue's political stand and for his helping Jews who wanted to leave Germany – in fact, he had helped them to take some money, over the allowed limit, out of Germany when they left in 1933.

London sent von Laue a proof about the positivity of the heat energy which could only be proven approximately. To answer von Laue's objections partially, the Londons were ready to sacrifice the spatial charges, since they were convinced that they did not exist. Ideas similar to von Laue's proposal were discussed in a paper by Fritz and Heinz London in *Physica*.[66] Von Laue did not agree with the proof and told London that he intended to publish a short note to this effect, sending him a copy of the manuscript. Eventually, London agreed that he was wrong and the positivity of the energy could not be proven with only the assumptions of the theory.[67] Von Laue, then, suggested that, instead of publishing the note alone, the three of them, Fritz, Heinz and von Laue, should publish a short article. Fritz felt that he was cornered and he was annoyed. He did not want to give von Laue the impression that, after the success of their theory, the Londons were now as famous as their peer, or that they were snobbish because of a joint publication with a Nobel prize winner. London thanked von Laue for suggesting the common publication and said that the Londons would prefer not to publish in a German journal, but von Laue insisted that they should publish the note in *Zeitschrift für Physik*.[68]

The Londons had claimed that the quasi-Joule heat would occur at those parts of the surface of the superconductor where the current went in and out. Von Laue disagreed with this conclusion, since the potential in the Londons' theory was related to the spatial current density which was not the same as the surface density. He suggested that there should be a footnote indicating their disagreement, unless they all adopted von Laue's position.[69] London disagreed with the inclusion of such a footnote, especially since it was not such an important point and because he thought that such matters were sufficiently objective to be settled by discussion. He also implied that von Laue often changed the focus of his objections.[70]

Von Laue continued to disagree about the possibility of surface heat production, but when London proposed a new form of the boundary conditions, von Laue decided to withdraw his objection and to allow the suggestion to be included in a footnote.[71] London admitted that there had been a mistake in their original paper on superconductivity. The change from spatial to surface charges was related to heat production. The thermodynamically stable state was the one with only the surface charges. In the original paper, they had wrongly thought that they were obliged to exclude surface charges at the boundary surfaces with the normal conductor.[72] Nevertheless, London was emphatic that the novel aspects of the theory were untouched by those problems: they were able to find a way out of the Bloch–Landau dilemma; now, the supercurrent could be viewed as a diamagnetic current in the experiments of Kamerlingh Onnes and Meissner; the magnetic field which sustained the diamagnetic current in the case of the experiment with continuous currents was shown to be affected by the supercurrent itself. Their joint paper, which included all these clarifications, was eventually published in 1935.

It is quite tempting to comment that von Laue never came to terms with the idea that the Londons could find a way to circumvent Bloch's impossibility theorem. It is quite clear that von Laue engineered the publication of the joint paper by putting rather strong psychological pressure on the Londons, after first expressing some doubts about the validity of the theory and, then, telling Fritz London that he was planning to publish a short note with his objections. And when London tried to answer the objections, von Laue immediately proposed that the three of them should publish a paper jointly. Despite his annoyance, London must have thought that a joint publication would have restrained von Laue from making public his objections. London was fully aware of what it meant to have a successful answer to the problem of superconductivity – especially during a time when he was so desperate about his professional prospects. And he also knew the damage that would be done to a newly formulated theory by a public challenge from von Laue. So a common publication suited all of them. When von Laue proposed that a footnote expressing his objections should be included in the joint paper, he was surely aware that such a format violated the rules of the game. But it was a good negotiating tactic by von Laue and, as we will see later, a good investment for the future. Evidently, von Laue would withdraw such a proposal after London's attempt to provide an answer to his objections. Von Laue appeared to have been successful on both fronts. Their joint publication gave the impression to the community that von Laue was in the game all along, or, at least, that the Londons needed von Laue's assistance to get over some difficulties and to strengthen the theory further; von Laue, on the other hand, did not appear to the Londons to have surrendered all his objections for the sake of a joint publication, but, on the contrary, appeared to have conceded half-heartedly without being fully satisfied with London's answers to his objections. In all this, London was determined to retain control of the situation – unlike the way he had dealt with the problem of the chemical bond. Of course, von Laue must have, justifiably, felt that, in his interventions concerning the de Haas–Voogd experiments, he had more or less foreseen the Meissner–Ochsenfeld result. The complete lack of acknowledgement of such a role by the Londons did not make the situation any better either. But the Meissner–Ochsenfeld result was so incontestable that for a publication in physics there was no need to refer to the history of the experiment.

London had miscalculated the situation once again. Less than a couple of months after the appearance of their joint paper, von Laue published a short note in the *Physikalische Berichte* containing his objections to the Londons' theory. London was fuming. In a long letter to Herr Professor, he told von Laue that he was annoyed about the note, which seemed to him to contradict the position taken in their joint paper. He was under the impression that the explanations provided were mutually satisfactory and, hence, did not warrant von Laue's comment in his article that he hoped the theory would be changed to avoid any shortcomings. London noted that he did not believe that a difference of opinion about scientific matters with von Laue was anything serious and considered: 'a scientific duel with you as an honourable thing'. But in this particular case the situation was made very difficult by the fact that von Laue had implicitly criticized their joint paper and gave the impression that it had the Londons' approval. London urged von Laue that if he wanted to distance himself from their common work, he should say so explicitly.

> Otherwise, it seems inevitable to me that misunderstandings will appear and this is separate from the fact that our scientific credit is damaged at its most sensitive period, if the opinion is established that the considerations upon which we are based, are to be seen as 'proven unacceptable' even if there is not even a trace for justifying such an opinion.[73]

Von Laue's answer to London came a month later and only after he had been sent a reminder. He told London that he had not answered his letter because he felt 'a bit angry'. The short notes were not all that important, and that anyone who was interested in their work after reading about it in the *Physikalische Berichte* would undoubtedly go and read the original papers.[74] London thought that such an argument was quite naive: if he himself read that a particular theory had unacceptable points, he would not make the effort to find the original papers.[75] Von Laue wanted to minimize the incident and thought that it was London's 'precarious position and the mood caused by it that [showed] these things in an exaggerated magnification'.[76] Von Laue's unwillingness to discuss the issues raised in London's letter made the whole situation worse. An embittered London felt a strong need to defend himself. He felt shocked that his 'undoubtedly precarious position' could be used as an argument against his scientific position and that a psychological explanation was given for the points he had raised in his letter. 'The "undoubtedly precarious position" in which I find myself has in no way affected my feeling for right and wrong.'[77]

Meissner became unwittingly involved with the whole mess. He planned to publish an article where, among other things, he noted that the Londons' theory needed certain changes, taking into consideration a number of impossible results which von Laue had described in his short article. But, of course, these changes had already been incorporated by Fritz and Heinz London before von Laue's publication. London pointed out to Meissner that von Laue's objections did not affect the basic assumptions of their theory. He reminded him of the conversation they had had at the Royal Society in London when he had told Meissner that 'it was agreed with Laue to have a common neutral publication. I wonder how come you have neglected this knowledge'.[78] Meissner was very sorry that he had not read all the subsequent papers and he said he was influenced by a letter sent to him by von Laue where he expressed his objections. He promised to insert the correction onto the preprints.[79] London thanked Meissner and told him it was not necessary to do this, since, in the future, he would have ample opportunity to discuss these points again.[80]

The discussion at the Royal Society

Fritz London was invited to take part in the discussion to be held at the Royal Society on *Superconductivity and other low temperature phenomena* on May 30, 1935. Among the other speakers were McLennan, Bohr, de Kronig, Brillouin, Meissner, Keesom, de Haas and Mendelssohn. London's primary aim during the meeting was to convince his audience about how their theory avoided the Bloch–Landau dilemma. Bloch and Landau 'have formulated the theoretical programme which seems to be indicated by the facts . . . We shall see that [their] "insoluble" problem has really never been set by nature and the interpretation of the facts has been in some way premature'.[81] London intended to show that superconductivity should not be interpreted as infinite conductivity and that it was possible to formulate the phenomenon in such a manner as to be consistent with the 'recognized conceptions of physics'. Hence, his program had a dual purpose: to (re)formulate the phenomenon and then to provide an explanatory schema. Such a theoretical approach was a reminder that London had not forgotten the rules he had set so systematically in his philosophical essays of the early 1920s. Understanding, London had insisted in his school essays, proceeded by two different transformations: T_1 took us from reality to experience; T_2 took us from experience to language.

According to London, to be insensitive to such differentiations led to premature interpretations of phenomena which may indeed seem to be indicated by the facts, but, which, may have never been set by nature.

London's reasoning relied on the analogy with diamagnetism rather than ferromagnetism. The most stable state of a diamagnetic atom was one where there was a flow of permanent current in the presence of a magnetic field. Such a phenomenon became immune from the primordial curse. Bloch's theorem dealt with situations with no external field. A small crack had, at long last, appeared in a framework which seemed to be so thoroughly sealed by Bloch's theorem. It was obviously the case that diamagnetism could not be considered to be the result of infinite conductivity. But surely nothing prevented the examination of the inverse situation. If diamagnetism was the fundamental feature, then perhaps superconductivity could be understood.

In attempting to 'sketch the programme which seems to be set by our equations for a future microscopical analysis',[82] London regarded 'the total supraconductor ... as a single big diamagnetic atom'.[83] He took the general expression for the electric current density to be

$$J = \frac{he}{4\pi im}(\psi \text{grad}\psi^* - \psi^* \text{grad}\psi) - \frac{e^2}{mc^2}\psi\psi^* A$$

where ψ is the wave function of a single electron in the self-consistent field of the others. The $\psi\psi^*$ gave the value of the statistical expectation for this electron at every point in space. The magnetic field was described by the vector potential A. In the absence of a magnetic field ($A = 0$), $\psi = \psi_0$ and the current density vanished. But, in a normal metal, the wave function responded to the magnetic field in such a way that there was cancellation between the paramagnetic contribution involving the gradient terms and the diamagnetic contribution proportional to the vector potential, A, leaving only a very weak diamagnetism. London proposed that, in a superconductor, the wave function displayed a rigidity such that it was essentially unmodified by the magnetic field. If the wave function was unchanged, the paramagnetic contribution vanished, even in the presence of the field. This left the diamagnetic term which gave an equation similar to the Londons' phenomenological equation:

$$J = -\frac{ne^2}{mc}A$$

This was a particularly significant result. London suggested that the average momentum did not change in a superconductor when the field was applied because of a long range order which maintained the local average value of the momentum constant over large distances in space. This order would be maintained even in the presence of the magnetic field. The ordered ground state was regarded as a single quantum state extending throughout the metal. It was these considerations which led London to present for the first time his views about superconductivity as a macroscopic quantum phenomenon.

The opening paper of the discussion was given by J. C. McLennan from the University of Toronto, Canada. McLennan had played an important role in the early days of low temperature research and, after the First World War, Toronto had become the first place outside Leiden where helium could be liquefied.[84] In his opening address, McLennan concentrated mainly on the experimental results, and described in very general terms the (few) theoretical developments. Immediately after the Royal Society meeting, McLennan asked London to give him a short account of superconductivity 'as free as possible from mathematics'.[85] London's answer was, indeed, non-mathematical and a succinct expression of the state of the art at the time. After discussing the significance of the Meissner effect in overcoming Bloch's dilemma, London pointed out the peculiar status of their theory:

> The progress I claim is mainly a logical one: by a new and more cautious interpretation of the facts I tried to avoid a fundamental difficulty (the so called theorem of Bloch) which stood in the way of explaining superconductivity by the customary theory of electrons in metals and which could not be overcome as long as one has considered this phenomenon as a limiting case of ordinary conductivity. Usually from a new theory one expects some predictions for new experiments. But as far as I can see this very probably does not happen in our theory.[86]

Termination of the ICI fellowship

At the end of April 1935, Fritz London was informed by ICI that they could extend the fellowship for only one more year and that their original offer would expire on August 1, 1936.[87] He immediately wrote to his colleagues in Paris, and Edmond Bauer suggested the possibility of a job opening there. London wrote: 'If this happens in Paris, I will consider it great luck'.[88] The prospect of the termination of the ICI fellowship took London by surprise. Both Lindemann and ICI had planned their whole enterprise on the assumption that the responsibilities of government would either oblige the Nazis to change their policies or force them to resign; then, all the scientists would return to Germany. The subsequent developments, though, did not leave much room for entertaining hopes for the return of the scientists.

ICI could no longer subscribe to the argument that short term help would have long lasting benefits and decided to dissociate itself from a role it was quite unwilling to play.

Soon after their joint paper was published in *Zeitschrift für Physik*, London informed von Laue about the termination of the ICI funding. Von Laue left for a lecture tour in the USA – his son was studying at Princeton – and promised London that he would talk to contacts in the USA about London's plight regarding a job.[89] 'I do not have any illusions and my situation will not be an easy one to be dealt with.'[90] The prospect of settling in England was becoming more and more unrealistic. 'In Oxford, as everyone agrees, there is a general resistance. This means that from the beginning assurances were given to limit our interaction with the normal university life. The personal behavior of the English does not give anyone the smallest hope.'[90] London felt that his future lay in the USA, where professional prospects appeared to improve. Von Laue wrote to London during his trip from Columbus, Ohio, to Buffalo, New York. The advice of many people with whom von Laue discussed London's case was that London should actively consider the position in theoretical physics being offered by the Hebrew University in Jerusalem. During the fall of 1935, a few months after London received the upsetting news about the termination of the ICI fellowship, Polanyi informed him that Chaim Weizmann, who was going to visit England, would like to meet him. They met in London during the latter part of October, and Weizmann asked him whether he would be interested in contributing in any way towards the attempts to establish a strong Science Faculty at the Hebrew University in Jerusalem.[91] Edith was quite enthusiastic about such a prospect, because, since going to school, she had been a Zionist. London expressed his sympathy with such an undertaking and told Weizmann that he would first like to visit the place. Nothing happened for a long time, but word about a possible offer to Fritz London got around. Von Laue did not know the details, but warned London to be cautious against exaggerated expectations. He advised that if London decided not to go to Jerusalem, it would be best for him to go to the USA, because the chances of finding a job were quite high, especially since there was considerable interest in superconductivity. Von Laue informed London that not even the non-Aryans who had fought during the Great War were exempt any longer from the stipulations of the civil service law.

> I must tell you a joke which was told to me here. What is the relation between Marxism and National Socialism? Marxism is the non-Aryan grandmother of the latter ... That the *Weltanschauung* of National Socialism means 'The World as Will without ideas' you already knew for a long time.[92]

Although Lindemann had not said anything to London about ICI's decision, London believed he knew something about it. London met Lindemann on September 17, 1935, and discussed the matter with him. Lindemann professed ignorance of the whole affair. All along, London's feeling was that he was wanted at Oxford University in order to establish theoretical physics there. He asked whether Lindemann was still interested in his doing physics at Oxford and whether he could expect any new possibilities to emerge, but Lindemann was non-committal.[93] He said he had not talked to ICI and unless ICI would reverse their decision, Lindemann could not see any other way out; also, he discouraged London from looking for a regular job in any of the British universities. After talking with Lindemann, London was convinced that they wanted to keep Simon and Schrödinger at Oxford University, and for this reason they could not commit themselves to any other cases.

London wrote to Lande, who was at Ohio State University, to inquire whether there might be an opening there, since, in 1929, Alpheus Smith had talked about two positions for theoretical physicists. In fact, any other offer from the USA would have been very welcome: 'This permanent uncertainty undermines one's creativity, and I would be glad to be able to do some physics with you.'[94] Lande did not have any encouraging news for London: Ohio University did not give him a definite answer, and other places like the University of Michigan and Purdue University did not appear too eager, either. Nordheim had not been given tenure and had to leave Purdue.[95] London could not follow Lande's advice to seek a job in the Soviet Union, not because of any prejudices, but because it would have been problematic for his mother and for Edith's parents. Furthermore, he would have had to put in too much time and energy to learn the language. Nevertheless, he was hopeful, since he said: 'one should have patience and something will turn up'.[96]

One of the despairing letters from Oxford was to Wigner. London told him how sorry he was to have to write about his situation, but he was forced to do it, since most of the people who came from Germany contemplated changing countries again. He said there was no prospect of his finding anything in Oxford as long as Schrödinger decided to stay, and any hope that Lindemann would do anything was totally illusionary. 'So everything points towards the USA, and I should try to find something there ... I cannot describe how bad and powerless we feel. And even if one accepts that one should suffer, this cannot be considered a permanent situation, especially when help is given as a kind of charity and not for other reasons.'[97] To add to their troubles, in the fall of 1935 Edith had a stillbirth. His mother, who had come over from Germany to help Edith after the birth, was severely shocked with the stillbirth, and had a mild stroke.

In mid-December of 1935, before going to Bonn to see his mother,[98] London arranged a meeting with the secretary of the AAC, Walter Adams. There was an advertisement for a junior position at Hull University College, England, and Szilard had suggested that London should see Adams to discuss the possibility of applying for the post. The AAC's position on that issue was quite inflexible, and was expressed by Adams in strong language. He told London that such an application would be most unwise. These junior positions were intended for promising graduates and were their first entry into the academic world. If German scientists with senior qualifications competed for these junior positions, it would be perceived by British graduates and other members of staff as unfair competition and would weaken the position of German scientists in general. 'Even if the local committee at Hull were so unwise as to try to get a senior man cheaply merely because he were a refugee, we ought to oppose such action.'[99] Szilard admitted that he had encouraged London to apply: 'chiefly because I was surprised and glad about the enterprising spirit which he unexpectedly displayed'.[100] Eventually, London agreed to follow the advice of Adams and Blackett and not to apply to Hull University College.

At the end of 1935, the possibility of a job emerged at the University of Manchester. Hartree and Bragg were doing their best to find something for London at Manchester. Polanyi was hopeful, but the whole situation remained very uncertain.[101] In early May 1936, London heard from Szilard that there was not much for him at Manchester. He decided to ask Polanyi to tell him exactly what the situation was, especially since Manchester appeared to be a good and realistic prospect, but Polanyi politely refused to pressure Hartree to learn how the situation with London was developing.[102] London wanted desperately to get into something permanent and to start working on a project. He was, thus, not willing to apply for a scholarship from the AAC, as Bragg and Hartree had suggested, since again it would be something temporary. He wondered whether Donnan knew of his case. Donnan was indeed interested in London, but Polanyi thought that already there were many German-Jews at University College, London, which made the prospect of an offer from Donnan not very realistic.

In the spring of 1936, Fritz London was in a miserable state: he was too senior for junior posts in British universities; there was no offer from any other place; he fell out of a railway carriage and broke his ribs; compared with other emigrés, he had great difficulties with the English language; and Heinz was heading for Bristol. There was a job offer from the Peking National University, but he was quite unwilling to accept it. He was hoping, as he wrote to Born, to be offered some kind of a research position from the Royal Society or any other institution in England. 'In the meantime it is totally incomprehensible to me the fact that Lindemann, who three years ago prompted me to stop my negotiations in Paris in order to bring me here, says that there is nothing more he can do for me here except to leave me to the mercy of philanthropy.'[103]

AAC's President, Lord Rutherford, had been aware of the new situation. Adams had informed him that at least fourteen of the consultants were affected by the ICI decision, and that almost all these scientists had quickly appealed to the AAC for help in finding alternative positions. Adams added that not all the recipients had grasped fully the temporary nature of the ICI arrangement.[104] Unknown to London, the AAC was doing its best to secure funds for him. Adams confidentially informed Lindemann that he had heard that Schrödinger had decided to accept the position at Graz. The question was whether there would be any chance of transferring the balance of Schrödinger's funds from ICI to London, even though it would be more satisfactory if a grant for a longer period could be obtained.[105] Meanwhile, Schrödinger had been informed about this possibility by Blackett and he sent a letter supporting such a prospect. He suggested that they should, somehow, find a way to involve London more in university life.

> Would it not be a pity not to make use of his extraordinary abilities for some or other purpose!... Being an active and assiduous man, he would prefer to be allowed to yield some services in return for what he receives, instead of just receiving it in the way of charity. Fritz London is really one of our foremost theoreticians. I am absolutely convinced that something good will come of it.[106]

Adams felt that any other arrangement, except the receipt of an extra amount, would create problems with the Home Office. Finally, it was decided that London would receive £20 a month for six months after the termination of his fellowship.[107]

During the second week of June 1936, London went to Paris, where he tried to secure a job. Upon his return, he met Adams on July 1, and told him of the developments. In a handwritten memo concerning his meeting with London, Adams wrote that London had been offered a research post with Langevin. London did not appear to like the idea of moving again. Adams thought that the offer had a lot to do with politics and the openings of the new government of France which was 'anxious to make a show of freedom'.[108]

It was almost a permanent position at the Institut Henri Poincaré, starting from October 1, 1936. The first year he would receive a grant from the recently formed Comité Français d'Accueil aux Savants

Figure 8. Fritz London (second row, third from the right) at a meeting in Copenhagen, June 2, 1936. Front row from left: Pauli, Jordan, Heisenberg, Born, Lise Meitner, Stern, Franck. Bohr is on the left. Heitler is behind London. (Courtesy of Edith London.)

Etrangers and from October 1, 1937, he would belong to the Caisse National des Sciences having, in effect, a state position.

Eventually, Hartree wrote to London offering him the position in Bragg's group at Manchester University. But London had already made up his mind and told Hartree about the offer made to him in France under very favorable conditions and that he must accept it, even though he would have liked to work with Bragg, since, at the time, he was very interested in the order–disorder researches of Bragg and Williams.[109] Adams communicated with Duncan-Jones who was at Birmingham University and urged him to offer a recently created position there to London, specifying that Oliphant knew London's work and thought highly of him. Nevertheless, both Simon and Adams were not very hopeful that there would be an opening for London in England and encouraged him to accept the position in France.[110] On July 1936, Magat assured London that everything was fine with his appointment in Paris and that all the necessary bureaucratic processes had been completed – including the financial arrangements. He advised him to go to Paris earlier than the beginning of October when his contract started, in order to improve his French.[111]

The end of London's three years at Oxford was rather uneventful. He dutifully thanked ICI for the grant which enabled him to continue his research 'stopped by political events in Germany'.[112] Although many colleagues offered their best wishes for his new job, Sutton, from Oxford, believed that it was a disgrace that Oxford had not been able to find a position for London and he felt that, after his departure, theoretical physics would become insignificant again.[113]

The three years at Oxford were among the unhappiest in London's life. He had been forced to leave Germany too quickly, and, though he felt lucky to have found a job at Oxford, he found himself in an impenetrable environment. As time went on, prospects in England became dimmer, and Lindemann became increasingly unreceptive to his agony. The job in Paris appeared to be the beginning of an exciting period. 'What I got is the best I could dream of; though I cannot yet quite get over the idea that I shall have to leave England again and shall go to another foreign country.'[114]

Notes to Chapter 3

1. *Daily Telegraph*, London issue (April 11, 1933).
2. People dismissed under the *Law for restoration of the career civil service* were not guaranteed pensions unless in office for at least ten years. Although the law was put into effect by November 30, 1933, there was a delay in enforcing provisions touching on organizational matters (March 31, 1934).
3. Unpublished decree, *Tages-Nummer*, 1644/34. During the summer semester of 1933, the structure of the University Administration was basically the same as that defined by the Richter–Peters statutes of 1924–1929 (*Die Statuten der preussischen Universitäten und Technischen Hochschulen*, edited by Werner Richter and Hans Peters, Berlin 1930). The new changes were instituted by a series of laws and decrees during the winter semester of 1933–1934. The faculties exercized their three traditional rights: authorizing a person who had received his doctorate as an instructor (conferment of the *via legendi*, usually followed by the title of *Privatdozent*); the proposal for the promotion of instructors to higher ranks; and the proposal for secondment of scholars from other universities.
4. The book burning at the Unter den Linden Strasse at Opernplatz included books by Thomas Mann, Heinrich Mann, Leon Feuchtwanger, Jacob Wasserman, Arnold Zweig, Stefan Zweig, Walter Rathenau, Albert Einstein, Alfred Kerr, Hugo Preuss, Jack London, Upton Sinclair, Helen Keller, Margaret Sanger, Marcel Proust, H. G. Wells, Sigmund Freud and Emile Zola.
5. Beyerchen (1977), 200.
6. London to his mother (March 1933, exact date not specified).
7. London to Ewald (April 15, 1933).
8. Von Laue to London (April 21, 1933).
9. London to Simon (May 15, 1933). E. must be Einstein, H. is Haber, F. is Franck. Franck had resigned from his post in Göttingen, protesting the Nazi policies. He had fought in the Great War and the decrees did not affect him. He indicated that he was unwilling to leave Germany. His resignation was not accepted, but at the same time he could not have an Institute Directorship. Forty-two members of the University signed a letter declaring that Franck's resignation was

'equivalent to an act of sabotage' and urging the Government to proceed to the necessary 'cleansing measures'. Two days later, the Göttingen press announced that on April 25 the Prussian Ministry of Education had placed on leave six university professors, among them Born, Courant and Noether. That meant that their positions were not officially vacant and that they would be paid their salaries. Franck was one of the last to leave at the end of 1933. He had a guest lecturership at Johns Hopkins University, USA, and a promise of a job from Bohr in Copenhagen afterwards.

10. Helene Solomon-Langevin to London (May 26, 1933).
11. From Moore (1989), as related to Thomas Kuhn by Schrödinger's wife, Anny.
12. When Schrödinger mentioned this to Lindemann during the latter's visit to Berlin in April of 1933, Lindemann was very surprised, since Schrödinger was not a Jew and had such a prestigious job (Moore (1989), Chapter 8).
13. Beyerchen (1977), 65. The reactions and relations of Planck and Heisenberg to the Nazi regime have been extensively discussed (Beyerchen (1977); Cassidy (1991); Heilbron (1986)). London did not have much contact with Heisenberg who was at the University of Leipzig. There is only one letter by Heisenberg to London (June 6, 1928). He invited London to a lecture at Leipzig and was very enthusiastic about the chemical theories of London. They did meet later at some conferences between 1933 and 1939. According to Edith London, in one of those meetings, London was surprised when Heisenberg told him that 'one has to look at what Nazism does for German youth. They now have a purpose in life'. London interrupted him, said good-bye and that was the end of their relationship. They last met at the Kamerlingh Onnes-Lorentz Congress at Leiden in 1953, but did not talk to one another.
14. I agree with Alan Beyerchen's note in the London Archive (Duke University, USA) that this draft may be a letter to Max Planck.
15. London to Ehrenfest (May 23, 1933); Ehrenfest to London (May 26, 1933).
16. Rosenfeld to London (May 10, 1933; May 23, 1933; June 4, 1933).
17. London to Raoul Lambert (May 23, 1933). The Academic Assistance Council (AAC) was founded on May 22, 1933, and in 1936 it was renamed the Society for the Protection of Science and Learning (SPSL). The American counterpart, the Emergency Committee in Aid of Displaced German (later Foreign) Scholars was founded a few weeks later. The AAC's staff was Walter S. Adams, general secretary (later replaced by David Cleghorn Thomson), C. M. Skepper, assistant secretary and Esther Simpson 'whose sensitivity to her charges' needs earned her abiding respect and affection' (Rider (1985), 137).
18. London to Mendelssohn (May 23, 1933).
19. Mendelssohn to London (May 25, 1933). This account differs somewhat from Moore's (1989) report in his biography of Schrödinger. Firstly, there is an inconsistency with the dates. Moore claims the incident occurred in mid-April, but the letter to Mendelssohn was written later. However, mid-April was too early for Lindemann to make any definite offer for an ICI fellowship, since final decisions had not been taken by them. So the story with Schrödinger should be placed about a month later. In fact, Edith London remembered that the first offer to London was when he was in Paris and Lindemann called him there. That was on May 28–29, 1933. Secondly, it is not clear whether or not a firm offer was made. Moore claims that, although Lindemann offered him an ICI fellowship, London wanted to think about it. When Lindemann was talking to Schrödinger and told him about London's attitude, Schrodinger's response took Lindemann by surprise: 'That I cannot understand. Offer it to me, if he does not go, I'll take the position'. However, in the letter to Mendelssohn, quoted in the text, there was no implication that London was offered anything concrete – quite the opposite was true.
20. Einstein to Lindemann (May 4, 1933). Lindemann papers D57. In Moore (1989), 269.
21. Lindemann (1932), vi; Lindemann (1933), 29. Lindemann expressed similar views in his correspondence with Born.
22. London to ICI (August 14, 1933).
23. See Rider (1985) and Morrell (1992).
24. January 28, 1933.
25. February 11, 1933, 191–2. In the same issue there was the announcement of the ceremonies for the formal opening of the Mond Laboratory in Cambridge, England.
26. Heinrich Kuhn, the spectroscopist, was also offered a stipend and started working with Derek Jackson, with whom he later developed the atomic beam method of high resolution spectroscopy. By the

fall of 1933, Simon, Mendelssohn and Kurti were hard at work setting up their liquefiers at the Clarendon Laboratory. Simon had designed an 'expansion' liquefier where helium was first subjected to very large pressures in cylinders immersed in liquid hydrogen and, then, it was allowed to expand and, thus, liquefy. While in Oxford, he perfected the techniques of magnetic cooling and started a collaboration with the Laboratoire du Grand Electroaimant at Bellevue near Paris. Cooling by adiabatic demagnetization had been first suggested in 1926 by Debye and Giauque, with whom Simon had worked closely while he was a visiting professor in Berkeley, USA, in the spring term of 1932. In 1935, when the ICI fellowships were terminated, Simon was appointed Reader in Thermodynamics. During the years before the Second World War, Simon was offered a professorship in Istanbul, Turkey, and at the Hebrew University in Jerusalem.

27. Mendelssohn (1964), 10.
28. Another person was Kurt Guggenheimer, who took his PhD oral examination at the University of Berlin on July 7, 1933, and received his degree on February 2, 1934. (I am indebted to the anonymous referee for this information.)
29. Cailletet & Bouty (1885), 104.
30. Dewar & Fleming (1893).
31. Dewar & Fleming (1904), 247.
32. Kamerlingh Onnes (1913*a*), 21.
33. See Gavroglu & Goudaroulis (1989).
34. Bloch (1966), 27.
35. De Haas & Bremmer (1931). However, accurate experiments carried out shortly afterwards on indium showed that even without a magnetic field the thermal conductivity did indeed suffer a slight discontinuity at the normal transition point (de Haas & Bremmer, 1932).
36. For a detailed presentation of this experiment, see Dahl (1993).
37. De Haas (1933).
38. Casimir (1973), 486.
39. Casimir to London (April 2, 1937).
40. London to Casimir (April 17, 1937).
41. Gorter to van Vleck (November 10, 1933).
42. Gorter (1964), 5.
43. Gorter & Casimir (1934), 307.
44. Ibid., 315.
45. London to Hartree (June 3, 1934).
46. London to Hartree (June 8, 1934).
47. Peierls and Bethe to London (June 8, 1934).
48. London to Peierls and Bethe (June 10, 1934).
49. Hartree to London (June 12, 1934).
50. London & London (1935*a*), 71.
51. Ibid., 74.
52. Ibid., 87.
53. London to Casimir (April 17, 1937).
54. London to Mott (October 10, 1934).
55. Mott to London (October 13, 1934).
56. London to Gorter (November 8, 1934).
57. Gorter to London (November 18, 1934).
58. London to Gorter (November 20, 1934).
59. Gorter to London (November 23, 1934).
60. Gorter to London (December 6, 1934).
61. Von Laue to London (March 25, 1935).
62. London to von Laue (April 30, 1935).
63. Von Laue to London (March 30, 1935).
64. Von Laue to London (April 9, 1935).
65. Von Laue to London (April 24, 1935).
66. London to von Laue (April 30, 1935); London & London (1935*b*).
67. Von Laue to London (May 18, 1935); London to von Laue (May 22, 1935).
68. Von Laue to London (May 25, 1935); London to von Laue (May 28, 1935); von Laue to London (May 31, 1935); London to von Laue (June 1, 1935); and von Laue to London (June 3, 1935). Heinz, of course, was in Oxford with Fritz all this time and we have no evidence of his reactions. There are no letters to Heinz from von Laue, who preferred to deal with the more senior of the brothers.
69. London to von Laue (June 3, 1935); von Laue to London (June 5, 1935; June 7, 1935).
70. London to von Laue (June 12, 1935).
71. Von Laue to London (June 17, 1935; June 20, 1935).
72. London to von Laue (June 25, 1935).
73. London to von Laue (February 25, 1936).
74. Von Laue to London (March 22, 1936).
75. London to Kallmann (March 26, 1936).
76. London to von Laue (March 17, 1936); von Laue to London (March 22, 1936).
77. London to von Laue (March 26, 1936).
78. London to Meissner (January 25, 1936).
79. Meissner to London (February 1, 1936).
80. London to Meissner (February 10, 1936).
81. London (1935*d*), 25.
82. Ibid., 31.
83. Ibid., 27.
84. Helium was liquefied for the first time in 1908 by Heike Kamerlingh Onnes at Leiden University. After the Great War, there were only 300 m^3 of helium remaining in Leiden. In 1919, the US Government presented Kamerlingh Onnes with 30 m^3 of helium and, in 1921, McLennan visited Leiden personally and brought 2 m^3 of helium. Kamerlingh Onnes showed McLennan the details of the

liquefier and, by 1923, liquid helium had become available at the University of Toronto. See Gavroglu & Goudaroulis (1991).

85. McLennan to London (June 18, 1935).
86. London to McLennan (June 21, 1935).
87. ICI to London (August 14, 1935).
88. London to Bauer (May, 1935).
89. Von Laue to London (September 10, 1935).
90. London to von Laue (September 20, 1935).
91. London to Weizmann (October 15, 1935; October 23, 1935).
92. Von Laue to London (November 13, 1935).
93. It appears that London also did some tutorials. 'The examination went off happily and I was given a special distinction in quantum mechanics, for which I thank you gratefully' (Robert Jackson to London (July 15, 1935)).
94. London to Lande (October 23, 1935).
95. Lande to London (February 6, 1936).
96. London to Lande (April 14, 1936). In mid-1934, a family friend, Cantorovich, who was a dentist and for a short period the Shah's dentist, sent a letter to Fritz's mother informing her that there would be a new University in Teheran, and inquired whether Fritz would be willing to consider applying for the Chair of Theoretical Physics. 'For those physicists who understand something, your son belongs to the very top.' Nevertheless, he warned Fritz that the difficulties over there were considerable and that one should think twice before resigning a position. At about the same time, the AAC informed London about the possibility of employment in the USA for $4000. He expressed his willingness to accept only a tenured position. London to C. M. Skepper, assistant secretary to the AAC (August 23, 1934).
97. London to Wigner (December 1935; exact date not specified).
98. There is no indication that he met with von Laue, but most probably von Laue had not returned from his trip in the USA.
99. Walter Adams to Szilard (December 19, 1935). Files of Society for the Protection of Science and Learning (SPSL), Bodleian Library, Oxford.
100. Szilard to Adams (December 23, 1935. SPSL files). It had been already realized that 'a great many of the English Universities will, from nationalist feelings, [probably] hesitate very considerably about introducing a foreigner into their staff' (C. M. Skepper to Thomas Hardy (March 8, 1935)). In the beginning of the summer of 1936, London was informed about a lecturership at the Herriot-Watt College in Edinburgh and of the possibility of becoming an Assistant Physicist at the Royal Free Cancer Hospital in London. Mott dissuaded him from applying, thinking that London had no chance of getting a job in theoretical physics, and suggested that he should apply for a job in mathematical physics, applied mathematics or natural philosophy, should such an opportunity arise. 'Because these subjects are much more related to what is called "theoretical physics" in Germany', Mott noted to Miss E. Simpson of the AAC (June 1, 1936. SPSL files).
101. Polanyi to Simon (December 17, 1935).
102. London to Polanyi (May 3, 1936); Polanyi to London (May 4, 1936).
103. London to Born (May 3, 1936).
104. Adams to Rutherford (January 30, 1936. SPSL files).
105. 'ICI made us a general donation of £2500 to assist German scientists, so that we could help London for a few months' (Adams to Lindemann (June 16, 1936. SPSL files)).
106. Schrödinger to Gibson (June 12, 1936. SPSL files).
107. Adams to Schrödinger (June 22, 1936); Adams to London (June 30, 1936. SPSL files). Until 1940, the SPSL had no funds from the British Government. Most donations came from British university teachers and the Central British Fund for German Jewry.
108. Fritz London File; SPSL files.
109. Hartree to London (June 1, 1936); London to Hartree (July 13, 1936).
110. Adams to Duncan-Jones (July 1, 1936); Adams to London (July 3, 1936. SPSL files).
111. Magat to London (July 11, 1936).
112. London to ICI (July 31, 1936).
113. L. E. Sutton to London (July 15, 1936). In another letter, London was told: 'I have heard from a very reliable source that Oxford was preparing to offer you another position. And I know they feel the loss keenly... Oxford will no doubt make you an offer next year'. From Corpus Christi College, but writer's signature is illegible (September 7, 1936). No such offer was made.
114. London to Donnan (late June/early July, 1936.

4

Paris and superfluidity

In 1936, London moved to the Institut Henri Poincaré in Paris. It had become evident that liquid helium had two different phases and, below 2.19 K, it defied all classical expectations for the behavior of a liquid. From the very beginning, London was convinced that the transition to the superfluid phase could not be understood as an order–disorder transition.

In the fall of 1937, he attended the Congress for the Centenary of van der Waals at Amsterdam, where he was impressed by Joseph Mayer's attempt to formulate a statistical theory of condensation for real gases. In the same Congress, Uhlenbeck retracted his criticism of the Bose–Einstein condensation which he had expressed in his thesis of 1925. In 1924, Bose (and later Einstein, after receiving Bose's paper) had discovered that, below a certain temperature, an ideal gas of integer electron spin will start condensing and that, with each condensing atom, there will be an increased probability that the next atom will find itself in the condensed state. This condensation was a purely quantum mechanical effect which was derived from the kind of statistics the atoms obeyed.

Upon London's return to Paris, he started working frantically and discovered a 'crazy thing' – as he duly informed Heinz London who was now at Bristol University. He proposed that the onset of superfluidity could be regarded as being the start of this peculiar condensation, since helium is a gas obeying Bose–Einstein statistics. His initial calculations gave results that were surprisingly close to the experimentally measured values. London was overjoyed not only because it was possible to explain the phenomenon of superfluidity, but also because the proposed schema was another instance that corroborated the macroscopic quantum phenomena.

He was offered the Chair of Mathematical Physics at the Hebrew

University and, in April 1938, he and Edith visited Jerusalem. Eventually, he declined the offer because of the uncertainties associated with the war and, instead, accepted the Professorship in Theoretical Chemistry at Duke University, North Carolina, USA, arriving there in late 1939.

During his stay in Paris, London wrote a short monograph with Edmond Bauer on the problem of measurement in quantum mechanics. It was the only time he had publicly, or privately, expressed his views on this problem. The booklet was a simplified version of von Neumann's mathematical analysis of the Copenhagen interpretation of quantum mechanics, with a further analysis of the observer's role in measurement.

The Popular Front

The years that London spent in Paris were the years of the Popular Front. They were the years of radical and lasting social changes, the years when the intellectuals were at center stage. Unlike London's experience during the period of the Weimar Republic, the intellectuals in France during the years of the Popular Front were preoccupied neither with voicing their worries about the loss of a glorious past nor with expressing by their works any 'hunger for modernity'. In the whole of Europe – including Nazi Germany and Fascist Italy – there was a redefinition of the relationship between the arts, politics and leisure. As Julian Jackson noted in his study of the Popular Front: 'if the Popular Front aimed to reinvigorate republicanism in France, it had to show that democracy could accommodate the imperatives of mass politics no less successfully than totalitarianism: culture and leisure lay at the heart of this endeavour'.[1] The masses, it was asserted, had a right to culture and the cultured had an obligation towards the masses. It was not a matter of a moral stand to alleviate the guilt of the intellectuals for being so far removed from the real problems of the less fortunate, but it was rather the expression of the new politics. To a large extent, London's own experience at the University of Berlin was about the denial of democracy through disputes about the procedural aspects of democracy. In Paris he lived through a period during which everyone tried to transcend those procedural issues and to seek the deeper political potentialities of democracy. It was no coincidence that, for Fritz and Edith, their most intimate connections were with Langevin, Joliot and Perrin – all very much involved with the politics of the Popular Front and with the Communist Party. Again, this was no coincidence, since the Communist Party had the central role in this process of redefining the relationship between intellectuals and the masses, and that between culture and politics.

Another of their close friends in Paris was the philosopher Aaron Gurwitsch, himself an immigrant from Germany and the person who pioneered the formation of American phenomenology after the Second World War.[2] London liked what was happening in Paris. 'It gives me great happiness to be in a position to participate in such events', he wrote to Becker who was in Göttingen. He thought that one of the significant changes brought about by the Popular Front was its recognition of the damages inflicted on science by the 'autocracy which reigned till the present', and its initiatives to amend it by organizing as he wrote: 'dynamically sparkling things ... even though I am convinced that this enthusiasm will be hit hard again and the French *Nationalindividualismus* [London's neologism] will not change much'.[3]

Frederic Joliot had deeply impressed Fritz. Quite uncharacteristically, Fritz said that it was the first time he had met such an intellectually rich and intelligent person. He wrote to Heinz about a weekly gathering organized by Joliot, where they discussed nuclear physics and where London was considered to be a theoretical showpiece. He was not doing anything new, but was learning many new things. He could not understand all the details of nuclear physics and it was 'like when one goes to a different family and hears gossip about relatives and uncles and aunts whom one does not know'. He was quite amazed at the progress in physics during the past years and he attributed it to the large number of people, especially in the USA, who had been doing physics. 'But there is a lot of useless work and I find the present situation in nuclear physics awful. But I think that in 2–3 years there will be serious theoretical work, and, then, I would have saved myself 5 years of work, since I believe that *nothing* of what is being done in theoretical nuclear physics will not last for ever.'[4]

London had a research position at the Institut Henri Poincaré and was not a professor at the Sorbonne as his mother thought. 'From the copies of the letters of our relatives that you send me, I see that they are not correctly informed. I am *neither* a professor *nor* am I at the Sorbonne. I am sorry that I have to disappoint them ... I would like to ask you with urgency not to overdo it. Flattery is really the thing I dislike most in people and *wherever* you wrote these stupid things please inform them accordingly. I would not want to do it myself.'[5]

But Paris was also a very enjoyable experience because Edith had, at last, the opportunity to pursue her interest in painting in a professional manner. While in Berlin she had attended classes on art and art history at the University and when they spent time in Rome she had painted quite a lot. Though she continued to paint while in Oxford, it was in Paris that she took lessons in studios, attended lectures at art schools and found other younger painters she could relate to. She had

Figure 9. Fritz London in Paris in 1936. (Courtesy of Edith London.)

planned to enroll in an art school, but when she listened to one of the talks by a teacher at the school, the Fauvist painter Othan Friesz, she was very disappointed. Friesz, in arguing about the growing influence of the Fauvists, had made a particularly nationalistic speech. She started going to galleries to look at the work of artists and to try to find a teacher. She came across the work of Marcel Gromaire and was immediately attracted to it. Gromaire had a teaching studio with about ten students. Fritz encouraged her to go to him, and she studied with him for more than six months. Often she was not satisfied with her own work, and wondered whether she was doing the right thing by making a career of painting. As Edith related, it was Gromaire who told her insistently that she should persevere and she would: '[soon]

find out that you are a painter. And you will do quite well'. She then studied with André Lothe, whose studio had more than fifty students and whose extroverted personality contributed to a much deeper communication among the students than in Gromaire's studio. She stayed with him till they left Paris in 1939.

The article in *Nature* 1937 and *Une Conception Nouvelle de la Supraconductibilité*

Though the first ideas about the macroscopic interpretation of superconductivity were expressed during the discussion at the Royal Society in 1935, London presented these ideas analytically in two publications a couple of years later. These articles included the thesis London submitted to obtain a doctorate from the University of Paris and titled *Une Conception Nouvelle de la Supraconductibilité* (or *Conception Nouvelle*) published at the end of 1937,[6] and an article in *Nature* titled *A new conception of supraconductivity*, published at the end of 1937.[7] For a long time, he had wanted to write an article in *Nature*, but he delayed it, since he did not have much self-confidence with the new approach he was developing. But in *Conception Nouvelle* he felt that he had developed: 'a totally different viewpoint, and a rather easily understandable form for anyone who is an outsider'.[8]

London, considering the diamagnetic character of superconductors as being their fundamental property, was able to show that it is, in fact, possible to show how infinite conductivity can be derived from diamagnetism. In a diamagnetic atom there are permanent currents. These currents appear when the system is in a stable state and in the presence of an external magnetic field – and, therefore, such cases cannot be excluded by Bloch's theorem, which assumed absence of an external field. He argued that a 'diamagnetic atom succeeds in representing an infinite number of currents by one single state' and that such mechanism of conduction was entirely different from that considered in the traditional theories of conductivity. In his theory, the transport of electricity was *not* based on progressive waves, but on stationary waves. 'By these a transport of electricity can only be effected in the presence of a magnetic field and this is precisely our assertion as to the nature of supracurrents.'[9]

When London talked of a 'macroscopic' interpretation he meant a phenomenological theory whose interpretation depended on a 'microscopic' mechanism which set it apart from that used to explain ordinary conduction. The differentiating characteristic of this new microscopic mechanism was the macroscopic dimensions of the stationary waves. London used, of course, the language of quantum mechanics, but, as yet, he did not regard superconductivity as a macroscopic quantum phenomenon – even though he knew that the explanation of the phenomenon of diamagnetism could be achieved only quantum mechanically. He talked more about the quantum mechanical description of superconductivity. During this period, London was thinking along quasi-classical lines and his use of 'macroscopic' was confined within a descriptive framework. In fact, the explicit introduction of the macroscopic current was, according to London, necessitated by Maxwell's theory, since the magnetic field of superconducting rings had a curl. What led London to adopt fully the concept of macroscopic quantum phenomena were the 'impossibility' of being able to deal successfully with the microscopic program he first enunciated in very general terms in these articles, and most importantly his proposed explanation of superfluidity in terms of Bose–Einstein condensation.

The development of London's notion of the macroscopic quantum phenomenon went through various stages. Initially, the concept was used to denote the new possibilities provided by the rigidity of the wave function. In its final formulation it involved looking at superfluids as systems ordered in momentum space. London's later writings, especially on superfluidity, and much of his later correspondence give the impression of a stubborn thinker. He was determined to convey his overall agenda. By no means did he regard his theoretical schemata as being phenomenological theories. He had told McLennan in 1935 that the progress he claimed was primarily a logical one and he avoided the difficulties of the traditional approaches by a new and more cautious interpretation of the facts. Though he granted that the success of a new theory often depended on its new predictions, he wrote: 'as far as I can see this very probably does not happen in our theory'. And in his speech when he received the Lorentz Medal – nearly twenty years after the letter to McLennan – he emphasized that the equations of superconductivity 'have often been called phenomenological, but actually they go beyond what is directly given by the phenomena. They express something like a *long range order of momentum vector*'. London had repeatedly stressed that his work in superconductivity and superfluidity was not only for the specialist, but was part of fundamental physics.[10]

Von Laue again

London and von Laue had not communicated for over a year. In the meantime, London had moved to Paris. London sent a copy of *Conception Nouvelle* to von Laue for comments. Von Laue agreed with everything relating to the preliminary discussions and the fundamental equations proposed by London. However, he objected to London's approach to the question of the change of phases as well as the whole question of the 'intermediate state'.[11] The latter was the name given to the state of a superconductor in the presence of an external magnetic field and it involved a series of alternating domains of normal and superconducting layers being formed as a result of the inhomogeneity of the magnetic field on the surface of an arbitrarily shaped superconductor.[12] London admitted that he would have preferred more formal proofs, but said that this was not possible yet. He also admitted that the whole treatment of the transition to the superconductive state as a process of phase transformation was a big worry for him too. He wanted to omit this paragraph after the galley proofs were sent to him, especially since the formulation of the theory of the intermediate state, made the discussion about the change of phases superfluous, but the publisher did not allow him to do it. London considered that the correct approach would be to show that the assumption of the total energy of the volume was not sufficient, but one needed an expression for the surface energy – which was calculated in a later chapter.[13] Towards the end of 1937, after having 'examined systematically the thermodynamics of superconductivity', von Laue pointed out a mistake in London's 'otherwise so good book'.[14] The difference between the free energy of the superconducting and normal phases should have been twice as much as that found by London. Despite some technical issues related to the definition of certain quantities, London agreed with von Laue.[15] Otherwise, von Laue was pessimistic, and wrote: 'personally things here are bearable, but I emphasize the first word'. They exchanged a few more letters before the Second World War broke out, and von Laue informed London about his and Pontius' experiments. The results, von Laue noted, did not verify London's theory, since such results could be derived from other theories. Of course, London did not agree.[16]

In the fall of 1936, Heinz London had already settled in Bristol[17] and kept Fritz informed of the various developments there. One of the remarkable aspects of Heinz's letters was that he wrote analytically his views and criticisms about various papers he may have read in the meantime – mostly about superconductivity. He did all the calculations; he had queries about the experimental set-ups; he stopped the letter and started it again a few days later when he had completed some derivation and had a proposal for a hitherto peculiar mechanism. Fritz was always his confidant. Heinz wrote to Fritz that Fröhlich planned to contact Weizmann through Bragg about a possible position in Palestine. Heinz rarely expressed his views on various social and political issues except in this one instance:

> Recently there was a Jew who was an acquaintance of Fuchs and who was there for a year without being able to realize his aims. He was cursing the Jews who, he said, with the British behave imperialistically towards the Arabs. The Arabs are not allowed to belong to the unions, they are paid less for the same work, the Jews lay them off from the big shops etc. Unfortunately, he seems to be right in most things, but it also seems that one cannot express these sentiments for the time being, since everything depends on the Anglo-Jewish capital. Concerning the 'great' policy, I do not understand how the Russians dare accuse the good Blum who does not give arms to the Spaniards while they themselves are neutral. Is your minister of the interior (or war) a radical or a socialist? I would advise him to unite the army.[18]

In the fall of 1936, a note by Schrödinger appeared in *Nature*. Schrödinger suggested that, as a result of the Londons' theory, there were now three different kinds of currents: the displacement current in an insulator; the ordinary conduction current; and the supercurrent. He wrote enthusiastically about the theory of superconductivity, but it embarrassed the brothers. Fröhlich told Heinz London that it was as if 'Schrödinger wanted to do us a favor'. He asked Heinz during the departmental colloquium what it all meant and Heitler burst into laughter.[19] It is not clear at all why Schrödinger wrote this note. There was, in effect, nothing new in the note and superconductivity was at no time high on Schrödinger's agenda. It may not be unreasonable to suppose that, in Oxford, London told Schrödinger about von Laue's tactics and that Schrödinger may have been coming to the Londons' 'rescue' – he had, after all, not done much for Fritz.

Fritz told Heinz about the initial reactions to his *Conception Nouvelle*. It was the first time he mentioned the problems with von Laue to Heinz. 'Laue starts to complain again. De Haas was equally dissatisfied with the citations, although I did try to include the work of his wife. But he wants to be included as one of the persons whose work foresaw the Meissner effect.'[20] Heinz did not respond to all this and his opinion was that Fritz should keep away from the complications involved with the intermediate state. The proposition that $H = H_T$ and the formulation of the non-zero electric field E lead to the intermediate state 'has been described by your theory, but since it is not our basic phenomenon and, in reality, a very confusing state of affairs, it may lead us to a general confusion'.[21] Still, Heinz talked analytically about the intermediate state to H. Jones who was going to the Mond Laboratory for a year so that, as he wrote: 'he can promote our teaching in Cambridge'.[22] Heitler had read the note by Fritz in *Nature* and liked it very much – even though he found the ending was grandiose and he thought that the difficulties which had to be resolved were ten times greater than the problems already resolved.[23]

Casimir and de Haas were among those to whom Fritz London sent his monograph on superconductivity. De Haas was one of the two directors of the Kamerlingh Onnes Laboratory at the University of Leiden (the other was Keesom). As discussed in Chapter 3, de Haas was dissatisfied with London's narrative of the Meissner–Ochsenfeld experiment. He granted that London had not undertaken to write a historical treatise nor to analyze exactly the significance of all the papers, but, even so, de Haas would have liked his work to have been properly acknowledged.[24] Not unexpectedly, London was quite upset by these remarks. He said that it was not his intention to underestimate the contributions of de Haas, that he did not plan to give a detailed historical exposition, but just a new version, of superconductivity. For this reason, he had to be very careful with the exposition, and had to simplify the historical developments very much, and especially the experimental developments. As for the development of $B = 0$, London told Casimir and de Haas that physicists in Leiden had played a very decisive role. Eventually, both de Haas and Casimir seemed to agree with London's version of the developments. However, Casimir insisted that the impetus for more exact research on the magnetic behavior of superconductors was given by von Laue's measurements of the resistance of wires in a transverse magnetic field as well as from the experiments of de Haas with wires with frozen-in fields.

The structure of solid helium

In Oxford, Fritz London's researches were not confined to superconductivity. During his last year there, he became interested in that other mystery of very low temperatures: the inability to solidify helium under its own pressure. Attempts to solidify helium had first been made by Kamerlingh Onnes himself on the same day he liquefied helium (July 10, 1908). The unsuccessful attempts were repeated on numerous occasions by Kamerlingh Onnes, who started thinking that helium might remain liquid under its own pressure all the way down to absolute zero. After Keesom's success, in 1926, in solidifying helium by applying pressure, it was deduced that the entropy difference between liquid and solid helium tended towards zero as the temperature was lowered, and this meant that the liquid phase passed into some kind of an ordered state. The nature of this ordered liquid state, which Simon called 'liquid degeneracy',[25] was one of the significant issues that preoccupied many physicists in the following years.[26]

'The expression "liquid degeneracy"', noted London many years later, 'has been suggested to give a name to this mysterious process. The mere naming of it, however, only served to emphasize the mystery'.[27] London decided to study this mystery.[28] Since liquid helium could be solidified only under pressure, the structure of the liquid at absolute zero became a problem of great significance. London proposed an interesting solution to the problem in 1936 and this also helped him to familiarize himself with liquid helium, whose superfluidity was discovered a couple of years later.

Simon had already shown that, at absolute zero, the liquid state of helium had less energy than the solid state. This was surmised to be due to the rather large zero-point energy and to the rapid variation of energy with the molecular volume. And, since solidification was due to applied pressure, there was no entropy change involved in the transition. Helium could not be treated with the usual methods of lattice theory, since its zero-point energy was of the same order of magnitude as the interaction energy. It was surely a peculiar kind of an ordering process whereby the transition below 2.19 K made liquid helium rapidly lose all its entropy of liquid disorder without going into the solid state. In fact, in 1927, Keesom and Wolfke had noted that there appeared to be a kind of transition at that temperature, and they suggested that at 2.19 K helium-I turned into helium-II. They could not observe any visual changes, but they were convinced that something peculiar was happening to liquid helium. In fact, they noted that it appeared that helium-II was very mobile. This remark, coming from the very reserved and conservative Keesom, is indicative of how much this highly unusual behavior of helium must have impressed the authors.

In mid-1934, Mott was studying problems related to liquids and was attempting to construct a theory where liquids were considered to be a limiting case of solids.[29] He asked London's opinion on his work, since, he felt, London had 'thought such a lot about liquids in connection with Simon's problems'.[30] London was quite enthusiastic about Mott's ideas. Mott recalled a conversation where London had insisted that helium-I, because of its large zero-point energy, had a simple cubic structure and, hence, no rigidity. If that were the case, Mott thought, such a lattice would melt with no latent heat.[31] But London changed his mind, and he now thought a diamond structure was more probable for helium-I. Still he could not calculate the zero-point energy precisely enough to make a final decision. A simple cubic structure seemed to be more probable for solid helium, whereas a close-packed structure was expected for higher pressures.

London wrote to Mott that in the diamond structure, while every atom has its four tetrahedral neighbors, not every tetrahedron of this size contains an atom. Tetrahedral neighbors cannot keep their atom in their midst, the latter may prefer the potential holes of the originally adjacent tetrahedra. There could be resonance, and, since the number of the empty tetrahedra is half the number of atoms, one would expect 50% probability for the occupation of each. The result would be the doubling of the original diamond lattice. This led to a space-centered cubic lattice. 'I cannot quite make up my mind whether or not it is all nonsense'.[32] Mott was not particularly convinced and thought that the model was rather indefinite, but it was satisfactory to know that the zero-point energy produced the right order of magnitude. He encouraged London to publish the idea that liquid helium had a diamond lattice.[33]

Simon, too, insisted that London should publish his results about the crystal structure of helium. Fritz was not enthusiastic at all, but he sent it for publication with, he wrote: 'the greatest scruples, for I am quite aware of its triviality'.[34] It was not possible to solve the problem properly by considering the simultaneous interaction of the many helium atoms and by taking into account their wave nature. London decided to start from another limiting case: first, he considered the zero-point energy of rigid spheres of finite diameter and, then, superimposed the interaction energy. Next, he showed that the zero-point energy could account for the expansion of the volume to three times its original value and that it became possible to understand why the energy content of helium at absolute zero was reduced to about one-tenth of the energy of the 'classical' equilibrium position.

The large zero-point energy of the substance, was, then, used to compare the properties of face-centered cubic, simple cubic, and diamond structures and also the values of their potential energies as a function of the atomic volume. He found that, under the actual conditions of helium, 'the diamond structure would have the lowest energy'.[35] He concluded that if helium-II tended to form any of these structures, the diamond lattice would be the one most favored. For London, the lattice points could give only an indication of the position for which the probability of finding an atom would be higher than elsewhere. Thus, the lattice structure must no longer be considered to be rigid, but was the preferred configuration of statistical distributions. Thus, London had shown that the closest packed structure was stable only under pressure and that this was the reason why liquid helium was solidified only under pressure. Liquid helium could now be conceptualized as being held together by the attractive forces balanced mainly by the zero-point repulsion.

Soon after the appearance of London's paper, Fröhlich suggested that, since two diamond lattices shifted by one spacing were equivalent to one body-centered cubic lattice, an order–disorder transition would be possible in which each point of the body-centered lattice is 'half occupied' at a temperature above the transition point, but, below, the atoms are in one of the diamond lattices. Fröhlich suggested that perhaps helium-II has diamond structure and helium-I is body-centered. This gave approximately the correct entropy change and a discontinuity in the behavior of the specific heat. Then, Fröhlich conjectured that the lambda-point transition at 2.19 K was an order–disorder transition between the holes and the atoms.

London did not favor such an order–disorder theory. He claimed that in a body-centered lattice there are $2n$ places for n atoms and, even at absolute zero, every place could be occupied with a probability of 50%. Thus, it was difficult to see why atoms should prefer to settle in one of the two available lattices assumed to exist in helium-II. He emphasized that these statistical structures were totally unrelated to disorder, due to thermodynamic considerations. The whole system of the suggested structure could be represented by a single quantum state, which, if it is in the lowest state, will describe the system at absolute zero. At that temperature the system has zero entropy in spite of the fact that it may involve a statistical distribution of n atoms over $2n$ places. Thus, Fröhlich's argument of an order–disorder transition could still be maintained; only the state of order was one in which $6n$ helium atoms settled down in $2n$ half-occupied states. Here was a situation indicating how a random distribution could be changed into an ordered distribution by quantum mechanical principles.

In none of his subsequent papers did London elaborate on these problems, not even in his review talk at the Cambridge Conference on low temperature phenomena, in England in 1946. Nevertheless, he discussed the problem quite extensively in the second volume of his *Superfluids*, not only to provide the necessary background to his main theme, but because, in this work, he appears to have discovered a series of plausible arguments for one of his most criticized assumptions: extending the methods used for ideal gases to the case of liquid helium.

In a letter from Leiden, Fröhlich informed London that Keesom and his collaborators, had found, in the Roentgen photographs of helium, wide rings and no difference between the structures of helium-I and helium-II. They thought that diamond structure should be excluded. 'I would like to know if you have thought out what to expect according to your theory.'[36] Fröhlich conceded that from London's theory it became possible to derive the lambda point and that the next step should be the calculation of the critical temperature.[37] He assumed that at absolute zero helium had a diamond structure. But the zero-point oscillations could not be considered to be 'small' and, he wrote: 'the whole discussion for the zero-point oscillations serves to clarify the problem that even in my proposal one cannot depend on the usual zero-point energy'. Fröhlich wanted to find the classical distribution of atoms in positions. Then, he suggested that it would be quite natural to put quantum states into a classical distribution. 'I think that if this is not possible, then it is meaningless to talk about crystalline structure. Are you totally opposed to this hypothesis?'[38] Unfortunately there is no trace of London's answer.[39]

The peculiar properties of helium

Ever since 1911, when Kamerlingh Onnes discovered that helium had a maximum density at about 2 K, there had been various indications that at that temperature 'something happens to helium'.[40] By the end of the 1930s, the phenomena associated with liquid helium below 2.19 K would defy all the attempts to understand the behavior of liquid helium by classical hydrodynamics.

During 1932, Keesom and Clusius reported that the specific heat curve had an extremely sharp maximum although there was no latent heat for the transition from helium-I to helium-II, but they could not understand the inner causes for such a transition. Keesom decided to repeat the same measurements more accurately and, in the paper he wrote with his daughter Anna Petronella, they proposed, after Ehrenfest's suggestion, for the first time the term 'lambda point' to indicate the transition from helium-I to helium-II. Then they attempted to measure the heat conduction in helium-II. They found that below the lambda point 'the heat conductivity is about 200 times that of copper at ordinary temperatures, or about 14 times that of very pure copper at liquid hydrogen temperatures. Hence liquid helium-II was by far the best heat conducting substance we know'.[41]

In 1930, Keesom and van der Ende, quite accidentally, observed that liquid helium-II passed with remarkable ease through extremely small leaks – something which was not possible at higher temperatures, not even for gaseous helium. This observation indicated that an enormous drop in viscosity occurred when helium was below 2.19 K. The viscosity of helium was measured in 1935 by Wilhelm, Misener and Clark in Toronto, Canada, and in 1938 by Keesom and MacWood in Leiden, using the method of rotating disks. It was found that the change in viscosity was continuous, and, even though it became less with the fall of temperature, it did not differ appreciably from that of helium-I. But, when compared to the results derived by the capillary method, the difference was a factor of one million. Such an enormous difference in viscosity by the two different, yet equivalent methods, could not be understood in the framework of classical hydrodynamics. More accurate viscosity measurements confirmed the earlier results and Kapitza used the term 'superfluidity' to characterize this strange behavior of helium.

'Perhaps the strangest of all the properties' was reported by Allen and Jones in February 1938. Allen and Jones wanted to extend the heat conductivity experiments to lower and lower temperature differences and to achieve this they used an apparatus consisting of a reservoir capillary. When they supplied heat to the inner vessel, they saw that the inner helium level, far from being depressed, seemed to rise above that of the reservoir. The rise increased with heat input and, for constant input, with falling temperature. This was the 'thermomechanical effect', a mass flow of helium opposing the heat current. In one of their experiments, they used a powder-filled bulb, open at the bottom and with a narrow orifice at the top. When they heated the powder by shining a light on it, they observed a jet of liquid helium rising from the upper end to a height of several centimeters. The phenomenon was named the 'fountain effect'. Extremely small temperature differences between reservoir and inner vessel were sufficient to produce a very large convection. It seemed, thus, impossible to treat the hydrodynamic and calorific properties of liquid helium-I independently.

In 1938, Daunt and Mendelssohn in Oxford and Kikoin and Lasarew in Kharkov, USSR, found that liquid helium flowed from one container to another inside it (or outside it depending on the relative height of the liquid helium surface) by means of a film which formed on the walls and whose thickness was of the order of 100 atoms. Such films, of course, are formed by any liquid which wets a solid surface, but the viscosity of an ordinary liquid is such that the film forms slowly and there is practically no movement. Helium-II is the only fluid which, owing to its superfluidity, forms a swiftly moving film.

As Sir William Bragg remarked, the situation with liquid helium was very much like the disorderly world of *Alice in Wonderland*.

Bose–Einstein condensation

In November 1937, the Congress for the Centenary of van der Waals took place in Amsterdam. It was a gratifying affair for London. The majority of the papers had as their starting point his work in molecular forces. He went to the conference with an open mind, since, in effect, he was not working on anything new at the time. Furthermore, in 1937, he had published a review of his work on molecular forces and superconductivity, and the fact that he did not present a paper at the conference is an additional indication that he had nothing new to say and was contemplating the direction of his future work. Casimir had invited London to give a paper at the conference. At first, London proposed to talk about aromatic molecules and superconductivity, but then he realized that it was outside the immediate theme of the conference, and, since he did not have anything new on the van der Waals forces, he did not give a paper.[42]

Beth and Uhlenbeck discussed the quantum theory of non-ideal gases; Lennard-Jones discussed the equation of state and critical phenomena; van Urk discussed the cohesive forces of liquids; de Boer and Heller discussed the anisotropic van der Waals forces; and a long paper by Hamaker discussed the London–van der Waals attraction between spherical molecules. Simon was also there and he gave a paper on the third law of thermodynamics. Schmidt and Keesom reported on new measurements at liquid helium temperatures. Born gave a paper on the statistical mechanics of condensing systems, where he developed some recent results derived by Mayer and which Born considered to be 'a most important contribution to the development of van der Waals theory'.[43] Mayer had attempted to solve the general problem for any law of central force between the molecules.[44] In this paper, Mayer used classical statistical mechanics, and tried to explain the phenomenon of condensation – a 'milestone in equilibrium statistical mechanics'.[45] Mayer's papers were discussed by others as well, and Kahn and Uhlenbeck showed that his treatment could be shown to be formally analogous to Einstein's equations for the ideal Bose gas – for which Einstein had predicted a condensation phenomenon. It was this work by Mayer which directed London's attention to the Einstein condensation paper. This is also corroborated by a letter to Fritz from Heinz London, who had heard that Groenewold, who had attended the conference, was saying that the Einstein condensation phenomenon might, after all, be correct. Uhlenbeck had talked about a transition point of 0.01 K – and the short letter he wrote with Kahn, where they retracted the reservations expressed in Uhlenbeck's thesis, was in the proceedings of the conference.[46] If condensation is considered to be a limiting property for $N \to \infty$, $V \to \infty$ (N/V is finite), then Einstein is right. 'All my wisdom comes from a letter by Groenewold to our Dutchman here who showed it to Fröhlich and he immediately translated it.'[47]

In a debate that took place during the conference and lasted all morning, the issue was whether the partition function contained the proper information to describe a sharp phase transition. It was not clear how it was possible to account for the existence of analytically distinct parts of isotherms that such a transition implied. 'The debate was inconclusive, and Kramers, the chairman, put the question to a vote. Uhlenbeck recalls that the ayes and nays were about evenly divided. Kramers' suggestion to go to the thermodynamic limit was recognized to be the correct answer. Uhlenbeck withdrew his objections'.[48]

After his return to Paris, London was very excited. The notion of a quasi-crystalline state 'which had never been taken seriously'[49] had already been given up by London, who turned to a new model. In this model, each helium atom moved nearly free in the self-consistent periodic field formed by the other atoms. According to Bloch's theory, this was similar to the way electrons move in a metal – but with a crucial difference. The helium atoms obeyed Bose–Einstein statistics, whereas the electrons in metals obeyed Fermi–Dirac statistics. As a first step, London disregarded the self-consistent field altogether and considered the ideal Bose–Einstein gas. In 1924, Einstein had already discussed a peculiar condensation phenomenon of this gas. Because of Uhlenbeck's observation in his doctoral thesis, London wrote: 'in the course of time the degeneracy of the Bose–Einstein gas has got the reputation of having only an imaginary existence'.[50] Uhlenbeck, with Ehrenfest, tried hard to reconcile the new statistical methods of Bose and Fermi with the classical theory of Boltzmann and Gibbs. That is how Uhlenbeck came across Einstein's paper. Einstein had replaced the partition function over the discrete energy levels of the particles by an integral which was not allowed near the condensation point. Ehrenfest agreed with Uhlenbeck; in fact, Ehrenfest and Uhlenbeck had started to prepare a short article for *Zeitschrift für Physik*, but it was never completed. The article started with a motto: 'When kings are building, then the sewer cleaners have work to do.'[51]

At absolute zero, liquid helium has characteristics reminiscent of the behavior of gases. There are no potential barriers to be overcome and the atoms can be carried over the barrier by the zero-point energy, without requiring thermal motion. The viscosity increases with increasing temperature. It was known that the viscosity of liquid helium was of the same order of magnitude as that of the helium gas, and, hence, it was possible to draw a parallel with the density independence of the viscosity of gases, as it was derived by Maxwell for ideal gases. London also noted that the temperature coefficient for liquid helium-I changes its sign under pressure. Such a property implied that liquid helium under low pressure 'might reasonably be contemplated from a point of view that emphasizes, more than is usual with other liquids, a similarity to a gas'. It was only under high pressure that liquid helium might be considered to behave like an ordinary liquid. Still, the explanation of the lambda-point transition and the strange properties of helium-II did not appear to be an easy job. Neither did it help the explanation that helium under zero pressure could not have a structure ordered in space. Nevertheless, it was found that the helium atoms could gain energy by dispersing in space. As a result, when liquid helium was cooled below the lambda point, it was expanding its volume just as it was losing most of its entropy.

> We may wonder whether this system, if it cannot gain stability by improving its space order, might not take advantage of this situation when approaching absolute zero and might, as a manifestation of quantum mechanical complementarity, settle in some kind of *order in momentum space* even at the expense of order in ordinary space.[52]

London pointed out that such a non-localized structure in condensed helium would, because of the high zero-point energy, be more favorable than a quasi-crystalline structure. Indeed, liquid helium-II, despite its high degree of order, instead of being close to a liquid or solid crystal, is, owing to its extremely large volume, much closer to a gas than to an ordinary liquid. London was prompted to look closely into the phenomenon of Bose–Einstein condensation for two reasons: first, because of the nature of liquid helium-II itself (its gas-like nature, combined with its high degree of order); secondly, because of the discussions at the conference in Amsterdam. But since all real gases had been condensed in temperatures higher than the temperature where the ideal Bose–Einstein gas started this condensation phenomenon, the mechanism appeared to be 'devoid of any practical significance'.[53] To the day of his death, London would continuously stress his view that such an apparent deadlock had, in effect, an immense potential and implied a new conceptualization which he had started articulating for superconductivity a few years earlier: quantum mechanics, whose triumph was hailed for the microcosm, could be used to define macroscopic quantum phenomena.

In January 1938, Kapitza, Allen and Jones published their papers in *Nature*. Later that same month, London called Tisza, who was at the College de France at the time, and said he wanted to visit him. '[London said] "Now listen to that." And then he showed me his Bose–Einstein condensation paper, and what he did during the few days, and I was enormously impressed and said, "Yes, it sounds wonderful". And then we said, "What should be done about it?"'[54]

In February 1938, Fritz wrote to Heinz that he had 'quit nuclear physics during the last three days because I have found "a crazy thing"'.[55] In the meantime, Heinz had learned of the new idea from Fröhlich. London had told Fröhlich some details about his new approach to the helium problem. Fröhlich's first reaction was that London's method gave a 'hump' rather than a jump for the specific heat. It seemed to him more possible that the higher energy state of the helium atom depended on the number of other helium atoms which were in a similar situation. And, therefore, one should calculate the increase in the zero-point energy when an atom is placed in a hole. In that case, London's assumption that the zero-point energy depended on the volume became doubtful.[56]

The very next day, however, Fröhlich wrote an enthusiastic letter to London. He was convinced that the behavior of the specific heat of an ideal Bose gas was really as London had described it.[57] Fröhlich's enthusiasm continued, and he communicated it to London again the next day. 'I like your paper more and more. Today I convinced Heitler and Mott that there is indeed a lambda point and I also think that it is possible to have zero point energy'. He thought that the entropy at the lambda point was quite large – something which could be corrected with a proper modification of the energy spectrum and this could mean that the condensed state near the zero-point energy increased faster than is the case with the free particles.[58]

Simon's first reaction to London's idea about applying the Bose–Einstein condensation to liquid helium was not enthusiastic, but not dismissive either. The calculated values for both the thermal energy at the lambda point and the entropy deviated appreciably from the experimental values. According to Simon, another difficulty was that the Bose–Einstein gas had no zero-point energy and this is why it had larger energy content in low temperatures as compared to the ordinary gases. He warned London that the connection between his model and that of real helium would not be so easy and that he should think about these difficulties, 'otherwise people will make fun of you'.[59] Simon thought that physicists from Leiden and some from Cambridge were silent on purpose about his own contributions, and he asked London, if possible, to mention his work, as he pointedly reminded London that: 'the term degenerate liquid that I proposed in my first paper for helium below the lambda point appears to me now to have been quite correct'.[60]

During a seminar in Bristol, Fröhlich derived the value for the entropy and found a value which was three times smaller than the one originally found by London, and so Heinz urged Fritz to redo the calculations.[61] At the end of February 1938, London realized his mistake: he had used units for the region above the lambda point which were different from those below it. When he did the calculations correctly, he found that the change of phase was of the third kind – contrary to the experimental results which implied a discontinuity of the specific heat at the transition point and, thus, a transition of the second kind. Still, he found that the calculated entropy was quite near the measured value and its magnitude was negligibly influenced by the form of the interaction.[62] London found the transition temperature to be 3.09 K. Interestingly, Simon thought that the experimental results showed that the transition at the lambda point was of the third kind.[63] Heinz was unsure how to deal with the new approach. 'Please tell me how secretive you want me to be about all this. My attempts at being secretive were all in vain this time, since Simon tells everyone about your Bose = He'.[64]

The note in *Nature*

Fritz London's short paper was published in *Nature* on April 9, 1938, having been submitted a little over a month before. He started with a critique of Fröhlich's scheme to account for the lambda transition as an order–disorder transition, and then promised to 'direct attention to an entirely different interpretation of this strange phenomenon'. He showed that those points which, according to Fröhlich, should become less favored for low temperatures had, in fact, quite a large binding energy and no cooperative phenomenon could appear.

For an ideal Bose–Einstein gas, the condensation phenomenon represented a discontinuity in the derivative of the specific heat and this was validated experimentally. Still on the one hand, London wrote in the note, 'it seems difficult not to imagine a connection with the Bose–Einstein statistics ... On the other hand, it is obvious that [we deal with] a model which is so far away from reality that it simplifies liquid helium to an ideal gas'. London's note in *Nature* dealt with the problem of the lambda-point transition and he only expressed his hopes that the proposed mechanism could, in the future, provide some kind of an explanation for the transport phenomena observed in helium-II. Only in his paper published in the *Physical Review* in December 1938 did London attempt to provide an explanation for these properties. Tisza's two-fluid model (discussed in the next section) which was published a month after London's original note, was, in fact, the first model to have accounted for those properties of superfluid helium.

Below a certain temperature that depends on their mass and density, a finite fraction of particles in an ideal gas begins to collect in the lowest energy state, that is they assume zero momentum. The remaining particles have a velocity distribution similar to a classical gas, flying about as individuals. Since both components (the 'condensed' and the 'excited') occupy the total volume of the container as if one is dissolved into the other, there is no condensation in the ordinary sense. There is no separation in space into two phases which can be distinguished by their density. 'But if one likes analogies, one may say that there is actually a condensation, but *only in momentum space and not in ordinary space*'.[65] There was, then, an equilibrium of two phases. One contained the molecules of zero momentum, which occupied a zero volume in the space of momenta. The second phase contained molecules with a distribution over all the momenta as it was found in temperatures higher than the transition temperature. No separation of phases was to be found in ordinary space.

The particles of the condensed phase could not form wave packets of the size of molecular dimensions by the superposition of the neighboring wave functions, since their wave functions were constant over the whole volume they occupied. Therefore, it was not permissible to use Ehrenfest's theorem (which showed the approximate validity of classical mechanics for small wave packets) in order to deal with the motion of these molecules in external fields. This was a rather serious problem, since this particular theorem had been the basis of the corpuscular treatment of transport phenomena. But London seized the opportunity to assert that there was a fundamental similarity between superfluidity and superconductivity, a similarity which transcended the statistics of the particles involved and was related to the macroscopic quantum nature of both phenomena. In the case of particles taking part in the Bose–Einstein condensation, the wave packets were of a very small extension in the space of momenta and in ordinary space they were spread over a region comparable with the extension of the inhomogeneities of macroscopic fields. 'The quantum mechanics of [such a case] has been investigated very little. Only one interesting special case has so far been discussed in connection with superconductivity'. And in order to discuss the phenomenon of superfluidity, London made a daring proposal to circumvent this difficulty. He wanted to investigate the problem of what would be the effect if, in Sommerfeld's theory of conductivity, the Fermi statistics were formally replaced by the Bose–Einstein statistics. He hoped that such an attempt was as justified as the classical treatment of diamagnetism on the basis of Larmor's theorem.[66] In order to do this, London depicted liquid helium as a metal where ions and electrons were replaced by particles of the same kind. Each such atom moved in the self-consistent field formed by the other atoms. The states of this system were divided into two classes. One was like a Bloch-type and corresponded to the electronic states of the metal and represented a transport of matter. The other class was more of the Debye-type quantized acoustical or elastic waves, corresponding to the vibrations of the ionic lattice. 'In this picture the *fluidity* of the liquid would correspond to the electrical *conductivity* of the electrons in a metal'.[67] London did acknowledge that Tisza had already discussed the transport properties; he remarked that his own proposal should not be considered more than a preliminary approach to the problem of liquid helium and that a molecular theory of liquids in general was still wanting.

Meanwhile, Uhlenbeck had heard from Goudsmit that London and Tisza were interested in the Bose–Einstein condensation. He sent them a paper, written with Kahn, where they had retracted Uhlenbeck's original objection. He did not agree with London in describing the transition as third type. If one insisted on Ehrenfest's nomenclature, then the transition should be of the first type.[68] Schrödinger found the proposed explanation quite successful, but he wanted to study the paper more fully, since he could not understand all the details.[69] London was informed by H. Jones, who was at Cambridge University, that, together with Fowler, they initially found it difficult to believe in the existence of a discontinuity, but then they succeeded in constructing a rigorous proof that a discontinuity did exist the way London had described it.[70] London was gratified. He was aware that his own quite elementary calculation, which went almost exactly along the same lines as Einstein's original proof, could not claim any particular originality.[71] Keesom asked one of his assistants, Groenewold, to send his comments to London.[72] London was not overly enthusiastic: when Groenewold pointed out that, for both the structure and the transport phenomena of helium-II, the correlations between particles were important and that, therefore, the Hartree method was not the proper method to use, London, uncharacteristically, wrote in the margin 'then let *him* create a better method!'.[73] But Groenewold, echoing Keesom's misgivings undoubtedly, questioned the validity of the extension of the Bose–Einstein condensation to real gases. London reiterated the strategy he would follow. He was confident that every method of physics was an approximation, although its area of applicability was not always precisely predictable. He was confident that physicists believed in its results until they constructed a better one or unless they had another reason to doubt the result.[74]

The two-fluid model

Laszlo Tisza, a Hungarian-born physicist, had been in Kharkov, USSR, with Landau from 1935 to 1937. Tisza considered himself to be Landau's student and, after he left Kharkov for Budapest, he was determined to pursue the thermodynamic ideas which he had acquired in Landau's group. In September 1937, he attended a conference in Paris, and London, who had been asked to help by Leo Szilard, recommended him for a scholarship at the College de France, which Tisza received. Less than a day after London had visited Tisza to tell him about his thoughts concerning Bose–Einstein condensation, Tisza got, roughly, the two-fluid idea. He concentrated on the greatest difficulty, that is on the fact that the results of the two measurements of the viscosity were absolutely irreconcilable. Right from the beginning he did not attempt to 'save the old theory', being convinced that 'something in the method had to be changed'.

> And that was my guiding idea. And I had a crazy night, but I do not remember the details, and probably there was nothing logical about it ... but by next morning, I had somehow the basic features of the two fluid model. But it was in an extremely immature and sloppy fashion.[75]

Tisza's first step was to examine the concept of viscosity in liquids and gases, in view of the discrepancy between the methods of measurement of viscosity, and he concluded that this was not a kinetic coefficient of an unusual value, but the breakdown of the viscosity concept. There was no Navier–Stokes equation with a viscosity parameter: 'I concluded there must be a "mixture" of a superfluid and viscous, or normal component'. A very narrow capillary (acting as an 'entropy filter') was permeable only to the superfluid flow, but not to the normal fluid. Such a division into interpenetrating, but distinct, entities was provided by the Bose–Einstein condensation, but the gas pressure was now more an osmotic pressure: 'When I presented all this to London the next morning, he was unimpressed, and his attitude was duplicated by every theorist I tried to convince for a number of years. It is true my ideas did not qualify as a "theory" by any admitted standard'. London became very disappointed and unpleasant about it. He said 'that is all nonsense' and he did not want to accept it at all. London's main objection was the impossibility of the independent flow of two interpenetrating systems. What was London's reaction, since Tisza's model could explain qualitatively practically everything?

> He did not address himself to this question. In fact, I had this experience in the years to come, that theorists with whom I have talked have certain standards, we should start from the usual Newtonian or the hydrodynamic equations or this or that, and making a completely new departure is simply not allowed. And in fact, to explain a phenomenon from factual assumptions was considered in a way sort of a dishonest way of going about it. I remember a few months later, when I desperately worked on the theory and tried to make it more understandable, we were at London's, and we played some game where you had to guess some sentence, and he posed a sentence to be guessed that 'I can explain all properties of liquid helium'. So in a certain sense, the fact that this explained everything made it very suspicious to him.[75]

London was much more conservative and he concentrated on understanding the lambda-point transition rather than the properties of helium-II. He wanted to give himself time to reflect on the new possibilities provided by the Bose–Einstein condensation. 'And in this sense, I can understand very well that he was angry at my jumping the gun, because he felt that he had a solid idea, and instead of pursuing it with care, I just rushed ahead, and I do not resent the fact that he did not accept it. He had good reasons'.[75]

Tisza's first note in *Nature* on May 21, 1938, received much attention.[76] In his own words, somewhat opportunistically, Tisza started the paper by noting London's paper on Bose–Einstein condensation, thus giving the impression that London's treatment of the phenomenon of condensation could provide some kind of a theoretical basis for his model. In this model, helium-II is regarded as a mixture of two (completely interpenetrating) components, the normal and the superfluid. These components or 'fluids' are distinguished by exhibiting different hydrodynamic behaviors, in addition to the difference in their heat contents. While the uncondensed normal fluid is supposed to retain the properties of an ordinary liquid (it is identical with helium-I), the condensed superfluid fraction of helium-II is meant to be incapable of taking part in dissipation processes. At absolute zero, the entire liquid is supposed to be a superfluid consisting of condensed atoms, while, at the transition temperature, this component vanishes.

The hydrodynamic properties of such a mixture were quite complex, but they were flexible enough to explain things which appeared to be paradoxical in ordinary hydrodynamics. Hence, an oscillating disk in helium-II experienced friction by the normal fluid, while a fine capillary allowed the superfluid to pass without experiencing friction. Similarly, an interpretation was provided for the thermomechanical effect. Since, in this model, the temperature of a volume of helium-II simply meant a relative concentration of the two fluids, a change in this concentration could be registered as causing either cooling or heating. Absorption of heat had the effect of increasing the concentration and osmotic pressure of the viscous component at the expense of the superfluid which was sucked into the cell. The obvious conclusion from this explanation was the prediction of the inverse effect, namely that helium forced through a fine capillary should be richer in superfluid and therefore should exhibit a drop in temperature. This effect known as the mechano-caloric effect was observed in 1939 by Daunt and Mendelssohn. The anomalously high heat transport in helium-II was also consistent with the assumptions of the two-fluid model. The superfluid and viscous components had different flow velocities, and this gave rise to an internal convection which was connected with an energy transfer without any mass transfer. This internal convection accounted for the super heat conductivity. Tisza predicted that the thermomechanical effect ought to have an inverse: a superfluid transfer from vessel A to B should lead to the heating of A and cooling of B. This was readily verified.

A few months later, in another short note presented to the Académie des Sciences in Paris, Tisza went much further; he recognized that his model implied a very strange feature, namely that, in liquid helium-I, the temperature would obey a wave equation. Tisza called these waves 'temperature waves' – later, Laudau called them 'second sound waves' (see Chapter 5) and the temperature dependence of their velocity would be a decisive test of the validity of the two-fluid model. This peculiar wave propagation ('perhaps the most interesting deduction from the Bose–Einstein theory')[77] was predicted in a paper which was published just before war broke out. Work on liquid helium in Europe stopped with the outbreak of World War II, and only in the Soviet Union were Kapitza and his group able to continue their researches in low temperature physics.

London did not think much about the thermal waves, partly because he tended to believe in Heinz's proposal that the fountain effect was a surface effect. He felt that Tisza was compromising the Bose–Einstein theory by carrying it to an improper extreme. When Simon and Kurti visited Paris in late 1938, Tisza discussed the model with them, especially Simon. 'They were amused. They considered it sort of ingenious, and sort of wild'. He visited Oxford and had long discussions with the physicists there, and he explored the possibility for detecting thermal waves with Mendelssohn. But they did not pursue the matter further. 'I think the official line is that the war came in between, but that was not really so. It was not the war that interferred, it was the war between London and myself'.[78]

London was not at all enthusiastic about these developments. 'Tisza's article in *Nature* has not convinced me – independent of whether it was morally correct for him to publish it. It was sent by him when I was in Palestine'.[79] Even after discussions with Tisza during walks in the Luxemburg Gardens in Paris,[80] London continued to disagree with his approach. He felt that their theories differed from each other in quite fundamental ways. 'I believe I have said something new about the statistics of Bose and all this is related to helium purely hypothetically'. Fritz strongly disagreed with Heinz when the latter referred to his paper as the basis for Tisza's paper. 'My paper is clearly related to Bose condensation, and whatever else I say there are only superficial references to helium. This could have been done differently'.

Fritz thought that the correct description of the situation would be something like

> Starting from the idea that the phase transition into helium-II might be interpreted as

> {the realization of
a manifestation of
due to}
>
> the condensation phenomenon of the Bose–Einstein statistics . . . and from certain kinetic speculations Tisza . . . has predicted the existence of reversible effects.

In connection with Fritz's observation about the possibilities being offered by the Bose–Einstein condensation, Tisza proposed a completely general and qualitative picture. The reversibility of the fountain effect was, according to Fritz, too inaccurate. Heinz London (1938) had deduced the thermodynamic formula and had also pointed towards the thermoelectric analogy as an example. Then, Fritz had shown how one can make a molecular-theoretic model for such a thermoelectric mechanism on the grounds of Bose statistics, and Heinz had 'laid the strongest claim to be viewed and recognized as the spiritual father of this child'. But it seemed to Fritz that Heinz thought that there were two proposed models which appeared to be equally general: the first was Heinz's and Tisza's; the second was the one by Fritz. Fritz believed that Heinz preferred the model by Tisza and that he did not want to consider theirs any more. 'I think you cannot turn your back like that. Such an underestimation of one of the alternatives, can be considered, by those who will study this work, only as a rejection of this standpoint, in a brotherly beautiful form, which, of course, cannot be your intention'.[81]

The trip to Jerusalem

At the beginning of 1937, Fraenkel informed London that, during his forthcoming trip to Europe, he would like to discuss with him the possibility of a Chair in Mathematical Physics at the Hebrew University in Jerusalem.[82] Fraenkel was Professor of Mathematics there and the Director of the Faculty of Science that was founded by Weizmann in 1935. But political and professional prospects seemed quite bright for London in Paris. He started having second thoughts about going to Jerusalem; he was not as enthusiastic and he wanted to see whether it would be possible to be invited for a short period before fully committing himself. If this was not possible, London suggested that Fröhlich might be interested in such an offer. Fraenkel promised to discuss London's proposal with the authorities.[83] It was obvious, though, that Fraenkel's associates in Jerusalem could not take too many chances and that if they wanted to go ahead in forming a strong faculty they could not afford to be too flexible to demands which,

Figure 10. Louise London before leaving Germany in 1938. (Courtesy of Lucie London.)

under different circumstances, would be part of normal practice in academia. At this stage, they had to have good scholars who would stay permanently – but at the same time they had to try not to discourage anyone. 'I am glad that you comprehended me. I will surely come to Palestine one day on my own initiative. I consider the fact that I do not know Palestine a great lack of my education'.[84]

London and Fraenkel met in Paris during August 1937. They agreed to a three-month visit by London, starting in December, and he received an invitation to visit the Hebrew University 'in order to be acquainted with its work'.[85] But London's heart was set on staying in Paris. He wrote to Fraenkel to tell him that he could not go without seriously severing his relations with colleagues in Paris. But it was not only his obligations towards his colleagues in Paris that made him change his mind. 'My only and decisive obligation results from the situation about the needs of the Jewish intellectuals'.[86] In France, under the new Government there were the first signs of a new and very positive attitude towards the intellectuals, which, London felt, had already led to quite specific developments. He was convinced that, if he had decided to leave, it could have created adverse effects and some could lose 'their appetite to continue the liberal policy, if the first person to whom this possibility is granted for staying in France, leaves immediately'.[86] He reiterated that the future of Palestine interested him greatly and that he hoped to be somehow involved in these developments. The letter was written in November, or December, 1937. All this, of course, was also relevant when he had met Fraenkel and when, a couple of months earlier, he had accepted the invitation. His change of mind was not independent of the immense importance for London of what had intervened, in the meantime. It was the van der Waals conference in Amsterdam and the ideas on Bose–Einstein condensation and liquid helium. He had not really done anything significant since the superconductivity paper in 1935 and he knew that, if he took the long trip to Jerusalem, he would have to postpone the work, which he could definitely not afford to do. There was, also, another aspect to London's half-heartedness about the whole enterprise. The offer made to him specified that he must be able to teach in Hebrew one year after he started teaching. It was another taxing condition, not only because of the difficulty of the language, but also because during the last years he first had to grapple with English and, then, with French. Learning languages was not among his talents, and he still recalled the embarrassing scene when after an English class in the Gymnasium, the teacher turned to him and told him not to come again unless he was serious about learning the language.

London agreed to take a short trip to Jerusalem after the end of

February 1938, but just as he and Edith were ready to leave, they cancelled their trip, having heard of serious troubles in Palestine.[86] Fraenkel wrote a polite letter to London that unless he went they would be obliged to offer the position to someone else. They could not afford to determine their course of action according to the developments in Europe and their insistence 'would show in practice that we can build a life outside the tensions of Europe'.[87] A few days later, the Londons changed their minds and were on their way to the 'precious Holyland',[88] taking the boat from Venice to Haifa.

In April 1938, London's name was mentioned in the correspondence between Jacques Hadamard and Einstein. Hadamard, a well known mathematician and champion of human rights, was a Professor at both the Ecole Polytechnique and the College de France. He was a member of the Franco-Palestine Committee and of the governing body of the Hebrew University. He wrote to Einstein that, after the information he had gathered, he agreed about 'London's superiority with respect to all the other candidates',[89] though he preferred Mathiesson or Infeld. Einstein was very keen that London should have the Chair at *notre université*. 'The best thing would be to obtain London's acceptance, who in every respect is a man of value. If this would not be possible, there will result a situation which appears to me rather dangerous'.[90] Einstein did not think that Mathiessen was the right choice because of the abstract nature of his physics.

They were in Palestine during April 1938 and it was the Passover vacation. They were taken around by Blumenfeld, one of the best known Zionists. Mrs Blumenfeld was the president of WIZO, the Women's International Zionist Organization. For London, Blumental proved to be too much of a propagandist, and he was not content by any of Blumental's answers to his questions. They also visited the Daniel Sieff Institute at Rehovot. On April 17, 1938, he met with Werner Senator from the administration of the Hebrew University, and was pressed to tell him his final decision. He officially declined the offer.[91] He was truly impressed with what he saw and felt grateful about the 'prospect of being able to cooperate in this dynamical life', despite his reservations about 'whether my work could bear fruits without the experimental background'.[92] While in Jerusalem the very conservative Fraenkel invited them to spend Passover together. To London it seemed to be an endless night. Later, Weizmann joked that London rejected the offer because he thought everyone in Israel observed religion like Fraenkel.

In December 1937, the Nobel Prize for Chemistry went to Sir Walter Haworth from England and Paul Karrer from Switzerland for their researches on vitamins. Fritz London had been one of the nominees for the Prize, and had been proposed by Roginskiy, who held the Chair of Physical Chemistry at the Leningrad Polytechnic Institute.[93] There are no indications that London ever learned about his nomination. In the same month, Paul Gross, Professor of Chemistry at Duke University, North Carolina, USA, communicated with London to discuss an offer from Duke. Gross had corresponded briefly with London earlier on some matters of molecular physics, and he was in Europe during 1936–1937. They had met briefly in Paris and had discussed the possibilities of a job offer and the duties involved in such a job. He asked London to accept the position of Associate Professor of Theoretical Chemistry at the Department of Chemistry at a salary of $4000. He would be expected to give a course of lectures and to act as advisor on theoretical problems arising in connection with the research in chemistry. The appointment was for one year starting from September 1, 1938, and subject to renewal.[94] London did not, of course, like the idea of being offered a non-tenured job. He expressed his strong interest, but proposed that he should go to Duke for a shorter period, since he felt that he could not really quit his present job which, for all practical purposes, was a tenured position.[95] Gross wanted to establish strong science departments at Duke and to make it one of the main centers for science in the South, so he convinced the administration to agree on an initially more flexible arrangement and to a visit of one semester. When Fritz and Edith returned to Paris from Jerusalem, they found the offer from Duke, 'an opportunity I should not let go'. Fraenkel wished him the best of luck. He told London that Weizmann could intervene to secure a legal stay in the USA, and urged him to establish some ties between the two universities. Fraenkel was hopeful that they would be able to build a good department 'with or without the Jewish State'.[96]

Leaving again

London was offered the post of Visiting Lecturer for the fall semester of 1938–1939, and was asked to start his lectures on October 1 and to complete his stay by the end of January. He gladly accepted the offer.[97] Gross' gentle insistence, versatility and initiative had convinced London to accept Duke's offer. Two other well known scientists had also accepted offers from Duke: Nordheim had an appointment in the Department of Physics and so had Hertha Sponer who eventually married James Franck. As a result, Franck visited Durham quite often from Chicago, every time a changed man, his eminent past repeatedly marred by unpleasant incidents. The presence of London, Nordheim and Sponer at Duke became Gross' great assets, while he successfully convinced many young and promising graduates from the North, especially from New York, to come to the South. Gross eventually became Vice President of Duke University.

During the academic year 1938–1939 Duke celebrated its centenary and contemplated inviting Irene Joliot-Curie and Frederic Joliot. Gross asked London to see whether their English was fluent enough for them to give an address in English, and whether they would be interested in taking part in the celebrations. London discussed this possibility with them and wrote to Gross that they would both be interested in going to Duke. 'I think they would certainly very much appreciate such an opportunity of coming to the States. I might venture to add my own opinion I would suggest that you could scarcely make a better choice than to get [such] a famous and attractive representative of French science as the Joliots to your celebrations'.[98] Duke decided, in the end, to invite only Irene Joliot-Curie to participate in a discussion during a symposium on women's interests.

About a month before his arrival in the USA, London asked the university to inform the immigration authorities, to avoid 'unpleasant difficulties'.[99] He planned to travel with Edith. Edith was pregnant and a couple of weeks before their departure the doctor advised her not to travel. Upon his arrival in New York, there ensued the unpleasant difficulties London wanted to avoid. To be admitted into the country, one had to have a passport which was valid for, at least, six months after the date of entry. The immigration authorities at Ellis Island, New York, noticed that London's German passport was due to expire in less than six months. So they did not clear him, and London had to spend five days at Ellis Island, waiting for the authorities to make the necessary inquiries. He asked the guards if he could be given a chair and a table so that he could prepare his lectures. The guards did everything they could for the professor. And after the intervention of many of his colleagues, especially Isidor Rabbi, Professor of Physics at the University of Columbia in New York City, London was granted entry into the USA.[100]

In November 1938, London visited the University of Chicago, where he gave a talk.[101] He was also invited to give a talk at Johns Hopkins University[102] and to present a paper at the Symposium on Intermolecular Forces during the Annual Meeting of the Division of Physical and Inorganic Chemistry of the American Chemical Society, which was held at Brown University on December 27–29, 1938.[103]

Early in 1938, London discussed his future in Paris with Joliot. He had told him that he was willing to risk the situation and was thankful to Joliot for finding some extra funds. As time went on, he started having second thoughts, particularly because of the increasing risks of war. But he also felt that there was 'another danger even more threatening than war. You will understand what I think, if I remind you that at a previous time I was obliged to leave my country for the same reason'.[104] He told Joliot about the good time they were having in France and hoped that a possible decision to go to the United States would not be regarded as a grave ingratitude on his part. While in the USA during the fall of 1938, London wrote to Joliot to thank him for his quick response to Wigner's telegram asking him to intervene so that London could be freed from Ellis Island. He felt that an Atlantic crossing could not get rid of the problems which he felt were now 'part of ourselves'. He was not particularly happy in his new environment. 'I am too European to be able to become too enthusiastic about life here, which even for those child-like adults here is too calm . . . It appears to me that people here are free of passions except bridge and football . . . Politics does not seem to be particularly interesting in Durham'.[105] On November 26, 1938, London received a telegram from Joliot asking him whether he wanted to accept the post of *Directeur des Recherches*. London did not feel able to commit himself by accepting a post which he might have to resign in a very short time.

Once back in Europe, the political developments reinforced his decision and served to make the offer from Duke – however isolated one felt and however far away it was from the main centers of research – the only viable alternative for his family with their newborn son, Frank.

The role of the observer in quantum mechanics

Reading London's works, one often gets the impression that at times he had the urge to clarify certain issues for himself and he proceeded by writing about them. Certainly his thesis in philosophy and his interventions in the dispute between Tolman and Ehrenfest (as well as some of the papers he wrote about quantum mechanics, superconductivity and superfluidity) bear this characteristic. After reading these papers and considering them within the larger context of his research programs, one can argue that these are papers expressing a peculiarly internal process of settling some accounts and closing a phase of long and tortuous thinking. His thesis in philosophy was his reconciliation with his Kantian tradition and his abrupt baptism into the overflowing streams of phenomenology. His paper about the possibilities provided by dimensional analysis was also an indirect statement about the inherent deadlock of all such attempts, independent of their mathematical sophistication. I do not think that the lack of any further paper discussing similar issues was related to the contingencies of his new professional interests. One can almost share his feeling that he had reached that point in those philosophical wanderings where he had clarified a series of issues for himself, and that any further preoccupation with similar problems required a decision concerning the character of his involvement with these problems. His *Conception Nouvelle* was another such case where he sought the opportunity to sum up his thoughts and to state the direction of his program in superconductivity. There was no such work of his about valence and the fact that he could never finish the book he had started on questions regarding the forces between molecules and atoms, was also suggestive of his specific style of work. Surely he could have written a book about the problems in molecular physics and quantum chemistry. But there are no indications that he had settled his accounts with the dominant issues there. His correspondence with various physicists and chemists, and especially the many letters he exchanged with Walter Heitler, bear witness to his uneasiness. In fact, what he perceived to be the charlatanism of the recently inaugurated approaches by the Americans, which were commented on so caustically in the correspondence with Heitler and Born, should have provided an additional incentive to write the book. This inability to finish his book on atomic physics should not be explained only in terms of his having to leave Germany and spending the three unhappy years at Oxford University. Not insignificantly, he was, at that time, in the midst of a methodological quagmire. On the one hand, he had renewed his contact with a trend he had neglected since completing his philosophy thesis. The antireductionism which was so pronounced in his philosophical thinking came into the fore and found a rather practical manifestation as a result of his work in low temperature physics. But, on the other hand, the work on valence, in the last analysis and after all the 'buts' and 'ifs', gave credence to a strictly reductionist program expressed so strongly by Dirac: chemistry is a problem of physics with immense calculational difficulties – but a problem of physics nonetheless. His work on the covalent hydrogen bond was a triumph of this program, and, though his work was in the mainstream of research, London refused to accept reductionism. He set out to convince the community to realize and to accept some radically new possibilities provided by quantum mechanics. Though quantum mechanics had provided the framework legitimizing Dirac's reductionist claim, that same framework also allowed the formulation of those concepts which could legitimize a non-reductionist claim by London, which was the description and explanation of macroscopic quantum phenomena.

In Paris, London wrote a monograph with Edmond Bauer titled *La Théorie de l'Observation en Mécanique Quantique*. The authors intended to present a text which would be more comprehensible than von Neumann's brilliant analysis of the problem. Furthermore, von Neumann's book was in German.

To write a 'simplified' version of von Neumann's difficult book, which could be understood by all who were not seeking such a sophisticated mathematical treatment of quantum mechanics and to analyze further the role of the observer which von Neumann had not fully elaborated, was all London and Bauer wanted to do. Neither they nor Langevin in his introductory note suggested that there was much else involved. In fact, the monograph was particularly interesting because it provides many insights into London's views about quantum mechanics generally. That Bauer and London did not want to get involved in the discussions about the epistemological and ontological issues which were being debated is supported by their neglecting even to mention either the Einstein–Rosen–Podolsky paper or Schrödinger's paper where he discussed the famous 'cat paradox'. Both papers had been published when London and Bauer were working on their monograph and, if anything, the criticism of quantum mechanics expressed in these papers may have had the opposite effect on London and Bauer. Convinced of the correctness of the 'orthodox' interpretation, they may have thought that all these difficulties were a misunderstanding and that a clearer exposition of von Neumann's work, which had summarized everything so brilliantly, was all that was needed to disperse the confusion among the physicists. Von Neumann did not include the consciousness of the observer to the measuring chain. The novelty of the London–Bauer treatment was the explicit claim that the reduction of the wave function was the result of the conscious activity of the human mind.

Any measurement in quantum mechanics changes the quantum mechanical state of a system. According to von Neumann, this change was a discontinuous, non-causal, instantaneous and irreversible process consisting of two stages: the interaction between the object and the apparatus, and the act of observation. Having shown that such an irreversible process was an inherent characteristic of a theory of measurement in quantum mechanics, von Neumann contended that there could not be any such theory without reference to human consciousness. London and Bauer's contribution was to discuss explicitly the issue about human consciousness introduced by von Neumann, in the context of his theory and without resolving (and not intending to resolve) any of the epistemological and ontological problems entailed by his analysis.

In quantum mechanics, the interaction between a system and the measuring apparatus does not result in a new pure state for the system, but, on the contrary, there is a statistical mixture that is uncertain, and incomplete knowledge. Such an uncertainty could not be reduced further, and their first conclusion was that 'a measurement is achieved only when the position of the pointer has been *observed*'.[106] The act of observing increases the knowledge about the system and it is because of this increase in knowledge that the observer can choose among the different components of the mixture predicted by the theory, and, of course, can reject those which are not observed. London and Bauer explicitly emphasized the 'essential role played by the consciousness of the observer' in this process.[107]

For the observer, the 'objective world' is the object and the apparatus. Because of his quite unique characteristic of having the faculty of introspection, he knows at any moment his own state.

> By virtue of his 'immanent knowledge' he attributes to himself the right to create his own objectivity [and thus] the observer establishes his own framework of objectivity and acquires a new piece of information about the object in question.[107]

A pure case is the result of an 'act of objectification' brought about by the intervention of the observer. If the measuring device is regarded as a filter that singles out a prescribed value of a certain physical quantity, then, it is stressed, such a filter does not put an individual object into a new pure state, but only into a mixed state. It is always possible to attribute the wave function of a pure case to, say, the states of atoms which have gone through a slit and have a particular property. But it is not possible to know in advance which individual atom has the particular property and, therefore, this assignment of the wave function undermines the individuality of the atoms. In order to find out whether a particular atom has gone through the slit or has been retained by it, an observer is required to proceed to a supplementary check.

> The filter alone thus produces pure cases, but in an *anonymous* form ... Anonymous objects are precisely the focus of one's interest in many experiments. The majority of the measurements in atomic physics really do not deal with an *individual system*; rather, they seek to find out the general properties of an *entire species* of atoms – or of molecules, or of elementary particles ... Quantum mechanics, truly a 'theory of species', is perfectly adapted to this experimental task. But given that every measurement contains a macroscopic process unique and separate, we can hardly escape asking ourselves to what extent and within what limits the everyday concept of an individual object is still recognizable in quantum mechanics.[108]

In the last section of their monograph, titled *Scientific community and objectivity*, London and Bauer discussed the notion of objectivity that 'appears to have been strongly shaken' by quantum mechanics. The reality and objectivity of a system in classical physics is a characteristic resulting from the fact that it is possible to picture the system at every instant in a unique and continuous manner by the set of all its measurable properties, even when this system is not subjected to observation. And the significant feature establishing the reality (that is, their existence independent of all observers) of the objects of classical physics is the fact that its theories allow for the continuity of the connection between properties and object. However, in quantum mechanics, the object itself is the carrier of a set of potential probability distributions, referring, of course, to measurable properties. But the mathematical structure of the theory renders meaningless any claim that these measurable properties could be realized independently of any well defined process of measurement. The coupling between the measuring apparatus and the observer is a macroscopic action and it is possible to neglect the effect of the observer on the apparatus. Furthermore, it is possible to be led to specific conclusions concerning the state of the apparatus, and, therefore, of the state of the object after the coupling is turned off and, hence, before the observation. Another observer can, then, look at the same apparatus and would observe the same results as the first. There is, at least, something like a community of scientific perception, an agreement on what constitutes the object of the investigation. The individuality of the observer can now be assessed only within this context of the collective perception of the scientific community.

London and Bauer attempted, finally, to adapt the notion of objectivity to the new state of affairs. Since the various properties of the object are present at the moment they are measured and since what is predicted by theory is in agreement with experiment, then one can, in fact, claim that the notion of objectivity is not undermined. They noted that the determination of the necessary and sufficient conditions for an object of thought to possess objectivity and to be an object of science was an important philosophical problem, first posed by Malebranche, Leibniz and Bolzano. Husserl's systematic investigations of such issues 'created a new method of investigation called "Phenomenology"'. The last paragraph of the text (written by London alone) is the most explicit statement by London about the relationship between physics and philosophy:

> Physics in so far as it is an empirical science cannot enter into such problems in all their generality. It is satisfied to use philosophical concepts sufficient for its needs; but on occasion it can recognize that some of the concepts that once served it have become quite unnecessary, that they contain elements that are useless and even incorrect, actual obstacles to progress. One can doubt the possibility of establishing philosophical truths by the methods of physics, but it is surely not outside the competence of physicists to demonstrate that certain state-

ments which pretend to have philosophical validity do not. And sometimes these 'negative' philosophical discoveries by physicists are no less important, no less revolutionary for philosophy than the discoveries of recognized philosophers.[109]

Abner Shimony has discussed some of the problems arising from this version of von Neumann's analysis. London and Bauer assumed that the states of an observer satisfy the quantum mechanical requirements only partially: they satisfy the vectorial relations and, yet, they do not evolve temporally. Although such an extension of quantum mechanics 'into the domain of psychology ... is a strange proposal, it contains no obvious inconsistency'.[110] To decide whether such a proposal could be factually correct, Shimony proceeded to do a brilliant analysis of two interrelated questions: whether mental states satisfy a superposition principle and whether there is a mental process for reducing a superposition. He found no empirical evidence to support such assumptions and concluded that the von Neumann–London–Bauer interpretation of quantum mechanics 'rests upon psychological presuppositions which are almost certainly false'.[111]

London never worked systematically on questions related to the measurement problem in quantum mechanics before, or after, the appearance of this monograph, nor did he ever correspond with anyone about these issues. In fact, there does not appear to be any evidence that questions about the problems of interpretation of quantum mechanics were of any particular interest to London. By the time London and Heitler arrived in Zürich in 1927, Schrödinger had already completed his papers on wave mechanics. He had already shown that the newly proposed mode was equivalent to Heisenberg's matrix mechanics, and he had started to express his objections to the interpretation of the wave function by Born and Bohr. There was not much talk about questions of interpretation of quantum mechanics, and Heitler and London were mainly preoccupied with understanding how to apply wave mechanics and to explore fully the contents of the new schema. Only after completing their paper on the hydrogen bond did they become interested in questions of interpretation, and, as Heitler recalled in his interview (see Chapter 2, note 5), they avoided discussing 'hot subjects with Schrödinger. We were youngsters and he was the great man, so even if we did not believe him we avoided telling him so'. By this time, Schrödinger was already preparing his intervention to the 1927 Solvay Congress which 'formalized' the Copenhagen–Göttingen interpretation of quantum mechanics. They read Born's papers, and, by the end of their stay in Zürich, they had adopted the probabilistic interpretation of quantum mechanics. When London moved to Berlin he was quite immune to the objections of Einstein and Schrödinger (and also partly of von Laue).

After the publication of the monograph, there is nothing in London's correspondence where one can find a discussion of some of the issues relating to the theory of measurement in quantum mechanics. The only evidence is provided by the letters he exchanged with Bauer when he was at Durham University for a semester in the fall of 1938. In section 14 of the monograph, London first suggested the word *spiritisme* to follow 'scientific community', Bauer thought *solipsisme* would be more appropriate, London answered by saying they should settle for *Communuté et realité scientifique*; eventually they decided to title the section *Scientific community and objectivity*.[112] When Paul Zilsel, in the late forties, asked London his opinion about all the disputes related to quantum mechanics, London did not think that there was much substance in those discussions and he felt that not much could be added to what he had already done with Bauer. Neither did Bauer write anything about these issues before or after his collaboration with London.

London received the galley proofs of the manuscript in the summer of 1939 – his last summer in Paris. It was very hard to find a passage to the USA and London had to bribe an agent to find them tickets. They were planning to sail with the *Athenian*, but at the very last moment, because of their German passports, they were not accepted on the *Athenian* and sailed with the *Île de France*. They left Southampton for New York on September 1, 1939 – the day Germany invaded Poland. On September 3, the *Athenian* was torpedoed by German U-boats.

Notes to Chapter 4

1. Jackson (1988), 114.
2. Gurwitsch was strongly antireductionistic in his book *Phenomenology and the Theory of Science*. Prof. I. B. Cohen recalled that Gurwitsch used to tell him about London's deep understanding of philosophical problems and how this was related to his understanding of problems in physics.
3. London to Becker (September or October, 1937).
4. Fritz to Heinz (November 4, 1937).
5. London to his mother (late fall, 1936).
6. London to Freymann (editor of Hermann Publications; April 12, 1936) informing him that the title would be *Théorie Macroscopique de la Supraconductibilité*. (See also letters of May 4 and May 6, 1936, with publishers.) Its publication was delayed and London was anxious to have it by June 1936 before the Low Temperature Conference in Holland. Magat had taken care of the language problems of the text. London insisted that the book should appear as quickly as possible, otherwise it might lose its intended value. By August 1936, the book was still not in circulation, and London felt that it had lost a lot of its value as an *actualité scientifique*. There had been quite a few theoretical as well as experimental developments over the summer and he asked the publishers to allow him to include some additional paragraphs and to guarantee him that it would be in circulation within two months from the date they received the corrections. The book circulated by the

end of the year (London, 1937*a*). His examination committee was Eugene Bloch (president), Louis de Broglie and Francis Perrin.

7. There was a rather unusual exchange with the Editor of *Nature*. London had sent the article titled *On a new conception of supraconductivity*. The Editor had misgivings about the word 'on' being 'very inappropriate in the title of an article appearing in *Nature*, although, of course, he was aware that it was frequently used in the titles of scientific papers'. The Editor of *Nature* to London (August 24, 1937). London answered: 'By using the word "on" I thought to indicate that the paper has not as subject a consistent representation of the new conception itself, but rather discusses some characteristic features of it and that the new theory is represented only so far as this is necessary for the understanding'. (In the draft letter, he also wrote: 'By the choice of the title I also wanted to avoid intimidating those readers who do not like mathematical theories'). The article was eventually published without the word 'on'.
8. London to Simon (undated, but probably September 1936).
9. London (1937*c*), 794. In addition, in 1937 and 1938, two papers by Slater were published, proposing a theory for superconductivity. In the first article, Slater did not mention any of the work by the Londons. Fritz decided to send a letter to the *Physical Review*, which did not attack Slater directly, but which expressed their theory in a more economical manner of which he said 'we should have done it all this time and I am amazed why we did not do it' (Fritz to Heinz, February 28, 1937). Fritz asked Heinz to decide whether he wanted to coauthor the paper, but warned him that Slater might get angry. 'I do not have anything to be afraid of, because he hates me anyway because of the chemical bond that he thinks he has done. But it can hurt you'. He told Heinz that the purpose of the letter was not to attack Slater, but to salvage their theory from being forgotten or from 'deathful silence', without, however, appearing that they were attempting such a thing. In his second article, Slater acknowledged that he had written it after Fritz London had 'objected quite properly to the earlier paper, on the ground that superconductivity has much closer resemblance to diamagnetism than to ordinary conduction'. Slater had shown that there could be discrete energy levels of the conduction electrons below the continuous bands, under certain circumstances. Slater's arguments were partly qualitative and it was not possible to solve the very difficult mathematical problem accurately enough to prove that the discrete energy states had the properties necessary to produce the phenomena of superconductivity. As Heinz noted, people at Bristol did not take Slater's theory seriously at all (Heinz to Fritz, February 17, 1937).
10. Interview with Professor Laszlo Tisza by K. G. (May 1988, conducted in English) which is deposited at the American Institute of Physics. See also London's talk at the Lorentz Medal ceremony (Chapter 5).
11. Von Laue to London (April 14, 1937).
12. London considered that the analytical solution of the problem was quite hopeless, since it involved a complicated boundary value problem. However, he found Landau's (1938, 1943) approach quite interesting. It was among the very few cases where London had a good word for Landau.
13. London to von Laue (April 17, 1937).
14. Von Laue to London (November 16, 1937).
15. London to von Laue (undated but most probably November 1937).
16. Von Laue to London (April 27, 1938; May 10, 1938); London to von Laue (May 22, 1938); von Laue to London (December 21, 1939).
17. At Bristol University, Heinz London resumed the experiments on high frequency currents and by 1940 he had observed the phenomenon he expected. At 1500 MHz, the alternating current resistivity was appreciably higher than the resistivity measured by direct current. At Bristol University, together with A. D. Misener and J. R. Bristow, Heinz was able to produce high quality thin films, and he could show the existence of superconducting penetration depth, which had been predicted by their theory. It was in Bristol that Heinz started studying the properties of liquid helium. In 1939, he suggested an explanation of the fountain effect (see section titled *The peculiar properties of helium*, this chapter) and predicted its inverse – the mechano-caloric effect. Tisza, using his two-fluid model, predicted the same effect, but Heinz's reasoning was purely thermodynamic. The thrust of the argument referred to questions of irreversibil-

ity: in the liquid helium problem, irreversibility due to viscous flow could be made negligible when the capillaries between the hot and cold reservoirs were made sufficiently fine. Characteristically, Heinz would say that the fountain effect was an example of the simplest heat engine and that it should be presented to students as a convincing example of the second law of thermodynamics.

18. Heinz to Fritz (October 21, 1936).
19. Heinz to Fritz (undated, but most probably fall 1936).
20. Fritz to Heinz (undated, but most probably spring 1937).
21. Heinz to Fritz (April 13, 1937).
22. Heinz to Fritz (October 20, 1937).
23. Heinz to Fritz (April 18, 1937).
24. Casimir, reporting de Haas' views, to London (March 22, 1937).
25. Simon (1927). In 1934, Simon pointed out that the high zero-point energy of helium was responsible for keeping the substance, under saturation pressure, in the liquid phase down to absolute zero. See also Simon (1934).
26. The first attempts to solve the problem were confined to describing liquid helium below the lambda point as the extreme case of a 'liquid crystal' (Keesom & Wolfke, 1927), since the notion of order was so closely related to coordinate space. In such a crystal, small crystalline regions of variable size and shape would account for a high degree of order, allowing at the same time the substance as a whole to retain its fluidity. Keesom (1932) called this state 'quasi-crystalline', and, in an unpublished paper read in Breslau in 1933, Clusius spoke of a 'crystalline state'.
27. London (1954), 17.
28. The zero-point energy may be estimated theoretically in two limiting cases: when the interatomic distance is less than an atomic diameter, and when the interatomic distance is much greater than the atomic diameter. London used an interpolation formula between these limiting cases. See Dingle (1952).
29. London to Mott (August 2, 1934).
30. Mott to London (July 16, 1934).
31. Mott to London (October 8, 1934).
32. London to Mott (October 10, 1934).
33. Mott to London (August 17, 1935).
34. London to Mott (August, 1935). London's paper titled *Remarks on condensed helium at absolute zero* was communicated to the Royal Society by Lindemann. The paper was accepted with some modifications suggested by the referee 'in order that the general reader may better understand the subject'. It is, nevertheless, quite remarkable to read the following in the report: 'As most in this country are not very familiar with cryoscopic work of this kind, it would be more readable if the author were asked to expand the introduction somewhat... For example, the referee had not heard of the lambda point and some description of the caloric properties near zero of liquid and solid would not take very long' (the Secretary of the Royal Society to Lindemann (October 1, 1935)).
35. London (1936*a*). Mendelssohn wrote: 'It is perhaps characteristic of the trend of thought at the time that F. London avoided the term "liquid" in the title of his paper, referring to "condensed" helium' (Mendelssohn, 1946, 376).
36. Fröhlich to London (February 18, 1937).
37. Fröhlich to London (April 30, 1937).
38. Fröhlich to London (May 12, 1937).
39. In 1990, Professor Fröhlich told me that he had not kept any of his correspondence from those years. No records of this correspondence have come to light as yet.
40. Dana & Kamerlingh Onnes (1926), 31, footnote 1.
41. Keesom & Keesom (1936), 360.
42. Casimir to London (June 18, 1937; September 29, 1937); London to Casimir (October 11, 1937).
43. Born (1937), 1034.
44. Mayer (1937); Mayer & Ackerman (1937), 74–86. There is no mention of London's work in these papers.
45. Yang (1983), 14.
46. Kahn & Uhlenbeck (1938*a, b*) published their results in two papers. Their paper of May 1938 was inspired by the discussions during the conference.
47. Heinz to Fritz (March 12, 1938).
48. Private communication from G. Uhlenbeck to Pais; in Pais (1982), 432–3.
49. Heinz London (1960), 34.
50. London (1938*a*), 644. London's first step was analogous to Sommerfeld's (1928) treatment of the free electron gas. Many years before London proposed his theory, M. C. Johnson discussed the degeneracy of the helium gas at a meeting of the Physical Society of London on February 14, 1930. Johnson used Fermi's correction to the pressure of the idea gas. Lennard-Jones pointed out that he should have used Bose–Einstein statistics: 'the author considers only Fermi–Dirac statistics, whereas the theory indicates that helium atoms should obey the

Bose–Einstein statistics. It would add to the value of his work if the author could consider the effect of the latter statistics on helium near the critical point'. However, this suggestion was not taken up.
51. Uhlenbeck (1980), 524–5.
52. London & Bauer (1939*a*), 643.
53. London (1938*c*), 947.
54. Interview with Tisza by K. G. (see note 10).
55. Fritz to Heinz (February 13, 1938).
56. Fröhlich to London (February 15, 1938).
57. Fröhlich to London (February 16, 1938).
58. Fröhlich to London (February 17, 1938).
59. Simon to London (February 2, 1938).
60. Simon to London (February 25, 1938).
61. Heinz to Fritz (February 28, 1938).
62. London to Simon (February or March 1938).
63. Simon to London (March 3, 1938).
64. Heinz to Fritz (March 12, 1938).
65. London (1938*c*), 951.
66. Ibid., 952.
67. Ibid., 953.
68. Uhlenbeck to London (May 3, 1938).
69. Schrödinger to London (May 15, 1938). In May 1939, London told Schrödinger that he could do the whole thing more elegantly.
70. H. Jones to London (May 4, 1938).
71. 'One has only to take care that summing over all states one is not allowed to replace the sum by an integral'. London to Jones (May 10, 1938).
72. Keesom to London (May 27, 1938).
73. Groenewold to London (July 8, 1938).
74. London to Groenewold (July 14, 1938).
75. Interview with Tisza by K. G. (see note 10). Tisza has provided a lot of material concerning the two-fluid model. He had repeatedly told me that he was reluctant to retell this 'old story, an intimate blend of success and failure. Moreover, the problem was complicated by controversies involving the "triangle Landau–London–Tisza". Being the last survivor I want to avoid even the appearance of making a partisan point'.
76. Tisza (1938*a*), 913. Tisza had sent the note on April 16, 1938, a week after the appearance of London's note.
77. Allen (1952), 90.
78. Interview with Tisza (see note 10).
79. Fritz to Heinz (June 12, 1938).
80. Tisza mentioned these conversations longingly in one of his last letters to London (August 10, 1951).
81. Fritz to Heinz (November 3, 1938; undated, but most probably during early 1939).
82. Fraenkel to London (January 3, 1937); London to Fraenkel (January 18, 1937); Fraenkel to London (January 29, 1937).
83. Fraenkel to London (February 28, 1937).
84. London to Fraenkel (March, 1937, exact date not specified).
85. Fraenkel to London (September 14, 1937); Werner Senator to London (September 16, 1937).
86. London to Fraenkel (March 18, 1938).
87. Fraenkel to London (March 25, 1938).
88. Fritz to Heinz (March 29, 1938, from Venice).
89. Hadamard to Einstein (April 14, 1938).
90. Einstein to Hadamard (April 26, 1938).
91. From a note in the files of Werner Senator about his conversation with Fritz London on April 17, 1938, Center of Archives of Hebrew University, Jerusalem. I thank Mrs Ety Alagem, of the Center for her help.
92. London to Fraenkel (May 1, 1938).
93. See Crawford, Heilbron & Ullrich (1988). Roginskiy's nomination was in the category of chairholders at invited universities. Roginskiy had worked in catalysis, kinetics of heterogeneous reactions, and on the chemistry and use of isotopes. There is no record of any meeting between them during London's stay in the Soviet Union. Until 1937, from among Slater, Pauling, Mulliken, van Vleck, Hund, Heitler and London, the only person to have been nominated was London. Hund had nominated Stern and Gerlach in 1932, and Slater had nominated Bridgman in 1933.
94. Gross to London (December 17, 1937).
95. London to Gross (December 29, 1937).
96. Fraenkel to London (May 22, 1938).
97. Gross to London (April 4, 1938); London to Gross (May 1, 1938); Wannamaker, the Dean of Faculty, to London (May 14, 1938).
98. London to Gross (June 7, 1938).
99. London to Gross (September 6, 1938).
100. George Pegram to London (October 4, 1938); London to Pegram (October 8, 1938).
101. 'I am sure you were delighted [from your visit to Chicago] to find *no* gangsters' (Charles Squire to London (November 1938, exact date not specified)).
102. Dieke to London (November 28, 1938).
103. George Scatchard to London (July 12, 1938).
104. Undated draft found among London's papers. There is no copy of it at the archive of Frederick Joliot in Paris. But almost certainly this draft had been written after London had decided, in the fall

of 1938, in his own protracted way, to go to Duke University.

105. London to Joliot (October 12, 1938).
106. London & Bauer (1939*a*), 251.
107. Ibid., 252. This quotation was not in the published text, but it is a translation of a typed addition inserted by Fritz London in his own copy of the printed book, and can be found in the translation of the monograph appearing in Wheeler & Zurek (1983). The translations were done independently by A. Shimony, and by J. A. Wheeler and W. H. Zurek, and by J. McGrath and S. McLean McGrath and were reconciled in 1982.
108. London & Bauer (1939*a*), 257.
109. Ibid., 259.
110. Shimony (1963), 759.
111. Ibid., 772.
112. Bauer to London (August 22, 1938); London to Bauer (September 29, 1938). Max Jammer (1974) argued that in the London–Bauer theory of measurement, one could discern influences by Lipps and Becher. Lipps' influence was traced to his student Pfänder. There is, however, no basis for such a claim, since the discussions between London and Pfänder were during a period when Pfänder had already criticized Lipps' psychologism and his *Logik* was a specific counter-example. Though London did attend Becher's lectures and mentioned his books in his thesis, there is no evidence that he had adopted Becher's interactionalism and the doctrine of psycho-physical parallelism. In fact, even though von Neumann explicitly mentioned the principle of psycho-physical parallelism (von Neumann (1955), 418–20), London and Bauer were conspicuously silent about it. Shimony (1963) has argued that von Neumann's explanation of intersubjective agreement, where he used the principle of psycho-physical parallelism, was rather different from that used by London and Bauer. At the concluding section of their monograph there was explicit reference to phenomenology. That would have been the 'proper' place to mention any other philosophical considerations London (who wrote this section) may have deemed to be necessary. Hence, Jammer's conclusion that 'London found in quantum mechanics a field where he could apply Lipps' and Becher's philosophy' does not appear to be borne out.

5

Tying up loose ends: London in the USA

After the unhappy years in Oxford and the hopeful years in Paris, North Carolina, USA, was a melancholy place. At Duke University, the very private London was lonely and scientifically isolated. He was not associated, even peripherally, with the atom bomb project, and he was far away from where the action took place after the end of the war.

He made several trips to Europe and he was the main speaker at the first International Conference in Low Temperature Physics in Cambridge in 1946. He enjoyed it immensely, even though he felt that 'the level of physics was that of 1939'. At the time, London was not sympathetic to Tisza's two-fluid model. Experimental results sent by the Soviet physicists who could not attend the conference, appeared to give credence to Tisza's model, and London changed his mind and adopted it, hoping to show that his proposed mechanism of Bose–Einstein condensation could be the theoretical rationale for the two-fluid model. But there was an additional reason for adopting the model. Lev Landau, the enfant terrible *of the Soviet Union, had proceeded along a different path, trying to explain superfluidity by quantum hydrodynamics. This particularly ambitious scheme had its weaknesses, but had to be reckoned with, especially since this author, highly regarded in the West, was ostentatiously ignoring London's work on superconductivity and superfluidity.*

In March 1953, London was informed that the Royal Netherlands Academy of Sciences had awarded him the prestigious Lorentz Medal, which was presented every five years. He was in good company: the only other recipients were Planck, Debye, Sommerfeld and Kramers. The first volume of his book Superfluids *had appeared and he was completing the second volume. He had developed a heart condition in*

1948 and his health was continuously deteriorating; in 1954, he died at the age of 54 years.

Duke University, North Carolina, USA

Leaving Berlin for Oxford, although sad, was also reassuring; leaving Oxford for Paris was promising; leaving Paris for North Carolina was painful. London loved Europe and he could even rationalize its unnecessary turmoils. He enjoyed his travels immensely and he felt that the cultural differences in Europe were indeed the coalescing factor of this most complex continent. The offer from Duke University was very welcome indeed: the war was imminent and the optimism associated with the initiatives of the Popular Front was followed by a deep sense of disappointment. The Democratic Front in the Spanish Civil War, which in the eyes of many (and especially of those closely associated with Fritz in Paris) seemed the only viable resistance to fascism, was by 1938 in great despair and tragically isolated. The USA was far away from the misery descending upon Europe and large enough to be able to provide shelter and a stable life to a person approaching forty, his wife and their newborn son Frank. When Fritz London arrived at Duke University in the fall of 1939, he was already a tired man. His continual moves and his numerous initiatives in-between to find a job – which at times he felt was an undignified pursuit – left him with a great yearning for quiet and for an opportunity to immerse himself in his work.

The private and lonely Fritz, with his humanist views so prevalent among European intellectuals whose respect for democracy was strengthened by their sympathies to the ideals of socialism, found himself in a desert land. To a European who had lived in Berlin and Paris, the American South of the pre-war years was the very seat of irrationality. Although London valued the democracy of the USA so much and expressed his approval on many occasions, he was witness to extreme forms of racial segregation and, later, had to live through the deleterious excesses of McCarthyism. He was continuously at odds with the stipulations of such aberrations in his adopted country. Fritz could not understand why Belle, their black maid, always looked so shocked when he greeted her by trying to shake her hand or why, despite Fritz's insistence, she always refused to eat with the family and ate in the kitchen. He was equally at a loss when the mania of anticommunism descended upon Paul Zilsel – the gifted young physicist who had come to work with London after receiving his doctorate. London remembered rampant anticommunism as being one of the ideological alibis of the Nazis in their methodical ascent to power and in their many attempts to legitimize antisemitism among republican Germans. London's reserved self and awe inspiring presence built slowly around him a net which was becoming progressively less and less penetrable. In Duke, he became like a treasured piece of art among an otherwise quite ordinary collection.

At Duke, London taught courses titled *Chemical Physics* and *Statistical Physics*. The first course was in effect about quantum mechanics for chemists. The students were impressed by the historical introduction to the subject, which was always detailed and never done in an indifferent manner. Though in many respects he was part of this history, the students had the distinct feeling that he did not want to capitalize on that nor on his personal acquaintance with nearly all the protagonists of these developments. He did not talk about personal reminiscences of Sommerfeld, Schrödinger and other well known physicists and chemists, and the only person about whom he divulged details of a somewhat more personal nature was Hönl – with whom London had written his first paper in theoretical physics. He rarely asked questions in class and students were especially appreciative of his attempts to convince them of the validity of the intuitive method and of not getting lost in the mathematical details whose somewhat obscure physical interpretations would have made a course on quantum mechanics for chemists a rather unattractive undertaking. In the early forties, most of the chemistry graduate students at Duke were from New York city. They really did not think that London was an oddity, since they themselves were considered to be peculiar and could not fit into the South. Henry Linschitz, a graduate student in chemistry during 1943–1944, remembers feeling a peculiar sense of communication with London: 'he was talking to Jewish kids from New York and they were not bothered by his heavy accent'.[1]

The Soviet Union, Kapitza and Landau

After being in the USA for more than three years, London had not published anything. Almost all experimental work at the European universities had stopped and so had the contacts among the physicists. In contrast to the sense of despair among the European scholars, their counterparts in the Soviet Union, at the end of the 1930s, were, on the whole, still living through a heroic period. There was a strong feeling that they were contributing to the realization of an egalitarian society and a peaceful country. The purges and the further curtailment of democratic rights had not made an appreciable effect in marring the vision of the 'new society'.

Figure 11. Fritz London with his son Frank in 1941. (Courtesy of Lucie London.)

The phenomenal development of low temperature physics in the Soviet Union is justifiably tied to the career of Pyotr Kapitza. Born in 1894, he graduated from the Petrograd Polytechnical Institute in 1919. He was part of the group of high-ranking Soviet scientists who travelled to Europe in 1921 after the end of the civil war to restore scientific contacts.[2] Kapitza accompanied A. Joffe, under whose supervision he had started his research. He stayed in England and worked with Ernest Rutherford at the Cavendish Laboratory, receiving his doctorate there in 1924. He was appointed Assistant Director of Magnetic Research at the Cavendish, became a Fellow of Trinity College and, in 1929, a Fellow of the Royal Society. The mutual respect and admiration between Rutherford and Kapitza is well documented. Rutherford was able to secure funds for building a new laboratory with Kapitza as its Director. The new Mond Laboratory in Cambridge was opened in 1933 and, by 1934, Kapitza was preparing to do a series of experiments with liquid helium. However, he never did those experiments at Cambridge because he went to the Soviet Union in August 1934 and he was not allowed by the Soviet authorities to return to England. After a prolonged period of being, in effect, a prisoner and totally isolated from his work, the two official newspapers *Pravda* and *Izvestiya* announced on January 3, 1935, that the government had decided to establish a new Institute of Physical Problems within the Academy of Sciences and that Kapitza would be its first director. By the end of 1936, the new Institute had started to function.

In fact, excluding some areas of applied physics, low temperature physics became the trademark of Soviet physics – especially during the war years, and Kapitza and Landau were the towering figures. Kapitza recognized the uniqueness of Landau's intellect – and accommodated his impossible behavior. The latter, fully conscious of his own brilliance, did his best to undermine his own social status and seemed to thrive by being unpleasant to nearly everyone. Landau was born in 1908 in Baku near the Caspian Sea. He was fourteen years old when he entered the University of Baku and two years later he entered the University of Leningrad, where he graduated in 1927. He had already formulated the notion of density matrix in quantum mechanics and, after his graduation, started working at the Leningrad Physicotechnical Institute, where he stayed till 1929. He was sent abroad for about eighteen months, where he visited Germany, Switzerland, Holland, England and, also, spent some time at Copenhagen with Niels Bohr – who became impressed 'with his power to penetrate into the root of physical problems'.[3] During this trip, he met Heisenberg, Pauli, Casimir and Rutherford. After his return, he developed the theory of diamagnetism in metals. He stayed at Leningrad University till 1932, when his relations with its Director, Joffe, started to deteriorate. Landau was asked to head the theoretical division at the Physicotechnical Institute at Kharkov; he accepted, and he stayed there till 1937. In Kharkov, he published his studies on phase transitions, formulating the detailed thermodynamic theory of the behavior of systems near the transition point. He also completed his first paper on superconductivity, introducing the notion of the intermediate state in superconductors and the surface tension between the normal and superconducting phases. In 1937, he resigned from Kharkov.

When Landau was arrested on April 28, 1938, on some unspecified spying charges, Kapitza immediately wrote a letter to Josef Stalin.

> I beg you, in view of his exceptional talent, to order the most careful attention to his case. I feel that his character should also be taken into account: he is, to put it bluntly, a horrid person. He is a troublemaker, he likes to look for shortcomings in other people, and when he finds them, particularly in high ranking elders such as Academicians, he starts mocking them most disrespectfully ... I forgave his antics in view of his exceptional gifts.[4]

He never received an answer from Stalin. Landau was in prison for a whole year without any investigation being conducted to bring charges against him. Kapitza wrote to Molotov protesting about Landau's imprisonment and asking him to release Landau, since it would not be possible to publish any of his newly completed work related to the discovery of the new phenomena 'in one of the most puzzling areas of modern physics' unless he had the aid of Landau.[5] On April 26, 1939, with Molotov's agreement no doubt, Kapitza wrote to authorities at the Butyrki prison:

> Please release from custody Professor of Physics Lev Davidovich Landau. I guarantee to the Commissariat for Internal Affairs that Landau will at my Institute conduct no counter-revolutionary activities against the Soviet Government, and I shall do everything in my power to ensure that he undertakes no counter-revolutionary action elsewhere. If I observe from him any statements to the detriment of the Soviet Government, I shall immediately report them to the Commissariat.

It was a courageous act. In order to have Landau released, he appeared to accept the initial charges against him – and, hence, to cover the Government, which had not conducted a proper investigation – and to be responsible for his proper behavior in the future. It is true that Kapitza had close connections with those high in the Party and Government hierarchy.[6] But he was experienced enough to know that the situation was very fluid because of the impending war and, most importantly, he also knew that rivalries and strong passions among scholars, which were the primary reason for Landau's arrest, could flare at any time because of Landau's quite uncontrollable contempt towards many of his colleagues and other, more respectable pillars of Soviet society.

What was common between London and Landau was the extreme dislike they felt, for one another. When Landau was visiting Europe in 1929, they had a long talk in London's house in Berlin and were together in his study for a long time. When Edith called Fritz to ask him whether he would ask the visitor to stay for dinner, as they commonly did with other visitors, London was not very enthusiastic about the idea. 'He is a good physicist, but too arrogant'.[7] Most probably this was their only encounter, since it does not appear that they met during London's stay in the Soviet Union. It was surely not a good omen for their future relationship.

Years later, when Landau started working on superconductivity, he never acknowledged the importance of the Londons' contribution, even though his overall theoretical approach was also phenomenological. Tisza always thought that it was strange that Landau had never mentioned the Londons' paper.

> His habit was to go to the library and search the literature every week. If he saw an article which seemed interesting to him, then he assigned it in his journal seminar . . . London's paper never came up, and I would have thought it was very much in the phenomenological quantum mechanical direction that Landau should have appreciated, and it is remaining a mystery to me why that wasn't the case.

Shoenberg recalled that when he was in Moscow in 1937, during a discussion about a paper by Heinz London, who had speculated on various alternative formulations, Landau claimed that the electrodynamic equations characteristic of a superconductor appeared to be identical with those of the vacuum. Despite arguments to the contrary, Landau would not hear of anything else and exclaimed 'he'd soon recognize the difference if I shot a superconducting lead bullet into his behind'.[8]

In 1941, Kapitza published the results of his extensive measurements on the behavior of the two kinds of helium. He put forward the hypothesis that the 'abnormal heat conductivity was not due to some exceptional thermal property of helium-II, but to heat transferred by convection currents whose presence could be anticipated owing to the exceptionally high fluidity of helium-II'.[9] It was calculated that to explain the values of thermal conductivity observed by Keesom and Keesom in 1936, the convection velocity must be assumed to be about 50 $m\,s^{-1}$. Kapitza decided to measure this velocity, but his experiments yielded a heat transfer at least twenty times greater than that measured by Keesom. Consequently, the convection velocity had to be of the order of 1000 $m\,s^{-1}$. It became quite obvious that the then accepted mechanisms for heat transfer could not be of much help in explaining such large convection velocities.

But it was 'an accidental observation'[10] which gave their work an impetus in a totally new direction. Kapitza found that the pressure pulsations transmitted from the helium pipeline of the laboratory into the helium in the capillary caused substantial changes in the thermal conductivity. In order to study this phenomenon as fully as possible, Kapitza set up new experiments, and thus he was able to establish the mechanism of the movement of liquid helium in the capillary as a result of a heat current. Kapitza's proposed explanation did not make use of the two-fluid model. He suggested the possibility of two spatially separated mass currents, flowing into the bulb on the surface layer of the inner perimeter of the tube and outflowing through the center of the tube. In order to explain the great thermal conductivity of helium-II on the basis of this pattern of movement, Kapitza suggested that there is a difference between the heat function of helium in this film and in the free state, and thus the difference in heat content between the two mass currents was accounted for by the van der Waals forces of the capillary wall on the surface of the layer of the liquid. This hypothesis proved to be fruitful: it led to the prediction that the thermal conductivity of helium would be strictly normal in the absence of surface phenomena. Subsequent experiments showed that the entropy of liquid helium flowing through the narrow tubes was indeed zero. This had already been predicted by both Tisza and London, but Kapitza thought that these schemata could not provide a 'rigid theoretical basis'[11] for his observations, and pointed to the theory of liquid helium proposed by Landau and published in the same year as his experiments.

Landau attempted to construct a quantum theory of liquids by direct quantization of the hydrodynamic variables, such as the density, the current and the velocity, without explicit reference to the interatomic forces. He considered the quantized states of the whole liquid instead of the single atoms, and started by considering the state of the fluid at absolute zero, which is its ground state. Excitation of vorticity would represent departure from the zero temperature states. Departure from the ground state could also arise from the excitation of one or more units of sound-wave energy or 'phonons', which gave rise to a Debye-type specific heat and could not account for the specific heat above 1 K. In this way, Landau constructed the energy spectrum of a liquid from two types of excitations; to the phonons of the solid body he added a spectrum of 'rotons' which defined the elementary excitations of the vortex spectrum. States near the ground state, therefore, were characterized by the numbers and energies of the phonons and rotons superimposed on the ground state.

Thus, in Landau's theory, helium became a background liquid in which excitations moved, and there existed only one fluid: liquid helium. In a way, the ground states and the excitations played the role of the superfluid and the normal state respectively. The excitations were the normal state because they might be scattered and reflected, and, hence, showed viscosity. The fluid associated with the ground state was superfluid because it could not absorb a phonon from the walls of the tube or a roton, unless it was flowing with a velocity greater than the velocity of sound or a 'critical velocity' respectively. Below the lesser of these two velocities the flowing helium would not interact with the walls and, hence, would be superfluid – unless, as Landau pointed out, some other mechanism, as yet undetermined, limited the flow. To use Andronikashvilii's expression, theories of superfluidity were attempts to understand the 'structure of heat'. When talking informally, Landau divided helium into *live* (normal) and *dead* liquids. This was inspired by a comment by Krylov, who related that sailors also talked about 'live' and 'dead' water: even if everything was fine with a ship, something would go wrong if a ship found herself in dead waters.

An experimental corroboration of Landau's version of the two-fluid model was Andronikashvilii's measurements in 1946. He had measured the change of the fraction of the superfluid as a function of temperature. The experiment consisted in measuring the moment of inertia of a pile of closely packed aluminum plates suspended in a bath of liquid helium-II. Since the superfluid component is capable only of irrotational flow, it had no effect on the rotation of the discs. By measuring the variation in the period of oscillation as the temperature was changed, it became possible to calculate the variation of the total moment of inertia of the oscillating system and, hence, to obtain the relative density of the normal component present at any given temperature. It was found that above the lambda point the liquid was entirely normal. Responding to the results of Andronikashvilii's experiment, Landau remarked to him 'you have been able to weigh thermal excitations'.[12] To relate the dramatic situation with superfluidity, Landau used to say that all theoretical physics is divided into two equal parts: theoretical physics itself and the theory of the viscosity of helium-II.

Landau's formalism led to two different equations for the propagation of sound, and, hence, to two velocities for sound. One of them was related to the usual velocity due to compressibility, while the other depended strongly on the temperature. This was the same phenomenon as Tisza's thermal or temperature waves (see Chapter 4) and Landau named them 'second sound waves'. The first and unsuccessful attempt to generate and to detect second sound waves was made with acoustic apparatus by Shalnikov and Sokolov. They failed and that was interpreted by London as meaning that Landau's theory was 'born refuted'.[13] The failure to observe second sound acoustically was explained in 1944 by Lifshitz, who made a more detailed theoretical analysis of second sound waves and showed that if the usual mechanical methods for generating sound were used then second sound was masked by ordinary sound. But a plate with a periodically varying temperature would radiate only the second sound.[14] Using such a 'radiator', Peshkov (1944) was able to demonstrate the existence of standing thermal waves for the first time.

It was in their dealings with superfluidity that London and Landau came to a headlong clash in the mid- to late-1940s. It was rather ironic that Landau's style was now closer to London's own style in his work in quantum chemistry, and London's approach in superfluidity was more reminiscent of the Americans' style in quantum chemistry that Heitler and London had been so critical about. Still, London was convinced that many-body problem techniques were not sufficiently developed to tackle the liquid helium problem. According to London, Landau introduced entities without properly specifying their ontological status, experimental results were implicitly considered as known while setting up the theory, and the energy spectrum of the excitations was arbitrarily adjusted when the experimental results did not fit the predicted results. Furthermore, Landau refused to acknowledge other theoretical work pursued in the West and he refused to give a theoretically legitimate reason for why he thought that statistics were irrelevant for understanding the onset of superfluidity. The feelings were reciprocal, and Landau used his beloved expression; 'pathology', to refer to London's work in superfluidity. The word pathological, according to Landau, described a piece of work which was not based on sound physical principles and therefore it was diseased. The Bose–Einstein condensation was a theoretical (indeed a mathematical) property of an ideal gas and it was by no means justifiable to use it as a property of real gases – and even less so to use it to make statements about the behavior of real liquids.

But their differences were not only visible in the style of their work. Landau was boisterous and extroverted, whereas London was quiet and intensely private. London was patient with his students, demanding yet encouraging, whereas Landau was outright intimidating. London's deep conviction that self-fulfillment can be achieved by giving happiness to a woman in a long-term relationship was at the antipodes of Landau's strong sentiments about the role of the opposite sex in his own life: as if to justify himself about the 'naturalness' of his views on women, he used to stress how natural it was for a teapot to be surrounded by many cups, but there were no known instances where a cup was surrounded by teapots.

The war years

During the war years, Fritz and Edith London's lives in Durham were rather uneventful. Slowly, of course, the impact of the realities of the war started being felt in the USA. Since the summer of 1940, there had been a considerable shift in public opinion concerning American involvement in the war. For London it became imperative that:

Figure 12. Fritz London with his son Frank in 1941 outside their house in Durham, North Carolina. (Courtesy of Edith London.)

'Hitlerism is crushed. Both candidates for the presidency have "more help for England" as their slogan. This is a *new* phenomenon and an important one ... Our life is under the influence of events in Europe and therefore there is little concentration on research. At present, physics is rather boring here'.[15] Progressively, the two brothers started writing to each other more and more about their personal feelings and less and less about scientific matters.

On June 25, 1940, together with other Jewish refugees from Germany, Heinz London was interned on the Isle of Man at an interim camp at Prees Heath. A number of them were from Bristol. Walter Heitler and his brother Hans, Fröhlich, Gross, Hoselitz, R. Sack, and G. Eicholz. The next day, Tyndall wrote a letter to the Royal Society, arguing that Heitler and Heinz London had already applied for citizenship in April 1939, and they would have become naturalized citizens in the autumn if war had not broken out. They had already turned in their passports and thus 'in the event of invasion and capture they would be among the first to be shot'.[16] On July 1, a letter signed by Heitler's sister, Annerose, and Heinz's first wife, Gertrude, was sent to the Society for the Protection of Science and Learning (SPSL), expressing their extreme concern for the two internees. All internees were released by the end of September 1940. It was also the time when direct mail was again resumed between the USA and Britain. The sense of separation and loneliness in a world at war is incongruously, yet bitterly, expressed in the first letter by Fritz to Heinz, right after the latter's release.

> I can preserve my hearty wishes now for more actual things, like health and happiness and successful work in a more peaceful future than the present times ... I felt that this interruption of mail must have been particularly unpleasant for you. But my most grateful thoughts go to mother; she must have suffered most of all by her loneliness and helplessness in those circumstances and I must say I never did admire her more than in those months. Her letters to us show a courage and a health in thinking in spite of her very difficult situation, only apprehension for the others, none for herself. In fact, all of us had wished and hoped for different circumstances for her when we made her come to England and we must try now to give her all possible compensation for the difficult times in which she has been alone in Bristol. If only we could do more from here![17]

Margenau, who was at Yale, told London about the Stirling Fellowship and suggested that, maybe, Heinz might be interested in applying. Fritz discouraged Heinz from applying and warned him that if he received it this would mean beginning again, 'whereas in England you are more known than here and people there have taken a definite responsibility for you, particularly since they have effectuated your release from internment. They will consider you as belonging to them now more than hitherto, since you have taken part in their sufferings'.[18] Fritz thought that after the war there would be a better chance for employment in England than in the USA and comforted Heinz that he would not have to wait for very long when people realized that he could be of valuable help in the aftermath of the war.

Some friends from the Paris years who had also emigrated to the USA visited them in Durham during March 1942 and London was elated. Edith was continually painting because one of the persons was a colleague of hers from Paris. 'European intensity is something one misses here, particularly in Durham. The few European colleagues here have either been assimilated or probably have always been somewhat dull'. Somewhat surprisingly, London sounded very happy with his teaching. He felt that it was the first time in his life that he had some rapport with the students, though he did not think that any one of them was fit to continue as a doctoral candidate. Many physicists and chemists had become quite interested in his 1942 paper on the van der Waals forces, even though he thought there was not much new material in it. He had read Landau's paper, but there were several points that he could not understand.

In 1942, London felt he was being excluded from working on Government projects, possibly because he was not yet a naturalized citizen. However, he hoped that he would find suitable employment with the Government, since, as he wrote, a distinction should be made between 'friendly' and 'enemy' aliens, as there had been in England. He was distressed about the situation with the refugees on the West coast. No laws had been passed about their future, there was a curfew and people could not even go to the hospital (as he wrote to Heinz) to give birth to 'an American citizen should the child arrive during the curfew'. What also worried him very much was the absence of any progressive criticism – unlike the situation in the House of Commons which was 'certainly very helpful in England'.

Heinz was naturalized as a British citizen in 1942 and Fritz hailed it as a great distinction. He was convinced that naturalization during the war meant that the British needed Heinz's talents. A short time later, Heinz wrote to Fritz that he had a job, but without giving any details. In fact, he had started working on the British atomic bomb project. Now that Heinz had a salaried position, they decided on a scheme of sharing expenses for their mother.[19] On a personal level, 1942 was a particularly taxing year for Heinz. His marriage to Gertrude Rosenthal, whom he had married just before the war, broke up, and in the fall their mother, who lived with Heinz, died.

The financial situation of Fritz and Edith was becoming quite restrictive. They were afraid that prices might start to rise, and that it would be impossible to make ends meet. Fritz asked Heinz whether they could share the expenses of their Aunt Clara, who lived in Berkeley, since Fritz was meeting about a third of her expenses and they also had Edith's parents to help. Fritz was particularly worried about Heinz who he felt was very lonely and withdrawn after his separation.

> I am very sorry to see that you have not yet found the right company. I believe that this needs time and there are always people who detest playing bridge etc., and who belong together, like those who enjoy playing bridge, only they do not find each other as easily. So I hope that by the time you get this letter, life will look better in Chester, otherwise you should make Trude come to you. It is not good in the present times to live separated from each other, if there is no major reason for it.[20]

It was during the war years that London articulated his views about biology. London's antireductionism was not as assertive in his thoughts concerning biology. He did not, of course, follow the traditional view that biology can be reduced to chemistry and chemistry to physics. He believed quite strongly that the possibilities offered by quantum mechanics for understanding macroscopic phenomena could, in the future, provide a link with biology.

In a detailed response to an article by Szent-Györgyi, London sketched, for the first time in 1941, his views about biology.[21] He commented on the notion of 'common energy levels', introduced by Szent-Györgyi, meaning those energy levels in which the electrons possessed a kind of 'omnipresence' as they did in metals. He confided that he had become interested in investigating some mechanisms of such 'omnipresent' electrons from a more physical viewpoint: 'with an *arrière pensée* that these mechanisms could one day be of interest for the understanding of biological mechanisms. What is the basis of this *arrière pensée*?'

According to London, in a living organism one dealt with one big inhomogeneous molecule (not a crystal) of macroscopic dimensions in some kind of stationary equilibrium surrounded by a series of ordinary homogeneous ensembles. This type of molecular structure was not well known in chemistry. For instance, a description of the single molecule by its thermodynamic variables alone could not be sufficient, and details of its physical construction and behavior would have to be considered. There was some indication that the nervous system of organisms typically had a kind of amplifier construction which might promote an enlargement of the quantum effects instead of their usual annihilation by averaging. These views could be accommodated with the 'pet idea, which circulates among some physicists and which may be quite prejudiced and wrong, but at least, rather instructive', the possibility that quantum mechanics could have manifestations in macroscopic dimensions. London shared Szent-Györgyi's opinion that there was a more specific interrelationship between quantum mechanics and biochemistry which had not been articulated yet.[22]

London returned to the problem of intermolecular forces in September 1941, at the Symposium on Surface Chemistry held as part of the festivities for the fiftieth anniversary of the University of Chicago. In this paper he criticised the use of point centers to represent the dispersion forces, because such an approach did not take into account the tensor nature of the molecular polarizability, and he proceeded to make some intriguing speculations about the possible use of these forces in the understanding of biological phenomena. The main thrust of his contribution was to find the dispersion forces in asymmetric molecules. He argued that his way of dealing with the van der Waals forces in long molecules could provide an explanation about the difference in elasticity between rubber and paraffin, something which could not be understood with other existing schemata. At the very end, he talked about a possible role for these forces in biochemical questions. He referred to some recent discussions by Pauling and Delbruck, who proposed explanations of serological phenomena in terms of molecular interactions.[23] He believed that, in most cases, chemical and van der Waals forces were inadequate for understanding the typical biological phenomena, especially since they lacked the specific selective character manifested by most of those phenomena.[24] The suggestion he timidly made was to see whether his new treatment of the van der Waals forces could play an essential role in biological mechanisms.

The war was forcing Fritz and Edith slowly to accept that events which started in the early 1930s in Germany, and which, at the time, very few people took seriously, were catalytic in bringing forth irreversible changes all over the world. The past was quickly fading and it was going to be a new world. And Fritz was a lost man in it – almost forgotten in the American South. Heinz was the only unchanging point of reference. Before going to the USA, Fritz had borrowed £50 from Heinz. He wanted to reimburse him as soon as possible, since no one could foresee the developments after the war. This was an excuse to write to Heinz about their financial situation in a uniquely sarcastic manner.

> I want to dispel any illusions you might have about our standing here. Life insurance is no form of saving for me as you seem to believe, at least not that form of insurance I actually could afford to take. It is what is called here 'term-insurance'. That means I have to die in a

certain period (20 years) in order to enjoy the protection of the insurance; if I do not die, the money of the premiums is completely lost. Most people insure themselves by an 'ordinary' life insurance, which protects also against 'the risk of survival'. [I give these details] only against your arguing that children, even promising ones, or life insurance are such a good investment . . . My physics is at its lowest standards, and comes partly from the lack of contact with the outer world, most other people being occupied with defense secrets. I envy you enormously that you can now do useful work with our science and hope only that you might not, like many of our equals here, with too much contempt look at those who have not yet reached the perfection, of which you can be rightly very proud.[25]

It was almost impossible for Fritz, Edith and Frank to take a vacation, and it was not because of the restrictions for 'enemy aliens' and the difficulties with transportation and fuel. Their financial situation did not permit such an extravagance, and it was also impossible for them 'to take part in the general defense boom. The latter is a very ugly trace, at present the most visible one, of this country being in the war'.[25] Among all these problems, there was a hopeful note 'I have ten students in my class. Twice as many as usual. 50% Non-Aryans!'.

By mid-summer of 1943, Heinz and Gertrude had decided to divorce.

I am glad to finally receive news from you, even though I could sense long before receiving it what it would be. Naturally, its content did not make me any happier. But I can at least write and express to you how close to me your destiny is and how painful it was for me to be unable to help and stand by your side when I knew you were alone. Certainly no advice could have helped you with respect to the experiences you went through. Life is now a science of experiences. I admire the indifference and clarity with which you were able to write about your bad luck, because I can sense what a burden it was for you . . . I hope . . . that you can really leave it behind you and that, in this way, you will be able to make decisions for your new start. Above all, please do not doubt that because of this failure one cannot have luck next time. You are still young and life can offer you a lot of beautiful things. Some people are not made for marriage, but I know you well enough to tell you that you are not one of them, that you fully possess the gift to make others happy and see in this your own happiness. To me it seems extremely important that you preferred to divorce rather than to make any compromises, so that the future of both of you does not stand in each other's way. I do not know what the legal situation is in England, but first be fully informed and then act.[26]

Fritz realized that he did not know the exact details of Heinz's work. He knew Heinz was going to work at Imperial College and advised him to find a house near Russell Square or the British Museum, even if it took a little longer to travel to Imperial College: one should not live only for work. He also informed Heinz that they were expecting their second child, and summer in Durham was especially unpleasant. 'The longer one is exposed to this climate, the less one gets used to it – one falls into a state of apathy and inactivity'.

They had already been in Durham for over five years, and Fritz hoped to get his American citizenship soon. Rosie (their second child) was born in 1944. They planned to send Frank to private school for the first year, and then to a public school, since the latter did not admit children before they were six years old. Fritz wrote: 'The schools are so bad here and the children learn so little'.

Most people at Duke were overburdened by teaching mature students from the Navy and the Army. London had not been asked to teach, even though he had told the authorities that he was ready to teach mathematics, physics or elementary chemistry. 'But the interdepartmental abysses seem to be too deep here and they preferred to hire a lot of people from outside the University for this teaching – the government pays them'. He was not unhappy, since the teaching would have been very frustrating: 'the boys have, of course, not the purest scientific interests in learning and in math – for instance – the teaching does not go beyond trigonometry'. In the meantime, Fritz had been 'cleared' by the War Department to do war research with two 'projects': one with the Chemistry Department and one with a group of physicists from Bell Telephone Company who had settled in the Physics Department. Again he confided to Heinz his sense of isolation.

> The character of my activities is designed to be that of a 'consultant'; however, to infer at least from the first three months of this 'activity' it seems that I shall not be overburdened by too much work. Though I really wished to do something useful. Durham has become less interesting than ever. Most of the few better brains have left and what remains is not even of third rate.[27]

A few days after the Victory of Europe day in 1945, Heinz met Lucie Meissner in Birmingham at a party given by a colleague. Some weeks later, Lucie arrived at Birmingham Town Hall for a concert, just after the first piece had started, and was allowed to stand just inside the curtain at the base of the steps leading to the gallery seats. In a few moments the curtain was opened again and Heinz appeared, he had come by himself and was also late. They sat together and then went for a walk. Before getting married to Lucie (in August 1946), Heinz warned her that he was already married to physics. Nonetheless, Lucie's feeling was that the 'the three of them lived in great harmony'. They stayed near Harwell for twelve years and all four of

Figure 13. Heinz London in the garden of his house at 9 South Drive, Harwell, in the early 1950s. (Courtesy of Lucie London.)

their children were born there. In 1958, the whole family moved to Oxford. Heinz's appearance left a lot of be desired and he always felt happy in large and old clothes. As Lucie lovingly reminiscences, the only time his vanity was hurt was when he came home with a new haircut and nobody noticed it. His pockets, more often than not, bulged with various gear. In 1945, Lucie, who had just met him, was surprised to notice that Heinz did not carry a watch, but had, instead a home made pocket sundial. He was always losing or breaking his watches and felt that he had enough. Not unexpectedly, he was not a person to enjoy small talk at parties and, much to Lucie's embarassment, be would withdraw even from social events which were going on in his own house and which did not directly concern him.[28]

The 1946 Cambridge Conference on Low Temperatures

In the spring of 1946 an opportunity arose for London to visit Europe. Simon invited London to be one of the main speakers at the Conference on Low Temperatures at the University of Cambridge. Kapitza and Landau were expected to speak at the conference as well. London accepted enthusiastically, even though the financial prospects were quite dim – the Rockefeller Foundation having promised only $500.[29] There were also bureaucratic difficulties about travelling abroad: the USA State Department approved only a limited number of passports for foreign travel.[30]

It was the first time since the war that physicists from all over the world had gathered to discuss developments in low temperature physics. London's invitation to a major conference in Europe and his being treated as one of the leaders of the field was a particularly welcome opportunity to break the isolation at Duke. He looked forward to this opportunity of reestablishing old friendships. He hoped that the physicists from the Soviet Union would be able to attend the meeting, since they had become 'most active and successful in this field'.[31] As it turned out, the scientists from the Soviet Union did not attend the conference, and this may have also been due to Kapitza's dismissal from his post as Director of the Institute of Physical Problems.[32]

London's passport application took a long time to be approved and, for this reason, London could not join the group of physicists who travelled on the *Queen Mary* to attend the conference. He informed the State Department that he also planned to visit the Institute of Advanced Study in Dublin, Eire, to 'confer with former colleagues' – meaning Schrödinger and Heitler, who were still there.[33] London flew to Dublin first and then continued his trip to England.

London immensely enjoyed the Cambridge Conference, which was attended by many European physicists, but very few Americans. (Joliot, Perrin, Casimir, de Boer, Michels, Pauli, Wentzel, Mott, Heitler, Schrödinger, Rosenfeld, Born, Bohr, Amaldi and Bernardini were there. From the American side the participants were Brickwedde, Boorse, Onsager, Andrew and Anderson.) Though he thought that it was not so interesting from the point of view of the physics discussed, he was glad to see so many of his colleagues from before the war years. It was also the first time he had met with Heinz London in about ten years, and they went to Wales together.

In all his papers up to 1946, London had provided a very short, almost dismissive, account of Tisza's two-fluid model. The 1946 Cambridge Conference was a turning point in London's views about

Figure 14. Fritz London at the house of Franz Simon in Oxford in 1946. (Courtesy of Lucie London.)

the liquid helium problem. He had now fully adopted Tisza's two-fluid model, abandoning his attempts both to construct a molecular theory and to explain superfluidity by analogy to superconductivity. In fact, his report to the conference was unusually assertive and considerably different in style from his other papers. He is nowhere as apologetic about the use of Bose–Einstein condensation. What had happened in the meantime to bring about this change of mood? The immense difficulties associated with constructing a molecular theory did not allow him to entertain any hopes for practical results in the foreseeable future. Kapitza's experiments and Peshkov's measurements of the second sound velocity had, in a way, legitimized, the two-fluid approach. London was also becoming aware that the notion of 'order in momentum space' from his theory of superconductivity was being complemented by the notion of 'condensation in momentum space' from his own approach to superfluidity. He began to think that 'macroscopic treatments' meant something more than phenomenological and that Bose–Einstein condensation might be considered to be the theoretical basis for the two-fluid model.[34]

London gave the opening paper titled *The present state of the theory of liquid helium*. The overall aim of the paper was to argue that 'there seems to be a good reason to suspect that [low temperature phenomena] are manifestations of quantum mechanisms on a macroscopic scale'.[35] Such a conjecture was supported by three developments. Firstly, the explanation that helium remains liquid down to 0 K involved the assumption of order in momentum space, and that only a description by a common wave function for the whole system has a well defined meaning. Secondly, the Bose–Einstein degeneracy of an ideal gas is an example of a state of order in momentum space, and it is a mechanism which can provide, at least, a qualitative account of the transitions at the lambda point. The third reason was the success of the two-fluid model. Overcoming the reservations he expressed when the two-fluid model was first proposed, London was now overtly enthusiastic about the model and the possibilities it offered for accounting for all the observed properties of helium-II, except for the problem with the critical velocity. He had nothing but praise for Tisza. He was also careful to imply that the two-fluid model was somehow based on the Bose–Einstein condensation phenomenon.[36]

Unsettled and unsettling issues in superfluidity and superconductivity

London discussed at length with Tisza many of the points he had included in his talk as well as in the published version of the report to the conference. During their correspondence about the conference, Tisza insisted on specifying those phenomena which could be regarded as a crucial test for the two-fluid approach and, at the same time, provided enough information for a choice to be made between the two-fluid model and Landau's quantum hydrodynamics.[37] London's answer was that there was no compelling evidence to justify one theory rather than the other. 'Nothing short of a macroscopic sample of the helium-3 [He^3] isotope would do the job'.[38] The proposed test was indeed crucial. A difference – so far as superfluidity was concerned – between helium-4 (He^4) and its rare isotope He^3, could be directly and explicitly ascribed to the role of statistics. But at this point, London thought that there was no need for these experiments, since Landau's theory already had enough shortcomings. The main disadvantage of Landau's treatment was the fact that his whole approach was quite independent of any considerations about the statistics involved. And thus, London argued, the absence of any criticism by Landau of the proposed mechanism of Bose–Einstein condensation and the fact that Landau developed his theory without basing it on statistics were quite sufficient for deciding the issue. He could not comprehend what Landau's rotons meant and, furthermore, he could not see how Landau's hydrodynamics were related to the assumption about the two fluids. London's criticism of the particularly problematic notion of rotons in Landau's theory led him to pay little attention to the way Landau was treating the phonons. London was always attracted by the possibility of developing quantum hydrodynamics. In fact, he confessed that, although he had tried, he had thoroughly failed to see how quantum hydrodynamics: 'taken alone, i.e. without considering something equivalent to Bose–Einstein condensation, leads to a *phase of zero entropy* (which, as we have learned from Tisza, is the quintessence of everything)'.[39]

In all his writings and correspondence, London expressed the belief that Landau's theory was a rationalization of Kapitza's experiments together with the insight provided by Tisza's first papers on the two-fluid model.

> He evidently tries to get rid of such a competition by obscuring the whole B–E [Bose–Einstein] business and by a post factum attempt to establish a theoretical rationalization which obstinately avoids B–E on the basis of his quantum hydrodynamics. Up to now I have found nobody who admits to having understood this rationalization.[39]

London disagreed with Tisza's view that all three theoretical schemata – Bose–Einstein condensation, two-fluid model and quantum hydrodynamics – appeared to be complementary to each other, and asked Tisza to write his views about the points London was raising. 'I do not quite see why you should be so keen to saw off the branch on which you were comfortably sitting for quite a while, unless you see something that apparently nobody outside Russia has recognized so far'. Tisza agreed with most of London's criticisms, especially with London's views concerning Landau's treatment of the phonons. He still thought that it might be possible to build a theory which was invariant with respect to replacement of excited atoms and vortices, and told London that he was trying to work out the details.[40] Though he promised to inform London about his results before the Cambridge Conference, the details were quite complex and Tisza could not complete the manuscript on time. Nevertheless, he did write to London that while attempting to rework the problem, he became more and more convinced that the Bose–Einstein theory appeared to have a better chance of being correct than Landau's vortex theory.[41]

London retained the same critical attitude towards Landau's approach even in the published report of the Cambridge Conference, despite Tisza's attempts to convince him that he was 'a little too rough on Landau'.[42] Tisza felt that even the use of the otherwise suggestive expression of 'ghostly rotons' might be avoided in print. The main thrust of Tisza's arguments was to find those crucial tests for which the two theories would yield substantially different results, since up to that point all experimental results were equally well explained by both approaches. In fact, Tisza believed that not only were the two approaches equally successful, but that they both had equally doubtful hypotheses as well.

London did not mention anything about He^3 in the Cambridge Conference report. To have admitted that He^3 constituted a crucial test of London's approach would have been to imply that the Bose–Einstein condensation program and the quantum hydrodynamics program would have to be dealt with on an equal footing. London was convinced that Landau's approach had serious theoretical deficiencies and it was reduced simply to a 'very interesting attempt'.[43] Peshkov had sent a manuscript, to be read at the conference, with his improved measurements on the variation of the second sound velocity with temperature. Peshkov's measurements for the second sound velocities between 1.36 K and 2.19 K, which were published just before the conference,[44] were not in agreement with Landau's prediction for this temperature range. Landau proceeded to modify the energy spectrum of the phonons after seeing the results.[45] London wrote to Tisza that Peshkov confirmed: 'your formula for this variation ... I made a point that you had predicted this in 1938, risking that you might perhaps again disavow [these former findings] and not agree with me!'.[46] Tisza was naturally gratified to hear that the measurements agreed with his formula, but he wanted to know more about the low temperature behavior of the second sound velocity, since, as London failed to appreciate, the predictions of Tisza and Landau were approximately similar down to 1 K, but sharply diverged below 1 K. Is it possible, he asked London, to use Peshkov's measurements 'to decide whether my formula is correct in *contrast* to Landau's or [are the measurements] only in the region where the two are identical'?[47] In contrast to Landau, Tisza did not include the phonons in the normal fluid. Tisza was well aware that the diverging predictions for low temperatures were due to the different role given to the phonons in the two approaches, and that the experimental results for the low temperature range would be quite decisive in resolving an issue related to the more fundamental aspects of each approach. In fact, referring to London's (nearly completed) draft of the published report, Tisza wrote: 'if I criticize anything, it is what you omitted rather than what you put in the paper. In fact, I think you do not mention a very important point, referring to the role of the phonons. Actually my formula for the velocity of second sound is not identical to that of Landau's'.[48]

This comment motivated a discussion about what a macroscopic theory should be. London replied that he concentrated on those points which seemed to him to be essential, and that he believed:

> It would only contribute to clarifying and to strengthening this point of view if one could build it up as a 'macroscopic' theory independent of the phonon – or other microscopic concepts maybe as an abstraction from certain molecular theoretical images, but not as based on them.[49]

London made a further interesting point in the draft of this letter, but he omitted it from the actual letter he sent to Tisza.

> Any macroscopic theory has in general to go beyond the strictly phenomenological data – the same is for instance the case in my 'macroscopic' theory of superconductivity which also was suggested by certain molecular ideas, but, of course, not based on them.

Since one was confronted with a macroscopic quantum phenomenon and since it was nearly impossible to solve analytically the many-body problem for a quantum liquid with Bose–Einstein statistics and interacting molecules, London adopted the view that a successful theoretical schema should be formulated in as macroscopic a manner as possible. It was conceivable, he wrote, that: 'a macroscopic theory can be a valuable achievement, even if one cannot yet carry everything rigorously back to first principles; one would say this with particular emphasis if, from the theory in question, one could predict the existence of previously unknown phenomena'.[50]

By the end of 1946, Tisza had started working on a paper which would epitomize his insight into the two-fluid model. It was an attempt to derive rigorously some of the results of his earlier papers and to provide a sound theoretical foundation for some of the assertions involved. For Tisza, the main problem of physical interest was the understanding of the nature of phases. He was in agreement with London that 'at present the theory of helium should be as macroscopic as possible'. And such was, in fact, the explicitly stated program in his *Journal de Physique et Radium* articles in 1940. But by 1946 he realized that he 'did not quite succeed. At present, I think I know how it should be done. It is not in conflict with anything you wrote, but I think one can go somewhat further'.[51]

In 1947, Tisza completed his work where he attempted to develop a phenomenological theory of the various thermodynamic effects in helium-II. His fundamental point was that helium-II could not be characterized by assuming extreme values for the kinetic coefficients, but that one should attempt to find the new hydrodynamic differential equations which would be suggestive of the novel mechanisms of helium-II. A quasi-thermodynamic theory of helium-II was developed, based on four postulates regarding the energy spectrum of the liquid. The first two postulates defined a quantum liquid by specifying the conditions obeyed by the lowest energy of a macroscopic system of helium atoms enclosed within a particular volume so that this system at absolute zero (and under vanishing external pressure) was in a condensed and not a gaseous state, and that, secondly, this substance was defined as being a liquid rather than solid. The third postulate, introduced 'elementary excitations' which had the main attributes of molecules in the kinetic theory of gases, but only when a definite quantum state was considered. At absolute zero, no such 'molecules' were present, but they could be created through thermal excitations. Unlike Landau's theory where the normal component was defined as a 'gas of phonons and rotons', Tisza's phonons did not contribute to the normal component mass. Tisza's last postulate defined the critical temperature for the lambda-transition in terms of the two densities corresponding to the normal and superfluid states. Then, by considering helium-II as being a mixture of two fluids capable of two velocity fields below the lambda point, the macroscopic hydrodynamic equations of the system were derived and nearly all the experimental results were explained. Tisza concluded by stating that: '[the] difficulties of the quantum mechanical many-body problem are not solved, but by-passed in the present theory'.[52] Tisza was under the impression that whatever went beyond straightforward thermodynamics was considered by London to be a kind of an aberration. But surprisingly, London seemed to agree with such an approach and thought that

Tisza had done an excellent job. He sent him a letter with detailed comments and pointed out that the two-fluid model had not been the result of a hypothesis: 'What we discussed in Paris was quite a network of ideas, not a hypothesis'.[53]

There was still a point of serious disagreement between London and Tisza. It was Tisza's third postulate for which London had 'little enthusiasm' thinking that is was an unnecessary commitment.[54] Shear and compressional modes in liquids were quite problematic aspects anyway and Tisza did not make his position any easier by letting the shear mode 'correspond' to the translatory motions of Bloch-like states. London insisted that Tisza should clarify the meaning of 'correspond' in this particular context. Tisza's response was '[to] refuse to be so phenomenological. I will not impress people by endlessly repeating that I knew that 8 years ago . . . The statement you dislike is my strongest card against [Landau]'. According to Tisza, the third postulate was crucial in justifying why the Bose–Einstein condensation should be applied to a quantum liquid. 'I challenge you to build up a complete theory without introducing it.' Tisza thought that it was meaningless to state anything about the macroscopic velocity field regarding an elementary Bloch state. Nevertheless, an enormous number of elementary excitations could cooperate in order to provide the macroscopic shear mode. 'The roles have suddenly been reversed. You, Bogolubov and Landau seem to think that there is no essential difference between the elementary excitations of the shear and compressional modes.'[55] He believed that the difference was quite essential. A Bloch wave was associated with a particle current, whereas a phonon was not. Nevertheless, Tisza acknowledged the intuitive character of his third and fourth postulates and hoped that he would be able to develop a quantum mechanical theory where he could start from the first two postulates and prove the last two; however, London was not convinced.[56]

It was Tisza who first informed London of the unexpectedly good news about the experiments concerning He^3 during the fall of 1947.

> I heard through the grapevine that at Ohio State they obtained a considerable increase of the He^3 concentration by letting helium flow through capillaries, indicating the importance of statistics. So this is a complete victory! Hurray![57]

London received the letter the day he read the published results. He was rather sceptical even though he wondered whether it really meant victory. He felt that, because of the great dilution of He^3, the same results would have been received even if superfluidity were a quantum effect *à la* Landau and not dependent on statistics. However, there was a much lower concentration of He^3 at the end of the experiment than at the beginning (2.2×10^{-7} compared with 1.2×10^{-6}). London wrote:

> Where did the He^3 go? Has it stuck in the slit? Again I do not understand that [they] seem to be astonished to have found a He^3 enrichment rather than a loss in container A! They should be glad if they had found an enrichment in A and surprised to have actually found a decrease . . . I think the merits of the B–E theory are at least to have motivated them to propose the experiment, a matter which is rather obscured in the paper – even if this experiment should not prove entirely correct.[58]

After his initial enthusiam, Tisza was equally disappointed: 'If I want to be conservative, I must admit that this is not yet victory'.[59] Yet, the rumors were that those who performed the experiments were convinced that He^3 did not take part in the superflow. London now regarded the newly performed measurements on He^3, rather than the second sound velocity determinations, as constituting the crucial test for the Bose–Einstein approach. Interestingly, the prevailing attitude of American experimentalists who measured the second sound velocity below 1 K was that only by studying the behavior of He^3 was it possible to provide a criterion for choosing between the two approaches.[60] The early measurements produced a short lived enthusiasm, since the experimental results (to the extent that they bore some relevance to the question of the dependence of superfluid behavior on the statistics) appeared to be inconclusive.[61]

Heisenberg's theory and London's program for a microscopic theory

In 1947, Heisenberg published a paper on superconductivity. The part of the electron–electron interaction responsible for superconductivity continued to be quite elusive. Heisenberg suggested that the singular part of the Coulomb interaction could lead to superconductivity. He assumed that, in an electrically neutral metal, the first-order perturbation caused by this interaction vanished and only the second-order perturbation was significant. For the lowest temperatures, Heisenberg suggested that there might be a very large number of 'current threads' which were randomly distributed and did not produce a macroscopic current. However, if these current threads formed a monocrystal by freezing, then the macrocurrent would persist in such a system and it could not be destroyed by collisions with the lattice of the ions. From such considerations, Heisenberg was able to derive the basic equations of the Londons.

For London, Heisenberg's paper was a 'welcome occasion to publish a few remarks concerning some related ideas I have nourished for several years but had thought to withhold until I could make a well substantiated contribution to this subject'.[62] The main emphasis of London's criticism was that the proposed mechanism could not yield superconductivity. Heisenberg's equations were derived by assuming true thermodynamic equilibrium, which was not a justified assumption at all. An isolated superconducting sphere in thermal equilibrium and in the absence of external magnetic fields, had no currents. Heisenberg's theory implied that a state of spontaneous current should be stable even in the absence of any such field. He felt that Heisenberg's two temperature intervals of different stability within the superconducting state were a 'strange feature' of superconductors.

In a second paper, Heisenberg proceeded to explain the diamagnetism of superconductors as being the result of the equilibrium between the Lorentz force of the magnetic field upon the supercurrents and the action of the stresses caused by the orientation of the spontaneous current domains. The Meissner effect became a 'secondary consequence' of such a balance of forces. London, who had received the manuscript of Heisenberg's second paper before his own article appeared, was still not satisfied, thinking that Heisenberg's theory depended on an assumption which presupposed what had to be proved.[63]

Heisenberg presented an expanded version of his work at a lecture he gave at Cambridge University in 1948. He explored the implications of assuming that electrons condensed in a metal at low temperatures. This condensation was assumed to be caused by Coulomb interactions, rather than by the magnetic interactions between the atoms. The condensation was visualized as having a very low electron density. The condensation gave rise to spontaneous currents consisting of moving electron lattices, and such an aggregation of electrons would lead to a moving ordered state which should, in some way, resemble the state of a ferromagnet. Hence, it became possible to build a picture where such a condensation caused electric currents, but not heat transfer or any resistance. The action of an electron field consisted in changing the momentum distribution of the free electrons which, in turn, transferred momentum to the electron lattices. Thus, Heisenberg was able to establish, though indirectly, a connection between the supercurrent and the electric field, much in the same way as was done by the first of the London equations. His outlook was, nevertheless, completely different from that of London. In comparing his proposed mechanism to that of other theories, he concluded that there was, indeed, an essential difference, since he regarded perfect conductivity, rather than diamagnetism, to be the 'primary feature of

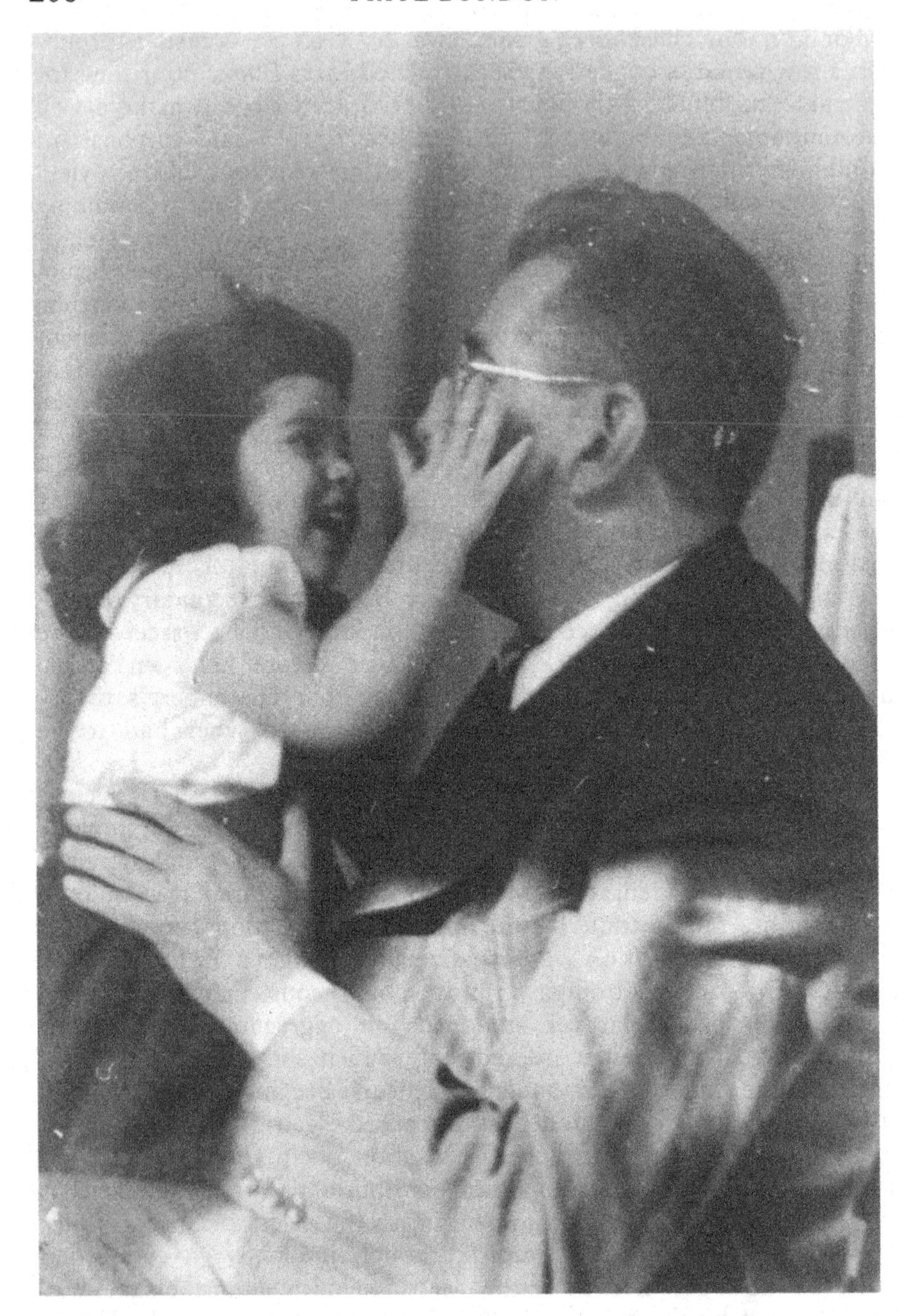

Figure 15. Fritz London with his daughter Rosie in 1948. (Courtesy of Lucie London.)

the phenomenon'.[64] Being aware of the various difficulties, Heisenberg concluded by noting that the theory should be considered to be an attempt to reconcile different features of previous theories, rather than being an entirely new description of superconductivity.

While in Cambridge, Heisenberg was Brian Pippard's guest and they had numerous discussions about the new theory. Pippard remarked that 'Each doubt and objection I raised was turned around by approaching the matter from a different standpoint, about which I was ignorant; I am not the only one to have found him a resourceful and ingenious opponent, so that although I remained sceptical I could not penetrate his defenses'.[65] In his doctoral thesis, Pippard, who later became the Cavendish Professor of Experimental Physics at the University of Cambridge, focused his criticism of Heisenberg's theory on three issues: he argued that neither zero resistance nor the Meissner effect could be derived from the theory, and that the transport of charge by persistent current threads through the thermoelectric effects, led to a violation of the second law of thermodynamics.[66] Heisenberg's theory, according to Pippard, helped a lot of physicists to think about a number of questions. In 1953, by combining Heisenberg's ideas with the non-local anomalous skin effect, Pippard derived an equation with which he could account for his measurements of the penetration depth in alloys. During the Kamerlingh Onnes–Lorentz Conference of 1953, Heisenberg, when asked whether he still believed in his theory of superconductivity, responded that he did not.

Heisenberg's theory, though, did have a rather beneficial effect on London. It was through his attempts to formulate his objections to it that he articulated his 'program for a microscopic theory of superconductivity' which a few years later became a valuable heuristic for John Bardeen. London took Bloch's constraint as the starting point rather than as a forbidding commandment. He saw it as an anathema only against those who could still entertain thoughts that superconductivity was primarily a phenomenon of infinite conductivity. He wrote that Bloch's constraint was an omen such that people should not try: 'to solve a problem which cannot be solved and, which, fortunately, is not the one set by the phenomena'.[67] He was convinced that the electrodynamics of the superconductors, which he had established with Heinz, could give clues about the directions for research for a microscopic theory of superconductivity. He first showed that if one treated the problem classically, then one always got zero for the local mean value of the momentum of the conducting electrons with or without a magnetic field. The equations for a superconductor gave a finite value in terms of the electromagnetic potential. Therefore, something prevented the momentum from assuming a finite local value and it could not be any mechanism which could be constructed by classical mechanics. This non-zero value implied, of course, that in a superconductor it was the momentum of the superconducting electrons rather than the current, which seemed to be held in a kind of long range order and, thus, the problem of superconductivity was 'accordingly reduced to that of finding a mechanism which at a sufficiently low temperature enforces the establishment of this peculiar kind of order'.[68] He formulated a plausible argument that the mechanism could, in fact, be accounted for only as a purely quantum phenomenon. The gist of the argument was that the momentum vector could be related to space coordinates through the de Broglie wavelength and the uncertainty relation. The superconducting electrons could not be free in the same way that the normal conducting electrons were considered to be free. This difference suggested that superconducting electrons were 'exposed to some particular inner field, presumably a result of electronic cooperation'.[69] Thus, wave functions representing the same momentum distribution throughout the whole metal would acquire a wide extension in space, irrespective of the presence or absence of a magnetic field. This would be a quantum structure on a macroscopic scale, rather than a mobile electronic lattice.

> To be sure this view is to be taken with due reserve. It is a possibility, a program, sufficient but perhaps not necessary for explaining superconductivity. At any rate we may say that by this program the requirements for the future molecular theory are considerably reduced. In previous attempts it was felt necessary to construct electronic superlattices which were freely movable within the metal. *We would require just the opposite: a kind of solidification or condensation of the average momentum distribution* . . . But it might well be that these conditions are unnecessarily restrictive, and it remains to be seen to what extend this program will be substantiated by the molecular theory of the superconducting state. [Emphasis in the original.][70]

More problems with von Laue

In 1948, there was renewed friction between London and von Laue. Von Laue had sent London a copy of his book *Theory of Superconductivity* which was first published in German in 1947.[71] Von Laue's aim was to accommodate the persistent currents of superconductivity within the framework of Maxwell's electrodynamics. He made no mention of the purely quantum mechanical character of superconductivity, and attempted to extend the Londons' formalism by including tensor and non-linear generalizations of their theory. London immediately replied with a very friendly note: 'I am glad that it is possible again to exchange thoughts between us after it was interrupted for so many years and I hope that the many wounds that were left behind by this tremendous catastrophe will heal. Many of the things destroyed, of course, do not come back'.[72] London was glad that his theory had found such a beautiful and clear exposition, and informed von Laue that he was occupied with writing a book on superconductivity. Von Laue had already learned from Kurt Jacoby of Academic Press that London was preparing a book.

Von Laue, now, was interested in understanding why it was necessary to introduce a surface energy for the boundary between superconductors and normal conductors. He wanted to know more, in case he decided to include it in the second edition of his book.[73] Concerning the surface energy at the boundary of superconductor–normal conductor, London referred von Laue to his *Nouvelle Conception* (see Chapter 4). If a longitudinal superconductor in a longitudinal magnetic field did not acquire the single phase structure, and, on the contrary, if it changed the structure to a kind of intermediate situation with an indefinite number of normally conducting layers, then it would be in an energetically advantageous position.[74] Von Laue could not follow the proof of the existence of surface energy as it was expressed in *Nouvelle Conception*: 'Some time ago I thought that the problem was me, but now I have a different opinion'. Von Laue thought that it was not particularly easy to find a relation between the free energy and the work produced in a cyclic process that took place in different temperatures. The relation between free energy and work could be found only in isothermal processes.

> These are the reasons that I was never able to reconcile myself with your thermodynamics. Thus your book has many mistakes next to some truly successful ideas. I had to say this for some time now, so that I could break the neck of this fully mistaken idea of surface energy. Please do not take this criticism to heart. You know very well that I consider your theory of superconductivity worthy up to a point.[75]

London tried to explain the necessity of something like the free energy in the boundary between the normal and superconductive phase. According to the Gorter–Casimir theory, if one accepted an energy difference in volume between the two phases, then a longitudinal superconductor in a longitudinal magnetic field would have been much better off, from the point of view of energy, if it did not create a one phase system, but changed into a kind of mosaic structure with an indefinite number of indefinitely small layers of normal conductivity. Such a mosaic structure would have let the magnetic field pass through because of the finite penetration depth and it would have been contrary to the Meissner effect. 'I cannot see how we can accept this catastrophe without new assumptions, such as that of a free energy for the boundary surfaces ... I hope to be able in the future to take back my revenge ... You cannot so easily break the neck of surface energy ... Even Nobelists sometimes perpetrate such sins'.[76] There was no response from von Laue.

Between May 31 and June 2, 1948, the Shelter Island Conference on Low Temperature Physics was held in the USA.[77] Both London and Tisza had been invited to attend. A short while before, Tisza had sent a short note to *Nature*[78] in answer to a claim by Green[79] that the theory of liquids developed by him and Born could, without further assumptions, explain the properties of helium-II. In this theory, the statistics were not taken into consideration, and the transition from

Figure 16. Fritz and Edith London in front of their house at Oakland Avenue, Durham, North Carolina. The photo was taken by Kurt Mendelssohn. (Courtesy of Edith London.)

helium-I to helium-II was attributed to a rapid drop in the potential viscosity of the fluid. Tisza argued that such an interpretation could not be reconciled with the experimental measurements of the viscosity of hydrogen (which is a classical liquid) and those of helium-I (which is a quantum liquid). But most notable in Tisza's argument was his insistence that the transition from helium-I to helium-II had to be described by a change in the macroscopic hydrodynamic equations, rather than by the assumption of a small value for the viscosity. Such a change could be manifested if one went from the ordinary one-fluid hydrodynamics to a two-fluid hydrodynamics. It was stressed that Bose–Einstein condensation was indispensable for understanding the 'crucial point in any theory of liquid helium'[80] which was the interpretation of the lambda point. There was no mention in the note about the second sound velocity measurements, and the behavior of He^3 was put forward as the decisive test. Concerning this note of Tisza's, London felt that the experiments with He^3 (despite their inconclusiveness) justified a stronger argument in favor of Bose–Einstein condensation.

> I would say to all other attempts like Green [and] Landau rest on *grounds still more hypothetical* than our theory, in so far as they have to *assume* that they will be able *in the future* to derive the discontinuities at the lambda-point. Presumably, they will be able to do so at best also only under greatly simplifying assumptions.[81]

A few months after the Shelter Island Conference, London received a letter from von Laue, who was visiting the USA. 'It is so nice to be welcomed in such a warm manner at a time when there is so much hatred in the world'.[82] Von Laue had attended the International Conference of Crystallography, where the main topic of discussion was the determination of the structures by X-rays and the bending of the neutrons. He found it remarkable that all significant advances in that specific branch were taking place in the USA. He wrote that he would be at Princeton University in mid-September; he was also going to be at the Institute and he wanted an invitation to give a talk at Duke University about London's theory of superconductivity. Von Laue reminded London of their disagreement: 'If there were no surface tension then the destruction would have taken place even in weak fields, with values much lower than the threshold value. And this was the only point on which the two of us disagreed'.[83] London invited von Laue to Duke and asked him to choose the topic himself. He was also curious to learn whether von Laue agreed with his position about Heisenberg's theory of superconductivity. Von Laue called the talk *The Barlow wheel and the theory of superconductivity*.[84] Quite inappropriately, this title was the same as that of a paper von Laue had coauthored with Heisenberg for Becker's sixtieth birthday.

Von Laue visited Duke at the end of October 1948. As Fritz wrote to Heinz, no one except 'Nordheim and myself could follow his lecture', even though many people attended it and were impressed by the speaker. The von Laues stayed with Paul Gross, but were mostly with the Londons. Fritz tried to convince von Laue about the inevitability of a surface energy between the superconducting and normal phase. Something, he wrote to Heinz, 'about which we have had correspondence over some time'[85] – a further indication that he had kept Heinz out of the controversy. Von Laue had told Fritz that Heisenberg and Gerlach were upset about the atom bomb and von Laue believed it motivated Heisenberg to work on something in pure physics again, like superconductivity.

> He wished to prove to himself that he was still capable of something ... Among wisdom from the new Germany, Laue told me the saying: The value of a human being is best judged by how much he busies himself with matters that have nothing to do with him. Germany has brought forth profounder wisdoms, but at that moment I was itching to use it as the motto in [my answer to Heisenberg's theory], but had to restrain myself.[86]

After von Laue left Duke, London was very uneasy. He confided to Edith that he felt von Laue was not his old self, events had taken their toll, it was as if he was under the influence of 'forces beyond human recognition'.

Von Laue insisted on his objections, even after the visit to Duke. '*Summa summarum*: I continue to hold my view after reading all the papers I mentioned. It cannot be otherwise. And I hope you will be convinced by it. Write to me if you consider it necessary to continue this discussion'.[87] London, still unconvinced, wanted to put a stop to these discussions.[88]

Hopeful signs from He^3

About the same time that Tisza's note appeared in *Nature* in 1949, measurements on the viscosity of He^3 were reported from Argonne National Laboratory.[89] The viscosity was measured by letting He^3 pass through a fine slit, and it did not show any discontinuity down to 1.05 K. 'This seems to be sufficient proof that the statistics [are] decisive. The temperature 1.05 should be low enough to settle this'[90] wrote London to an equally enthusiastic Tisza. 'Congratulations! This seems like the end of a long, long fight'.[91]

Feeling that the question of the dependence of superfluidity on statistics had been settled, London decided to send a review article to

Nature. As London himself admitted, his attitude had changed since he had prepared his report for the Cambridge Conference. He no longer insisted on the second-sound velocity at temperatures around 1 K as being crucial, but emphasized the implications of the statistics. London wanted to concentrate his arguments on He^3. Tisza's attitude was that a coherent description of the situation would be: to present the idea of Bose–Einstein condensation; to note that liquid helium was not quite a normal liquid, since it remained liquid at absolute zero and, hence, to argue that it was justifiable to start from considerations involving an ideal gas; then, to mention the two-fluid concept (having been verified, according to London, by Kapitza's experiments) and to express the view that Landau's theory was a new molecular interpretation of the two-fluid concept. 'A priori this was neither better nor worse than the Bose–Einstein theory, and Landau's mistake was that he pretended that his theory follows from the principles of quantum mechanics'. Tisza asked London one more time to be 'less emotional about Landau – not that he deserves it, but it gives a better impression'.[92]

However, London did not think that what was at stake at that time was the two-fluid model. He believed that the reported absence of superfluidity of He^3 settled the issue concerning the necessity of the assumption of Bose–Einstein statistics for any theory professing to provide an explanation for the properties of helium-II. Once this was decided, then other considerations, most notably those associated with Tisza's two-fluid model, would follow quite unambiguously. London did eventually include the two-fluid model in his article to *Nature* in 1949, and for the first (and last) time referred to the model as the London–Tisza theory.

Second sound velocity measurements at very low temperatures

At exactly the same time that London's article was published in *Nature*, two letters appeared in the *Physical Review*. The first was by Landau (1949) criticizing the work of Tisza and the second was Tisza's (1949*b*) response. Landau insisted that his approach was the only one consistent with the first principles of quantum mechanics and that it was the correct microscopic theory. He stated that experiments on the entropy and specific heat of helium-II verified his thermodynamic equations. These equations had been derived from the general considerations of conservation laws, and had been combined with the requirement of Galilean invariance rather than the assumptions concerning the energy spectrum of the liquid whose validity had been questioned by both London and Tisza. London had stated that the experiments with He^4–He^3 mixtures could not bear on the question of statistics. But the main thrust of Landau's argument was concerned with Tisza's treatment of the phonons. Tisza, unlike Landau, did not include the phonons as part of the 'normal fluid' and assumed that they were associated with the liquid as a whole. This interpretation gave rise to the divergent predictions of the temperature dependence of the second sound velocity at temperatures below 1 K. In his answer, Tisza pointed to one fundamental difference in attitude between them on how to tackle the liquid helium problem. Starting from the conviction that, at the time, the techniques for handling the quantum mechanical many-body problem were quite inadequate to formulate a theory of liquid helium, he repeated the basic assertions of his 1947 paper. Landau's approach, even though it could be regarded as an attempt to formulate a fundamental theory from first principles, was not free of ambiguities, since 'he has introduced into his theory more or less disguised assumptions which cannot claim the same degree of certainty as the principles of quantum mechanics'.[93] Tisza was fully aware that the different role he ascribed to the phonons was expressed in his third postulate – which, as we have seen, was also criticized by London, who believed that it overcommitted Tisza and did not make the theory 'quite as macroscopic' as London would have liked. Appraising the overall situation, Tisza considered the macroscopic aspects of the two theories as being on the whole, equivalent and complementary to one another; even though they both accounted equally well for a series of properties, each approach provided a better understanding of different elements in the behavior of liquid helium. London was quite enthusiastic about Tisza's answer to Landau, agreeing that there could not be a claim for one correct approach, since there were assumptions on both sides which went beyond first principles. He repeated his view that the two-fluid model should really be considered to be part of a program initiated by the idea of Bose–Einstein condensation.[94]

The second sound velocity measurements at very low temperatures continued to haunt London and Tisza. There were new results by Peshkov on the velocity of second sound. The velocity first went through a maximum for which both theories by Landau and Tisza gave identical results, and then reached a minimum, rising sharply as the temperature approached absolute zero. Despite the fact that at the 1946 Cambridge Conference the measurement of the second sound velocity down to 1 K agreed with the predictions of the two-fluid model and was considered by London to be the strongest corroborating evidence for the validity of the model, his attitude changed after Peshkov's improved measurements: 'I do no longer insist on the difference between the two curves for the second sound velocity ... since this difference between you and Landau is really accidental or rather concerns the secondary question, how to dispose of the phonon entropy'.[95]

In 1949, Pellam's measurements below 1.4 K showed an increase in velocity and differed considerably from Peshkov's. Thinking that this effect might be due to temperature discrepancies in the determination of the very low temperatures, London chose not to consider them to be as reliable as the measurements above 1.4 K. Also, he suggested that Peshkov's reported flattening and slight increase in the velocity might have been due to the same reasons. The experiments of Maurer and Herlin in 1949 settled the issue of the temperature dependence of the second sound velocity below 1 K. Using the pulse method initiated by Peshkov, they lowered the temperature to 0.85 K and observed an increased velocity starting at about 1.1 K. The results were quite unambiguous and could have been used to corroborate Landau's approach – except that new experiments, completed at the same time, tried to detect superfluidity in a pure liquid sample of He^3, and found negative indications down to 1.05 K. Maurer and Herlin believed that the results were not necessarily contradicting the predictions of the Bose–Einstein hypothesis, but they felt that further refinements should be introduced in the model to account for the second sound velocity results. A few months later, Pellam and Scott (1949) also observed the increase of the second sound velocity in the very low temperature range, and believed these measurements distinguished between the two competing theories. 'As you see it is not what I expected and I think I should both "explain" it and eat my words' wrote a disappointed Tisza to London.[96]

Ironically, the measurements that finally gave credence to Landau's theory came from the same laboratory where it had all begun in 1938 – the Mond Laboratory in Cambridge. Atkins & Osborne (1950), using two different demagnetizations, were able to measure velocities down to 0.17 K. They found that there was a marked increase in second sound velocity, which, when extrapolated to 0 K, was the same as that predicted by Landau. Their pronouncement on the two-fluid model was not encouraging at all. They thought that the London–Tisza approach could have led to the same result if it had included interatomic forces, and 'until such a theory has been developed . . ., the present experimental results must be considered strongly in favour of Landau's method'.[97]

Writing *Superfluids*

The first mention that London himself was planning to write a book about superconductivity and superfluidity was in his response to an inquiry from Academic Press concerning their plan to translate von

Laue's book on superconductivity. London thought it was a very good idea and that it was an excellent and very useful exposé. London felt that the book which he was planning to write would have a more general scope than von Laue's. Its main characteristic would be the elaboration of a set of ideas about the fourth state of aggregation, i.e. the quantum fluids. He promised a thorough discussion of 'superfluids'.[98] A few days later, Academic Press made an offer to publish *Superfluids*. London declined the offer, since there was a 'kind of an understanding' about publishing the book with Wiley, which was reached during the Washington meeting of the American Physical Society in April 1947.[99] At the beginning of 1948, Simon suggested that London should publish his book with Oxford University Press. London stuck to his original commitment and signed a contract with Wiley in March 1948. The two volumes which were eventually published epitomize London's most mature views on these subjects, had a rather long period of preparation and include, perhaps, the most articulate expressions of his obsessive belief in the macroscopic quantum character of superconductivity and superfluidity.

Towards the end of 1948, London asked Wiley to publish three different volumes: *Macroscopic Theory of Superconductivity; Macroscopic Theory of Superfluidity; Molecular Theory*. The first volume was ready and the second was being thoroughly reorganized in view of the experimental results with He^3. He thought that the available information as well as what was expected in the near future from measurements with this 'quite unique substance' would be absolutely decisive for the content of the second volume. Concerning the volume about the molecular theory, London could not make any commitments. He hoped that the problem might be solved during the period when he would be writing the two volumes, but even if it was not, the other volumes would justify themselves by 'bridging part of the gap which has so far prevented the development of such a molecular theory'.[100]

London's collaboration with Wiley was quite difficult from the very start. There was constant frustration, because of London's slow pace and the changes he wanted to make in almost all the stages of the manuscript. On October 1949, he was informed that: 'reluctantly we have concluded it will be impossible for us to work with you to your satisfaction and my fellow officers have asked me to grant you a release from your commitment to us for the publication of your manuscript'.[101] London called Jacoby at Academic Press to see whether they were interested in publishing his book. They were quite receptive to such an offer, but they were also worried about a possible overlap with von Laue's book, which they were translating. London did not deny that there was, indeed, such an overlap. But he felt that

Figure 17. Lucie London talking with Lise Meitner. On the right, Heinz London; on the left, Lady Cockroft. (Courtesy of Lucie London.)

the differences between the two books were significant enough to merit the publication of both. The differences could be understood in the same terms as the distinction between mathematical and theoretical physics. Von Laue's book was mostly mathematical physics, and his main purpose was to solve equations. 'My book is more concerned with finding the equations and in understanding their meaning ... in terms of molecular physics'.[102]

Realizing the difficulties involved in changing publishers and the less than enthusiastic reaction by Academic Press, London took the initiative to continue with Wiley. He apologized for his failure to 'arrive at an objective appreciation of facts' and for the 'unnecessary misunderstandings' and informed them that O. K. Rice from the University of North Carolina would help him with the manuscript. The final stages were plagued with all kinds of technical problems. It took London almost two months to correct the galley proofs and, apparently, he added quite a lot.[103] The first volume of *Superfluids* appeared on November 22, 1950.[104]

Though there were a number of reviews of the first volume, very few of them were anything more than a presentation of the contents of the book.[105] The most insightful of the reviews was that by Pippard: 'The fact that in fifteen years London has not seen fit to make any substantial alterations to the fundamental framework of his theory is to be regarded not as evidence of a conservative nature, but as testimony to the logical beauty of the theory in its original form'. Pippard's view was that there had not previously been a rigorous experimental test of the theory because of the technical obstacles to such an investigation and that the last word on these subjects was still to be said. In his book, London examined the relation between the macroscopic electrodynamics and the quantum mechanics of an assembly of electrons, in order to formulate the conditions under which, in his view, the superconducting state might exist. The major requirement, that the wave functions of the electrons should be rigid, in the sense that they were not readily perturbed by an applied magnetic field, was something which had not in the past been accepted by all who had attempted to construct theories of superconductivity. But it was significant that the most recent attempts had been along these lines, and, in other respects, showed a considerable improvement on their forerunners. 'The book is a personal account of an important aspect of the theory written by one who has contributed perhaps more than anyone else to our understanding of the phenomenon'.[106]

Mendelssohn's review was not so gracious. In the beginning he referred to Heinz's own contribution in his doctorate and the paper by Becker, Heller and Sauter. Then followed a description of the content of the book and the review ended with a criticism for not including Mendelssohn's verification of Tisza's prediction of the mechano-caloric effect, for not elaborating too much on the work of Becker, and for not emphasizing the fact that others (among them Mendelssohn) had also suggested the connection between superconductivity and superfluidity.[107]

The trip to Europe

In the summer of 1948, the first signs of Fritz's deteriorating health appeared. He and Edith were planning to go to the mountains, where it would be calm and cool and where he would be able to work on his book. He could not walk more than 100 meters, and he felt well only when he did not have to move at all. The problems with his circulation were such that he could hardly walk in cold weather without getting cramps in the chest and arms. In warm weather it was better, but he could walk uphill only step by step. 'We pondered for a long time whether to travel at all, but there is nothing to be gained by waiting, as long as I do not travel alone and avoid strain and cold'.

During his sabbatical leave in the spring and summer of 1949, London wanted to visit Leiden, Amsterdam, Paris, attend a conference in Florence, go to Oxford and spend a month in Cambridge at Shoenberg's invitation. Fritz suggested to Heinz and Lucie London that they should all rent a house in Cornwall or Devon for the summer vacations of 1949. Edith urged Heinz to be persuaded by Fritz to go on a men's tour: 'so that you men could get a rest from your women (and vice versa)... We talk of nothing but Europe for next summer'.[108] They planned to leave New York on March 30 and reach Le Havre, France, on April 8, via Southampton, England. Even an hour's meeting with Lucie and Heinz 'would be wonderful'. They planned to return to the USA on September 7 on board the *Queen Mary* from Southampton, though Fritz had to return earlier by plane to attend the Low Temperature Conference at MIT (6–10 September).

They wanted to show Paris, Provence and Northern Italy to Frank and Rosie. They rented a car and travelled for 2000 miles, stopping in many places along the way. The trunk was full of books, but Fritz never opened it during the month-long travel. They sent a postcard to Heinz and Lucie from Rapallo: 'We are now in Stazione Margaritta. Do you remember the beautiful days we spent here with mother in 1928? It is still as enchanting as twenty-one years ago'. They were looking forward very much to the holidays they were planning to have together.[109] London attended the conference in Florence and, after a few days in Paris, the children were sent to a boarding school in England and were 'most happy in England and the parents as happy on the continent without them!'.[110] Fritz and Edith spent part of May and June in Paris, and Fritz was in Cambridge during July. They had decided to take their vacations at Saint Malo, France, but, at the last moment, Lucie could not go with them, since her mother was very sick and she had to stay at Harwell. Only Heinz went. His health problems made mornings especially hard for Fritz and he had to sleep till very late. Despite the problems, the union of the two families was, indeed, a memorable experience, 'an incredible dream'.[111]

The International Conference on Statistical Mechanics took place in Florence from May 17 to May 20, 1949. Bauer, O. Klein, Pauli, Casimir, J. Mayer, Onsager and Kirkwood were there. Kramers, as the President of the International Union of Physics, opened the conference, and the first lecture was by Born on the foundations of quantum mechanics. London did not present a paper and participated very little in the discussions except after Gorter's paper which was on the two-fluid model.[112] During the discussion of the paper, one senses a defeated man. Never before was he so committed to the two-fluid model and to its relation to the condensation mechanism. Nor are there any other places where he was so dogmatic in dismissing

Figure 18. Rosie and Fritz London aboard the steamer *Queen Mary* on their way to Europe in 1953. (Courtesy of Lucie London.)

Landau. Gorter had referred to the two-fluid model as a phenomenological theory. But such a judgement did not 'give justice to the fact that this theory was developed (mainly by Tisza) at a time when the greater part of these properties was not yet known. They were originally expressed in the form of predictions, which only later were confirmed by experiment'. Of course, there were questions about the theoretical basis for Tisza's model and he agreed that there was some confusion about this, even though it was known that the theory of the degenerating ideal Bose–Einstein gas and its condensation mechanism played 'a suggestive role' in the formulation of the two-fluid model. Furthermore, no other attempt succeeded in deriving this model from any mechanism other than the Bose–Einstein condensation. The development of a molecular-kinetic theory was still very much a matter for the future and at this point it could be regarded only as a program for statistical mechanics. That such a program could be established became possible 'by a mental act which I believe might best be characterized as an act of abstraction'. None of the qualitative features of the ideal Bose–Einstein gas and of its condensation were altered by introducing sufficiently weak intermolecular forces. At the conference, Max Born raised the objection that, by neglecting the interaction among atoms, one would have no liquid to talk about. London's response was that the forces between helium atoms were sufficiently weak so that the qualitative features of the Bose–Einstein condensation were not affected – despite the fact that the forces have to be strong to achieve liquefaction. Onsager mentioned the possibility of quantized vortices – the first time the possibility of flux quantization was ever mentioned. Responding to a question by Casimir about how his vortex model was connected with the two-fluid model, Onsager said that, on the phenomenological level, the two-fluid model was 'probably fit to summarize any molecular theory of superfluidity'.[113] Throughout the discussion there was only a fleeting reference to Landau's theory.

Fritz was not sure whether he would be back from Europe in time to attend the Low Temperature Conference at MIT in September 1949. He had urged his collaborator, Zilsel, to attend the conference and to be the first to raise his voice in the discussion in case Born, Green or Heisenberg should present their views, even though he expected them to have changed.[114] He managed to be back on time and from New York went straight to Cambridge, Massachussetts. He talked about his program for a microscopic theory of superconductivity. His talk had substantially the same ideas as those in his answer to Heisenberg and was almost identical to the relevant chapter in the first volume of *Superfluids*. Tisza reviewed the present state of the helium problem. Despite the low temperature measurements of the second sound velocity that were in agreement with Landau's predictions, Tisza was confident: '[that] there seems to be no difficulty in incorporating this aspect into the Bose–Einstein theory'.[115]

There had been changes in Duke while Fritz was away, since the University had a new President. 'Last week he was initiated with much celebration and he introduced himself with words against the Communists who would have no place in the University.' There was an additional change as well. Paul Gross, who was the Chairman of the Department of Chemistry, had become Vice-President of the University and as London wrote to Heinz: '[he] can no longer be approached about minor matters of the Department: very regrettable as he was the only sensible person who had often given me valuable support'.

Fritz's health showed some signs of improvement. Fat had been found in his blood and he was on a very strict diet which helped him a lot. He took no salt and little fat with little meat, no eggs and no milk. His diet seemed to help him more than the medication he was taking. Nevertheless, his strength was diminishing and work progressed 'devilishly slow'. He was making numerous corrections to the second volume and he was irritated by all the problems associated with publishing it: 'The thing was stale before the first corrections came'.[116] He declined professional invitations, as they were too demanding on his health, but he did not give this as his reason.[117] By the summer of 1950, Fritz had serious problems in walking uphill and could manage the walks on even ground with some difficulty.

Some developments in the theory of superconductivity

In 1950, Landau and Ginzburg proposed a model in which the energy needed to produce a change in the superconducting state over any distance was explicitly included. They worked out the thermodynamics of their model by defining a parameter ω, which was a measure of order in the superconducting phase and which was zero above the transition temperature. Then they identified ω with the square of an effective wave function ψ, which they set equal to the concentration of the superconducting electrons. The wave function ψ did not describe a single particle, but the motion of the superconducting condensate as a whole. Their theory predicted correctly the dependence of critical field upon the temperature. When the effective wave function was considered constant, the Ginzburg–Landau theory produced the London equations.

Since the discovery of superconductivity, there had been a widely and firmly held belief that the ion masses, being so much larger than the electron masses, could not play an important role in the establishment of the superconductive state. In 1950, Fröhlich conceived the idea that just the 'opposite of the "dictum" contains the truth'.[118] The development of such an approach was closely connected with the introduction of new methods, those of field theory, into solid state physics.[119] Fröhlich applied the field theoretical methods to the interaction of the electrons in a metal with the lattice vibrations, and he found that the interaction would lead to an attraction between the electrons.

At the same time, and independently of what was predicted by these theoretical developments, experiments were undertaken to determine whether or not the critical temperature was dependent on isotopic mass.[120] These experiments showed, surprisingly at the time, that the critical temperature varied inversely with the square root of the isotopic mass. Thus, the mass became an important parameter when the motion of the ions was involved, and this, in turn, suggested that superconductivity could be derived from some sort of interaction between the electrons and zero-point vibrations of the lattice.

During March 1950, Fröhlich visited Duke and gave two lectures: one was on the theory of dielectrics and the other was on superconductivity. Fröhlich had just been informed of the experiments at Cambridge University concerning the isotope effect at critical magnetic field. By the fall, Fröhlich had not made much headway in his attempt to show that if one considered a homogeneous magnetic field and a simply connected conductor, then such an assumption led to a contradiction when the conductor became a superconductor. He thought that there could be much progress if he and London 'could work together for a few weeks. If you also think so, we may examine the possibility of finding funds for such a collaboration'.[121] He hoped to be able to discuss the various problems with London during the forthcoming Oxford Conference in Low Temperature Physics of 1951. By the end of October 1950, Fröhlich informed London that he had found a simple method which resulted in the London equations and that the method incorporated the extra energy required for a redistribution in momentum space.[122]

London was very enthusiastic about all these developments. 'As a matter of fact I have not been thinking much about superconductivity lately, partly because it is now Fröhlich's and Bardeen's turn.' He felt that the proposed interaction should contain the solution to the problem. Despite some criticisms about Fröhlich's approach[123] London was convinced that Fröhlich had cracked the nut:

> Perhaps I have not yet been able to completely follow your argument which is certainly far from being trivial, and I shall study it in greater length. Today I wanted only to write to you my very *first* impression

> which is that although I believe and wish that you have the solution I am not convinced that you have it yet.[124]

Fritz wrote immediately to Heinz praising Fröhlich's work and telling him that he was convinced that Fröhlich was 'capable of bringing this matter to a praiseworthy conclusion'.[125]

John Bardeen, who was at the Bell Telephone Laboratories, was thinking along similar lines. He had been working on semiconductors and his researches with William Shockley and Walter Brattain had led to the invention of the transistor. All three would share the Nobel Prize for Physics in 1956. Bardeen's interest in superconductivity dated back to his Harvard years when he had first read Shoenberg's *Superconductivity*. After learning about the experimental discovery of the isotope effect in May 1950, Bardeen was immediately able to suggest that superconductivity might arise from a new, attractive interaction between the electrons and the phonons, resulting from lattice vibrations. At the beginning of December 1950, he sent London copies of his work. In the accompanying letter, Bardeen asked how much his own proposal about superconductivity was based on London's approach.[126] London was exalted. Such an enthusiastic letter by London about something done by another person is nowhere else to be found:

> Many thanks for your kindness of sending me your two extremely interesting manuscripts, which seem indeed to give a brilliant solution to this old provocative problem and at the same time to confirm some conjectures I had advanced in this connection. After Fröhlich's discovery of the new interaction field there could be little doubt that this field must contain the so far missing electronic interaction. Yet it was not at all clear in which way this interaction would produce superconductivity. It seems to me that here you did the decisive step by making a small effective mass and Landau–Peierls diamagnetism responsible for the effect and I wish to congratulate you on this great achievement.[127]

Bardeen thanked London for his comments on the manuscripts and informed him about the manuscript he had just received from Fröhlich.[128]

There was an incident which displays London's gracious behavior whenever he felt that his work had been given the proper credit. John Pellam had suggested that London should give a talk in the Low Temperature Symposium to be held at the Bureau of Standards in March 1951 about 'your dual concept of superfluidity and superconductivity, perhaps on the basis of a condensed version of your book'.[129] London declined the invitation, and a disappointed Pellam tried to convince London to change his mind by conveying what

Bardeen had written to Pellam. 'I was surprised that F. London was not listed to talk at the Theoretical Session ... London has been right all along and probably has contributed more than anyone else to the theoretical side of the subject.' Pellam continued 'I am sure that this represents the general viewpoint and that everybody would be glad to hear from you'.[130] London continued to refuse to give a talk, although he agreed to chair a session at the symposium.

> I should feel only too happy if I could live up to your great expectations. However, I cannot always be on the front line and now it is Fröhlich's and Bardeen's turn to carry the torch. I very much appreciate Bardeen's reaction ... But 'to have been on the right track all along' is no reason for me to shout about it still now ... I shall certainly participate in the discussion if I can contribute something to it.[131]

An ugly finale

In late 1949, von Laue sent London the second edition of his book on superconductivity, published on the occasion of his seventieth birthday and, about a year later, London sent him the first volume of his book *Superfluids*. 'On browsing through it I saw that our opinions differ on certain points. So much more interesting for me' wrote von Laue to London.[132]

Upon a closer reading of the book, von Laue complained that London had failed to include various references to his work. He felt that London should have mentioned von Laue's interpretation, in 1932, of the difference between the limiting value of the magnetic field and the threshold value, which depended on temperature and the material. Such an interpretation followed from the general theory, but in 1932 there was no hint of such a development. In his book, Shoenberg had emphasized that the strength of the field on the boundary was the only significant parameter. 'But that had to be explicitly expressed sometime, and *I* had done it ... It was the beautiful papers of de Haas and his school that you briefly mention which first clarified for all physicists the correctness of my theory. I have the impression that my publication concerning the researches in superconductivity, that was clear from approximately 1932, opened a new and successful avenue.' Von Laue published a short note expressing his disagreements with London; London answered and there ensued a rather unexciting and inconclusive series of notes in *Annalen der Physik*, published in 1952.

The years had passed, but not all accounts had been settled. Now it was London's turn: he felt that nothing he had written was wrong, but the historical account was not as generous as some of the leaders of the German physics community expected. Von Laue, the truly courageous anti-Nazi, was wounded. He expressed his bitterness in a deeply shocking way.

> You know I am not an antisemite, but sometime I have to say it: I have the impression for some years now, that some Jews who are very gifted and indispensable for science, and who work on superconductivity, conceive of this research field as a kind of scientific Israel to where others do not have a permission to immigrate. The fact that you do not mention even once the attempt of Justi–Zickner for the induction in superconducting coils [. . .], together with what I tell you above, strengthens this impression inside me.
>
> I am sorry that I have to write to you such a letter.[133]

London was devastated. Edith London remembers how much he was hurt by this letter, and how he spent long hours trying to compose his answer to von Laue, not liking what he wrote and starting all over again. In his final answer, he first tried to refer to the technical points. He admitted his negligence, since he had included the references to von Laue in the original manuscript, but then the reference was lost when he decided to change the format of the bibliography. 'I should tell you that it was not, naturally, in my intentions and that it would have been paranoid to have consciously made such a breach of good manners or an inaccuracy in my book. Such things happen unfortunately even with the greatest of attention and I am sorry for this negligence.' For the remaining points, London's attitude was that he should have had a freedom of choice with respect to the approach he was going to adopt, otherwise:

> I would not have attempted to write a book on the same subject for which you have published such an excellent report.
>
> Your remaining observations are outside the field of scientific discussion and bring to the fore emotional generalizations and malevolent hypotheses that cannot be justified under any conditions, for an Israel of Jewish Physicists; they revive painful memories from the past and fit badly in the picture I have of you, so it is impossible for me to be engaged in such a discussion.[134]

Von Laue was sorry for expressing himself in such a manner and his only explanation was that he had: 'many troubles with my colleagues in physics that sometimes I become very angry'.[135] He asked London to contribute an article to an issue of *Zeitschrift für Physik* to be dedicated to the seventieth birthday of Franck and Born and that he hoped that: 'you have not taken personally my polemic about superconductivity. I have the impression that in this way there will be something new about the subject, and in this respect your book was useful'.[136] London was not so sure that he could contribute to the volume, since he had promised to send an article to a volume dedicated to de Broglie.[137] There was no other exchange between them, and when von Laue sent his condolences to Edith in 1954, he noted that Fritz London's ingenuity was best displayed when he was in Berlin.

For a long time, von Laue's colleagues had been aware that he often had depressive moods which 'in later years took the form of a feeling of being persecuted, be it by scientists or the military which he abhorred'.[138] In fact, a few days after the letter he sent to London accusing him of a conspiracy, von Laue sent a similar letter to Einstein, where he raised the same issue. He told him about a 'painful matter' which concerned London's new book on superconductivity where his own contributions were mentioned only briefly, and he noted that Shoenberg's book did not mention many similar things done by von Laue. He wrote to Einstein that the previous day he had overheard a conversation between two colleagues of his at Göttingen who were concerned about similar occurrences in astrophysics.

> These would all be trifles, if one did not get the impression that a plan stood behind them. In superconductor research, I unfortunately must comment, there is a group of Jewish researchers who want to make a scientific 'Israel' out of the field, to which no outsiders would gain entrance . . . Such occurrences do not improve the mutual understanding between peoples which is necessary to the preservation of peace. These people could, if they continue to accumulate, seriously endanger this understanding. Fortunately, you involve yourself in promoting peace. Couldn't you use your large influence in scientific circles to favor the institution of a relevant citation method? In the big picture we scientists unfortunately only have small influences on world events. But the big occurrences, without exception, are based on many small events. These we can – and should – try to influence.[139]

In 1952, Pippard visited Duke. He found London was still very hurt by von Laue's letter.[140] London was worried about the latent anti-semitism of North Carolina society, but this was more because he saw his children beginning to become aware of differences in color, and was afraid that they would imbibe from their schoolfellows something of the virulent racism that he saw and detested around him.

Could Landau be right?

The new measurements by Atkins and Osborne[141] on the second sound velocity down to 0.1 K, which showed agreement with Landau's predictions, became a source of deep disappointment for both London and Tisza. It is quite surprising to read that London, up to that time, had not really appreciated an absolutely essential aspect of Landau's theory, namely the effective phonon mass. 'I want now to understand the main point in Landau's argument which I have never understood.' London sought Tisza's help, since he had thought so much more about phonons and knew Landau's approach much better than London did.[142] A few days later, London sent another note to Tisza telling him that he had consulted the 'classics and learned the acoustics I should have learned at school. Excuse me that I bothered you with such a trivial question'. Tisza's view now was that the second sound measurements justified Landau's assumptions to consider the effective mass of the excitations rather than only the rest mass, which would result in a zero contribution for the phonons. London noted that helium-I was 'getting ripe for a decent theory where Bose–Einstein and phonon features are united'. He informed Tisza that he had been thinking of giving a better quantum mechanical foundation to his idea of Bose–Einstein *liquid* as distinct from the *non-ideal* Bose–Einstein *gas*, since everyone tried the latter with no success.[143]

Nevertheless, London continued to think that, despite the experimental results, Landau's handling of the effective mass of the phonons was theoretically inconsistent. Using Landau's approach, he found two different values for the effective mass of the phonons, depending on the particular method of calculation. If the momentum of a moving system was divided by the velocity of translation, then the effective mass m was equal to $(4/3)\ (E/c^2)$ which was the correct value. If, on the other hand, one calculated the transformation of the energy, then one needed to have an effective mass five times larger. This (apparent) inconsistency in Landau's theory prompted London to write that his impression was that Landau's argument was nonsense.[144] Tisza's response implied in no uncertain terms that they might have to proceed to drastic revisions of their ideas. Even though he felt as much puzzled by the phonon mass, he failed to see how it was possible to derive the superfluid properties from only the excitation energy. He regretted that things should be in such a mess. Then, Tisza proceeded to cast doubt on what was one of their strongest criticisms against Landau: the status of the (ghostly) rotons. 'To tell you frankly I can even imagine that the roton idea is correct... I hope this is all nonsense, but truth is presumably independent of our wishes.'[145] This change of attitude was further expressed by Tisza when, in suggesting some improvements on London's manuscript for his forthcoming book, he, again, advised London to go easy on Landau:

> It seems to me that you are not quite fair to him. Nothing very definite, rather an undertone. I think people have the tendency to give him more

> credit than justified. It's your natural reaction to give him less and even that somewhat grudgingly. I think our common aim would be better reached by leaning over backwards. In addition to the purely personal point, there is also a factual question. I think that Landau is not so wrong when he emphasizes the many-particle aspect of the helium problem. There is ample ground to criticize him on the grounds that he promises more than he actually fulfills.[146]

London was invited to the Low Temperature Conference of 1951 at Oxford University by Simon, and was asked to give the introductory lecture at the liquid helium session. Simon suggested that his talk be titled *Theories of liquid helium*. London had his misgivings. There were, at present, no acceptable theories of liquid helium, he said, and he proposed to speak on the *Limitations of the two-fluid model of liquid helium*.

There were difficulties in finding adequate funding to go to the conference. He informed Gorter of the dim possibilities for his trip and told him how much he wanted to go to Europe to 'get the ultimate blessings for the second volume [of *Superfluids*]'.[147] Eventually, he did secure financial support and the whole family travelled to Europe.

They travelled with the *Nieuw Amsterdam* and reached Rotterdam on June 16. Then, they hired a car and drove to Northern Italy via Bonn, the Black Forest, Zürich and the Gotthard pass. They went to Bonn because London wanted to show the children the place where he grew up. It was the only and last time he visited Germany after the war and in Bonn he met one of his childhood friends.

During the summer of 1951, Fritz, Edith, Frank and Rosie met with Heinz, Lucie and the children, and stayed at Harwell, England. It was, as always, a particularly joyful affair. But it was too brief a visit and Edith was sad to leave. 'The cruelly wide ocean lies now between us again, [and] the thoughts of our togetherness are bridging the wide span.'[148] Fritz had a wonderful time, feeling as though he was at home: '*Kunststueck* [hardly surprising] as Lucie would say. You have taken a lot of trouble to make it comfortable for me in every way. I can still hear Lucie's orders to the children: nothing is allowed to disturb Uncle Fritz's sleep'.[149] He worried, though, about Heinz, who seemed overtired and overstrained to Fritz. His habit was to stay in the laboratory till very late and he went to bed too late. Heinz was getting tired very frequently and he had to fight fatigue as part of his routine. He consumed vast quantities of glucose tablets, peppermints and his high powered vitamin pills.

At the Oxford Conference, London's was a pessimistic report. The low temperature second sound velocity measurements and the careful reconsideration of Landau's theory motivated by these measurements led to a talk that was in striking contrast to London's 1946 report. He felt that no satisfactory molecular theory of liquid helium had so far been produced, and there were limits to the validity of the macroscopic two-fluid model. He noted that the He^3 results could suggest that the Bose–Einstein statistics were essential for explaining the properties of helium-II, but the point was by no means emphasized. The impressive agreement of the measured second sound velocity with the predictions of Landau's version of the two-fluid theory was mentioned. According to London, the outstanding problem that a successful theory of liquid helium would have to resolve was to find a way of combining the properties of the Debye phonons which obeyed Bose–Einstein statistics with the particle (translational) properties which might obey either type of statistics. Furthermore, though there was no doubt about the basic correctness of the two-fluid concept,[150] it appeared to be limited in a number of ways. These were its inability to accommodate the existence of a critical velocity of the flow above which superfluidity was destroyed, the dependence of the normal viscosity on the slit width, and the possibility that at large velocities, in addition to ordinary viscosity, there might also be a dissipative process.

Despite all the difficulties involved, in his second volume of *Superfluids*, London continued to be quite enthusiastic about the two-fluid model. He repeated his arguments about why results derived for ideal gases could be applied to liquid helium, and noted that the temperature at which an ideal Bose–Einstein gas started condensing was of the same magnitude as the lambda-point temperature. He stressed that it was this 'correspondence' which prompted him to advance the hypothesis that the transition from helium-I to helium-II might be caused by such a condensation process. He was aware, of course, that the numerical agreement was quite meager, yet he could not resist the temptation to proceed with the Bose–Einstein condensation, which appeared to be sufficiently exotic to promise a qualitative interpretation of the unusual super properties of liquid helium. He considered Tisza to be the first to have recognized the possibility of evading the pitfalls of a rigorous molecular kinetic theory by employing the qualitative properties of a degenerating Bose–Einstein gas to develop a *macroscopic* theory.[151]

There was no compelling reason why London should have been persuaded to abandon his approach and to adopt that of Landau. After all, London's agenda included the understanding of all the peculiar phenomena associated with liquid helium: its being liquid at absolute zero under its own vapor pressure; the unexpected change in the behavior of many parameters at the transition temperature; the phenomena defying a strict description with the current concepts of hydrodynamics. Furthermore, Landau's theory was a difficult theory to understand. In 1945, Pauli, who was at the Institute for Advanced Studies in Princeton, USA, wrote to London that they had been studying Landau's papers on superfluidity, but, since Landau 'gives so often merely allusions instead of proofs',[152] his paper could not be understood. Uhlenbeck had informed Pauli of similar feelings: something was there, but it was by no means easy to make some sense out of it. Uhlenbeck had also written to Tisza that Landau's paper had impressed his colleagues at Michigan University immensely and they had seminars to discuss the paper – 'the Russians seem to be able to do good physics and beat Hitler at the same time, so perhaps we should try to follow their example!'[153] For London, Landau's achievement was based on a quite unacceptable circularity: the mechanism of superfluidity which appeared to be following from his theory was nothing other than a mathematical formulation of what was originally derived on purely intuitive grounds. What was originally a theoretically barren conception, was made 'respectable' by what appeared to be a quantum theory of hydrodynamics. The energy gap between the ground state which was interpreted as being the superfluid, and the normal state which included the excitations, was a crucial notion and absolutely necessary in order to make the whole treatment a credible enterprise. But the energy gap was inserted by hand, and therefore, according to London, what was projected as a microscopic quantum theory of superfluidity had a rather serious element of arbitrariness. In his own approach, the only element of arbitrariness was the analogy he introduced between Bose–Einstein condensation and liquid helium. But after all there was a theoretical basis which legitimized the analogy and convinced him that there had to be a connection. Nothing even remotely similar was true for Landau. The irony is that even though his own formulation of the theory of superconductivity was a case of a hypothetico-deductive procedure, London became adamantly opposed to the very same procedure when it was used by Landau. London failed to see that part of Landau's contribution was to develop a theoretical framework for reinterpreting what were considered to be different fluids in classical language in terms of different quantum states. Landau fully realized the serious problems involved in the interpretation of this newly (re)formulated concept of motion, and he was extremely careful while introducing these concepts to point out that they should not be taken literally. It is obvious from his writings that Landau was striving not only to develop a new theory, but also to develop a different descriptive language in parallel with the development of the new theory.

London's proposal of the Bose–Einstein condensation did not provide any predictions. It was rather a plausibility argument for the transition to the superfluid case and it was projected as a viable alternative for the order–disorder approach. The two-fluid model did, of course, have specific predictions. But London's adoption of the two-fluid model was in 1946, eight years after the initial formulation of the assumption about the Bose–Einstein condensation, when it was evident that one of its more interesting predictions, namely the one about the existence of second sound, had been corroborated. When, later on, it also became evident that the two-fluid model gave the wrong quantitative prediction for temperatures below 1 K, it was the two-fluid model which bore the brunt of the criticism and not the proposed Bose–Einstein mechanism. At that point, the absence of superfluidity in He^3 became the prediction for the Bose–Einstein approach. Thus, so far as He^3 was not found to be superfluid, London's faith in the Bose–Einstein condensation as a way of accounting for the transition to the superfluid state was strengthened further. He had not liked the two-fluid model to start with, anyway. London was so deeply convinced about the crucial role of statistics in explaining superfluidity, that he did not pay any attention to Landau, since the latter failed what London considered to be the litmus test of any correct theory of superfluidity. The fact that Landau's theory gave the correct results for the second sound, could have been accounted for by many reasons. But still, Landau's theory could not have been a fundamental one, since, according to London, it failed to make any connection between statistics and superfluidity.

London's attitude towards the need for statistics to explain superfluidity and the effectiveness of Landau's approach remained more or less unchanged until the end of his life. Three issues gave him misgivings. The first was a methodological issue: Landau's theory was developed in 1941 after Kapitza's experiments, whereas the two-fluid model of Tisza had been formulated three years earlier than Landau's without any of the new experimental results and it had, in fact, predicted new phenomena. 'Indeed, in view of the evidence of Kapitza's experiments the two fluid concept must have appeared almost obvious'.[154]

The second issue was a theoretical one and, in effect, reiterated questions which had been posed systematically since the latter part of the nineteenth century, and concerned the actual nature of liquids as well as the most effective ways for setting up a theoretical framework for dealing with liquids. Could a theory for liquids be developed by regarding liquids as condensed gases or rarefied solids? London thought it was the former, whereas Landau believed it was the latter. Landau extended the quantum theory of fields by representing the liquid as a quasi-continuum whose excitations are quantized in the same manner as is done in the Debye theory of the solid body. There was, nevertheless, a serious difficulty. It was not at all obvious how, for example, it became possible to enumerate the characteristic modes in a liquid taken as a continuum. In contrast to the situation in a gas and solid, the different modes of the sound waves, for example, did not exhaust all degrees of freedom. To be able to deal with this difficulty, Landau's proposal was to introduce two kinds of elementary excitations, the phonons and the rotons. But as it was repeatedly stressed by London, the definition of the rotons and, in the last analysis, their ontological status, left a lot to be desired.

The third issue raised by London concerned the possibility for reaching a decision on what would constitute a crucial test for each of the two schemata. Such a decision could not have been taken after the assessment of the quantitative differences between the predicted and the measured values of parameters. This is why he rejected the prediction about the behavior of the temperature dependence of the second sound measurements. In fact, this is the only point where there was a change in his attitude from the very beginning where he, himself, considered the test as verifying Tisza's two-fluid model. But when measurements below 1 K were in agreement with Landau's formula and in disagreement with Tisza's, London proposed a different phenomenon which would actually test the fundamental assumption about the role of the statistics. Thus, the presence or absense of superfluidity in He^3 became the *experimentum crucis* for the role of statistics in a theory of superfluidity. This was, of course, more in character with London's overall approach to such questions. He considered that yes–no type answers settled debates related to the fundamental issues of theories.

It was also psychologically difficult for London to accept Landau's approach. It had to do with the beginnings of his career. The great success of the Heitler–London paper was to have shown that the homopolar bond was a phenomenon which could only be understood in terms of quantum mechanics. In fact, London's career was spanned by his successful attempts to articulate the role of purely quantum mechanical notions: with the electron spin and the exclusion principle he explained the homopolar bond; with diamagnetism he proposed the macroscopic character of superconductivity; zero-point energy was decisive in understanding solid helium; with the uncertainty relation he resolved the problem of the intermolecular forces; by using the Bose–Einstein condensation he afforded a purely quantum mechanical mechanism for superfluidity. Somehow, accepting the correctness of Landau's approach would have been a serious drawback in London's attempts to complete his theoretical agenda.

London felt that superfluidity – more than any other phenomenon he had studied – gave him the opportunity to develop and explain in full his views about the theoretical aspects of macroscopic quantum mechanics. Perhaps this may be the best way to understand why in that remarkable introduction published in the first volume of *Superfluids*, nearly four times as much space was devoted to superfluidity as to superconductivity. He thought that his own approach to the two phenomena implied a 'striking parallelism apparent between the conceptual frameworks'[155] and he was more confident than previously about future developments. He believed that both phenomena were the result of a macroscopic quantum mechanism which produced a gradual condensation and induced a superflow which would be maintained: 'not by the absence of collisions, which is hard to accept, but would represent the thermodynamic equilibrium state established, under the given boundary conditions, by the very collisions'. Nevertheless, London felt that, in contrast to superconductivity (where the underlying mechanism was not yet known), in the case of superfluidity such a mechanism, at least qualitatively, was provided by a 'quite intriguing concept', namely the Bose–Einstein condensation. In the case of superfluidity, he thought that the main emphasis should be to devise effective techniques to deal with the quantum mechanical many-body problem. 'Yet this so tantalizingly simple idea is a conjecture which will have to be substantiated or else reduced to its true bounds by a workable analysis of the quantum mechanical many-body problem before we can say that we have penetrated the mystery of the superfluids.'[156]

The structure of London's arguments in defense of his approach to superfluidity was uniquely suggestive of his agenda concerning interpretative aspects of quantum mechanics. The notion of 'macroscopic' found its optimum substantiation in the phenomenon of superfluidity. His theory of superconductivity was never seriously challenged and he had even articulated a strategy for a molecular theory. But the theory of superconductivity could be accepted and applied without any commitment to the particular interpretation. True, the long range order in phase space was absolutely necessary for appreciating the theory, but it was by no means necessary for accepting it.

But with superfluidity there was a dramatically different situation. The notion of 'macroscopic' was inherent to the theoretical foundations of the schema. The application of the Bose–Einstein statistics to an ideal gas resulted in a (mathematical) oddity devoid of any possibility of testing it experimentally. Examining the possibilities provided by considering this idea as the basis for a theoretical treatment of liquid helium attributed the necessary ontology to the Bose–Einstein condensation. And the macroscopic character of such a pure quantum effect was no longer a matter of an interpretation imposed on the formalism, but imposed itself because of the formalism.

London realized that the study of liquid helium and superfluidity gave him the best opportunity to articulate his viewpoint fully. He was not just interested in proofs, but he wanted to persuade the community about the legitimacy of this viewpoint.

The researchers in low temperature physics in the Soviet Union have acknowledged London's approach very rarely, if at all.[157] Soviet physicists referred exclusively to the achievements of the Soviet physicists in low temperature physics; an attitude which had been formed quite independently of Landau's feelings towards London.[158] Soviet physicists did not have access to journals from the West and vice versa, which is why the West was not aware of Landau's work until after the war. But the end of the war did not bring only peace. The former allies found themselves in the scorching waters of the Cold War and relations among the scientists from the two camps, at least in the beginning, were not altogether free from the contingencies of politics. Work in low temperature physics, especially superfluidity, was the pride of the Soviet physicists and they considered it to be their own achievement during the most difficult years of the Soviet Union. They were proud and, at the same time, possessive of the origins of superfluidity. Kapitza's remarkable experiments and his astounding results functioned as a glorious corroborating instance of what could be achieved in this new society by starting from scratch and by working single-mindedly within an overall program. Landau's theory, proposed almost immediately after the experimental results, was perceived not only as a tribute to his genius, but also to the harmonious collaboration between experimentalists and theorists. The remarkable work of Lifshitz, Peshkov, Shalnikov, Andronikashvilii, Khalatnikov and others was proof that their particular organization of science was making the Soviet Union a world leader in a particularly difficult area of research. There was no doubt that the theory was correct and, furthermore, it manifested their way of doing correct physics: from experiment to theory to confirmation to changes in the theory as a result of new experimental results, resulting in further confirmation. Furthermore, it was the right kind of physical theory, since the phenomenon was reduced to the 'more' fundamental entities. That was the important thing. Their ontological status could be clarified along the way. This group had discovered a new phenomenon, had devised its theoretical explanation and had confirmed this explanation. To many physicists of the Soviet Union, this was a triumph due to the political decision to build the Institute of Physical Problems in Moscow and to let the scientists continue their work during the war. It expressed the success of the philosophical approach to the relationship between theory and experiment.

The worrisome realities of the postwar era

By the end of the 1940s, the 'front' against the communists had not only become the cornerstone of American foreign policy, but it was also determining, to a large extent, domestic affairs. Fritz was deeply troubled by these developments. He felt that there was a kind of a war psychosis which had influenced the economy as well. Food prices had risen, and scarcity and more price rises generally followed immediately. The news was not particularly comforting, and he felt that the people were no longer being given objective information. 'Madness has again gripped mankind. This time it seems that *we* are making the main contribution and that we are manoeuvering ourselves into a fine mess.' Somehow, he could not help seeing certain parallels with 1933. He saw bad times coming and felt that the USA was drifting to a point of no return. Not even in those unhappy months during the end of his stay in Oxford was he so pessimistic. 'Madness will defeat reason!', he wrote to Heinz.[159] When, two years later, there were presidential elections, he knew that their outcome would influence developments all over the world and would show 'the maturity or immaturity of the American people, and whether they do or do not prefer an unprincipled demagogue to an intelligent, responsible Democrat'.[160] Even though Dwight Eisenhower was proposed as the candidate for the moderate wing of the Republican Party, having defeated the conservative and extreme R. A. Taft at the Republican Party convention, London's sympathies were with the candidate for the Democratic Party, Adlai Stevenson.

In 1950, the USA Naval Ordnance Laboratory offered London a part time consultancy in its theoretical physics group. It would be for a two- or three-month period per year and he would be paid the same salary as he received in the university. The offer mentioned that London would be expected to work on problems related to the quantum mechanical explanation of the research directions of the Laboratory, which were being expanded to include solid state physics.[161] He declined the offer because such a commitment would have been very disruptive for his work.[162] The real reason for declining the offer was, of course, the state of his health.

The children were growing up and each needed special attention. The Londons lived in west Durham, near the city limits. Many poor people lived in the same neighborhood, and having intellectuals as parents was uncomfortable for Frank, whose peers came from more 'normal' environments. Fritz was at a loss with the everyday interests of his son who was developing happily in the new environment. Much to the dismay of his parents, Frank asked to have a hatchet so that he could join the rest of his friends and chop trees. In the summer of

1952, Fritz was invited to spend the summer at the Department of Physics, University of Wisconsin. When he discussed the matter with the family, Frank was deeply saddened by the news. How could he face the other children after coming back from up there, where people had 'fought on the wrong side' during the American Civil War? In the school that Frank attended there were no Jews and the rest of the boys did not identify him as a Jew. When there was talk among them about the church each attended, Frank was slightly at a loss, but his best friend Willis told him to say that he attended his own church. Nevertheless, Frank did not feel very comfortable when he joined the Jewish boys' group and started attending Hebrew School. 'Frankie is getting to the teen-age phase, absolutely normal and charming, but causing constant quarrelling compared with his previous obedience!'[163] Frank wanted a new bicycle and London wanted to get him an English Raleigh lightweight bike, but to no avail. He insisted on what London called an 'American monster. He is an American conformist and admitted eventually that he does not want to be different from the others'.[164]

Rosie (now six years old) developed an interest in piano playing. It was a particularly welcome surprise for Fritz. Of course, he had played the piano since childhood, and he played it very well. She practiced by herself exercises and scales which Fritz worked out for her, since he thought that in traditional piano teaching too much importance was given to note reading. 'I let her do everything by ear and intuition; only separately a little note reading.'[165] Fritz was teaching her Bach minuets by ear, but he realized that he would soon have to find an adequate teacher for her. He preached faith in intuition, whether in physics or piano playing, but he was, also, the first to acknowledge that nothing could be mastered by intuition alone. When Rosie started piano lessons in the fall of 1951 with a qualified teacher, Fritz started using her child-sized fiddle to teach himself the elements of violin playing. Many times, Fritz expressed his gratitude and marvelled at the way Edith coped with housework and her art. He was worried that she was overworked and thought that she ought to take a proper holiday away from them. He was really proud of her paintings and told Lucie and Heinz that one of her paintings was at the State art gallery at Rayleigh, in a most prominent place, near the entrance, where, he said: 'the experts find hers by far the best picture in the exhibition'.[166]

London's sentiments towards the new political situation were also displayed through his relationship with Paul Zilsel, who in 1947, having just received his doctorate from Yale, went to work with London and was London's first post-doctoral fellow. Zilsel was born in Vienna in 1923. His father, Edgar Zilsel, was the well known sociologist of science and a Professor at the University of Vienna. They had emigrated to the USA in 1939. He received his bachelor's degree in physics in 1943 from the University of Wisconsin. His father committed suicide in 1944 when he was 53 years old and his mother spent most of her time in the USA in a mental hospital. She went to Austria in 1948 and stayed there till she died. 'The USA was disastrous for my parents. It was London's experience exponentiated'.[167]

Zilsel met Gregory Breit, one of the dominant figures in nuclear physics, at the University of Wisconsin and started his graduate studies. Madison was the hot bed of radicalism and he joined the American Youth for Democracy, successor to the Young Communist League. The organization's main task was to help the war effort, but Zilsel was quite apprehensive, since he had formed a rather distrustful attitude towards the American establishment. When the American Youth for Democracy expelled Brower in 1946, he joined the Communist Party while he was a graduate student in Wisconsin. He had already received a master's degree in mathematics. In the meantime, Breit had gone to Los Alamos to work at the Manhattan Project and, when he came back, Zilsel started his doctorate with him. His doctorate was on nuclear scattering and he studied the effect of bonding on low energy nucleon–photon scattering. Breit wanted to have Zilsel as a member of junior faculty at Yale, but discovered that they could not get clearance for him. He asked Zilsel to leave Yale as soon as he could, but he made sure that he found him a job. Through Onsager, who was also at Yale, Breit contacted London, who agreed to have Zilsel at Duke. When he went to Duke, Zilsel had to change his field of research completely.

After Breit, to work with London was a liberating experience for Zilsel. He felt free, he could talk openly and argue with London. Zilsel remembered that London had an ironic way of commenting about what was going on, he was too much of a sceptic and not a whole-hearted political enthusiast. 'London was a famous man. But there was this sense of isolation and craziness in the South. This was probably one of the reasons why I felt so close to London, because in a way it was my own family, only less extreme'. Zilsel felt that, while he was at Duke, London was rather disdainful about his situation and Zilsel's view was that London could have been less isolated if he had succeeded in hiding his contempt. He remembered how much London longed for European culture. London reminded Zilsel of his father: 'He was very lonely. In a way he wanted to be accepted, yet he could not hide his contempt'.

They spent a considerable amount of time discussing various issues of their work and most of the time was spent in elucidating the physical meaning of mathematical terms. London always asked questions and he worked by talking things through.[168] Of course, London knew Zilsel's political views, but he did not know about his organizational affiliations, although when he tried to get him a research grant from the Navy he learned about them. Eventually, London did manage to get him the money to be at Duke during his absence in Europe in 1949, but Zilsel would not sign the non-communist affidavit. He had to leave. After a year of teaching at Fred Collins College, Breit got him a job at the University of Connecticut at Storrs. While at Duke he was expelled from the Communist Party, because of his disagreement over its policy concerning Henry Wallace – the 'left-wing capitalist for whom the Communist Party wrote his speeches'. In 1953, Zilsel was called to testify both at the House Committee for UnAmerican Activities and at the Senate's Internal Security Committee. They asked many questions about his activities at Yale, but none about what he did at Wisconsin. He learned that, because Wisconsin was McCarthy's home state, McCarthy did not want his record to look bad. Zilsel refused to name people, despite the insistence of the Committee. After he had appeared at the Committee, the University of Connecticut suspended him. He had tenure and was paid, but they would not let him teach or go to his office.

Throughout his life, London was sympathetic to the plight of the Left, to its ideals of equality, to its avowed humanism, internationalism and antinationalism. The following incident was quite characteristic. In 1946, James Franck had asked him to sign a statement calling for help for the victims of nazism. Of course, London signed the statement, but in his letter to Franck he urged for some changes to be made to it:

> I think it would be better for the opening sentence to read 'We the undersigned men and women, enemies and victims of Nazism' and leave out the 'of German and Austrian origin'. Otherwise there is the danger that the unsuspecting reader may think that the reason of this movement is only because of the German–Austrian patriotism due to the origins of the people involved, and that it is not an issue which should concern *every person.*[169]

Despite his strong convictions, London was never politically active. Having lived through the tempestuous years of the Great War and, later, the Weimar Republic, he was always distrustful of politics and ambivalent about suggestions of possible good which might come from it. The motivations for London's visit to the Soviet Union in 1930 were not exclusively scientific: not only was he curious about what was going on there, but also he was greatly interested in the details of those developments. His associations and friendships in Paris were with people who were deeply involved in politics. His friendship and respect for Joliot and Langevin was also an expression of his political affections. Linschitz has a vivid memory of the following incident. In 1942, the Londons had invited the graduate students of the Department of Chemistry to their home. At one point during their conversation, London said assertively: 'every intelligent man was a communist as a young person'. As it is invariably done, he did not add what is usually attributed to Churchill 'and a capitalist in his forties'. The words fell upon the ears of young people coming mainly from a Jewish environment in New York, where such ideas were often discussed and frequently adopted. One afternoon, Fritz and Frank had a discussion about Frank's views on a series of issues, and more particularly on the problem of equality. In the end, Fritz told Frank that if the things he believed in were to be realized, then that would be communism. But he did not agree with what was actually happening in the Soviet Union. He thought that the USA had all those institutions where democracy could flourish – even though the existing situation left a lot to be desired. During this time, when even the expression of personal sympathy towards the 'enemies of the people' was often considered to be an endorsement of their political views, London had no qualms in expressing his sentiments, and he wrote the following to Zilsel:

> Last week I read in the local paper a very impressive and, as far as I can guess, very fair article about your recent hearing. I was very proud of your testimony, and my son even said: that is a patriot. I brought the matter up here again, hoping they would reconsider their decision under the present circumstances: however, there was no reconsideration. The project is *indefinitely dropped*. I am now writing to the Office for Naval Research. Perhaps they will look at the matter differently and then the position of the University would be truly untenable.[170]

Right after London heard that he would receive the prestigious Lorentz Medal, Edith remembered that he turned to her and said that, with his increased prestige, he would immediately write a letter to the President of the University requesting some funds for Zilsel to work again at Duke – but the request was not granted.[171]

The second volume of *Superfluids*

London felt that it would be difficult for him to summon up the courage to get involved in the writing of the second volume of *Superfluids* – though Maria Goeppert Mayer strongly recommended Wiley to proceed with it.[172] Eventually, he was convinced to proceed with the second volume which he, provisionally, titled *Macroscopic Theory of Superfluid Helium*. The new manuscript was ready by May 1952, at which time Wiley informed London that they would need some time to do 'a little investigating regarding the potential market of the book'.[173] London was extremely upset by such developments, because he felt that nothing had happened since the time the original contract was signed to indicate any changes in the prospective market for the book.

Wiley answered with a rather terse letter. They did not think that they could make any profit on the first volume, even if the whole stock was sold[174] and they wanted to emphasize that Wiley was not running a 'philanthropic operation'.[175] London was very disappointed. When he received a letter from Lowdin, writing about the lively discussions of his work in the seminar of the physicists at the University of Stockholm, Sweden, London remarked that: 'it is difficult to work in such a vacuum and it is good to know that my work finds appreciation'.[176] A second letter by Mayer to Wiley changed the situation. Wiley wrote back and apologized for having offended London, saying that the choice of words in the previous letter was rather unfortunate.[177] Their only consideration was economic and 'if allowed to operate freely in this fashion would soon leave us no name at all, good or bad'.[178] There was a promise to honor the contract; London was asked to send the manuscript and Wiley promised to proceed directly with its production.

But London himself was facing problems with the content of the book. He told Paul Marcus that his manuscript was at present in a complete uproar. He had found out that the derivations of the two-fluid hydrodynamics from Hamilton's principle were all faulty and it seemed to him that it was not possible to escape the conclusion that there was a hidden assumption which could not be reduced to classical mechanics.[179] Marcus was convinced that only a statistical theory of (quantum) liquids would clarify such problems.[180] The final draft of the manuscript was sent to the publishers just before London was ready to leave for England, in May 1953, on his way to Leiden to receive the Lorentz Medal.

London urged Wiley to publish the monograph as soon as possible: 'I cannot reiterate often enough that these monographs have a short life time'.[181] He was assured that the volume would be published by the middle of June 1954. The last letter London wrote before his death on March 30, 1954, was to Wiley. He wanted the transposition of two paragraphs, since the new order would 'give a closer logical connection of things which belong together'.[182] After London's death, it was proposed by Hobbs that the proof-reading be taken over by T. C. Chen who was a graduate student working with London, William

Fairbank and Nordheim. Felix Bloch agreed to write a commemorative introduction to the volume. Eventually, Nordheim took all the responsibility for supervising the publication. The only new thing to be incorporated, which was also known to London, was the new data about nuclear spin paramagnetism of liquid He^3. Measurements made at Duke by Fairbank and others showed definite deviations in the direction of Fermi degeneracy below 1 K, while none had been found at higher temperatures. London wished to incorporate these developments, especially since he had predicted the effect. The book was published on November 4, 1954.

William Fairbank

Starting in the fall semester of 1952, London joined the Physics Department at Duke on a half-time basis, since Nordheim had gone to Los Alamos to work on the hydrogen bomb. As the prospects concerning low temperature work at Duke started to improve, London felt his strength to be progressively diminishing. There were now three people working with him: Chen, a doctoral candidate; R. Smith, who succeeded P. J. Price and was a student of J. G. Daunt and who worked in Duke for a year; and R. de L. Kronig, a Dutchman on a Fulbright scholarship. London felt the burden of these commitments, though he was overjoyed with the prospect of William Fairbank's appointment to Duke, where he had been offered and had accepted an Associate Professorship. Duke had bought a Collins Liquefier, so London needed a good experimentalist and Fairbank was among the most promising. He was very knowledgeable about microwave techniques and London wanted to have him measure the critical velocities of helium-II with wide capillaries. 'Fairbank is just the right man for me'.[183]

Fairbank had gone to Yale University in 1944, had worked in low temperature physics for his doctorate and became closely associated with the work of Heinz London. At the MIT Radiation Laboratory he had already worked on various aspects of perfecting radars and, at Yale, he decided to search for alternating current losses in superconducting cavities.[184] Immediately afterwards, he started working with C. T. Lane who, together with William's brother, Henry Fairbank, was engaged in measuring the velocity of second sound and studying He^3–He^4 mixtures. Duke made an offer to William Fairbank, who gladly accepted it and went there in the summer of 1952, very excited with the prospect of interacting with London. By the spring of 1953, there was liquid helium flowing at Duke and it became possible to produce temperatures of 1 K.

Of course, London had many questions he wanted to test experimentally, but, at that time, foremost in his mind was the dependence of superfluidity on statistics. He convinced Fairbank to examine the extent to which He^3 behaved as an ideal Fermi–Dirac gas. Fairbank had already discussed this issue with Onsager at Yale. It was proposed to measure the strengths of the He^3 nuclear magnetic resonance signals as the temperature of liquid He^3 was reduced. Particles of an ideal Fermi–Dirac gas at sufficiently low temperature were expected to have increasingly antiparallel alignment of their spins, which caused magnetic susceptibility to deviate from the classical $1/T$ Curie law and, thus, to become independent of temperature when all the spins became aligned. When, after a series of problems, measurements were resumed below 1.2 K there was a definite departure from the predictions of the Curie law and the liquid appeared to behave as an ideal Fermi–Dirac gas, having a degeneracy temperature of 0.45 K. London was very interested in the results of the experiment, but because of problems with the liquefier no data were produced before his visit to Leiden in June 1953. London used to pay frequent visits to the Laboratory, standing at the door and impatiently waiting for every new datum point. The measurements from 1.2 K to 0.23 K were completed in February 1954 and confirmed his expectations.

One of the best-known results derived by William Fairbank was the discovery of the flux quantization predicted by London. Fairbank discovered the phenomenon with Bascom Deaver, his first student at Stanford, by detecting macroscopic quantization of the magnetic field outside a superconductor. Walter Gordy wrote: 'I doubt that anything in Bill's scientific career would have given him more pleasure than showing these data to Professor London'.[185]

Further developments

The London equations were local equations because they related the current density at a point r to the vector potential at the same point. In 1953, Pippard proposed a generalization of the London equations in such a way as to give a non-local relation between the current density and the magnetic field.[186] The basis of his theory was his concept of coherence: that the range of order of the wave functions of the condensed superconducting phase extended over rather large regions of space, of the order of 10^{-4} cm in a pure material, and the coherence length τ was a measure of non-local effects. His coherence length was the shortest distance in which a significant change of electronic structure in the superconductor could occur. Although it seemed to be merely a modification of the London theory, Pippard's theory represented a significant change from the earlier theoretical attempts because of its treatment of the problem of how an applied field influenced the superconducting electrons. A basic point of the Londons' theory was the absolute rigidity of the superconducting wave function in the presence of a field. Pippard abandoned this point and suggested instead that a perturbing force acting at one point in the superconductor would be felt over a distance τ, and, conversely, the response at a point due to a spatially extended perturbation would be obtained by integration over a finite region surrounding the point. Pippard arrived at the concept of the range of coherence partly empirically and partly on the basis of his interpretation of the nature of the superconducting state.

At the Leiden Conference of 1953, London criticized Pippard's theory as being an attempt to 'demolish the whole picture of the long range momentum order', in favor of a much more involved description, where the concept of coherence represented a kind of mean free path mechanism analogous to the mean free path mechanism in ordinary conductivity. He was unwilling to accept 'such return to classical mean free concepts for superconductors', despite the fact that it stood the test of experiments. As it turned out, though, after the Bardeen–Cooper–Schrieffer theory of superconductivity, non-local equations were seen to be compatible with London's fundamental principles. Pippard's suggestions were attempts to explain experimental results and did not claim to present a new theory of superconductivity. In 1957, Pippard met Landau whose reaction was not unexpected: '[he] gave me the impression I had committed some unmentionable sin!'.[187]

In 1953–1954, the relevance of Bose–Einstein condensation for superconductivity was demonstrated independently by Ginzburg, Feynman and Schafroth. All three had shown that an ideal gas displaying Bose–Einstein condensation has the magnetic response characteristic of the Meissner effect. Schafroth was the only one who did not think that such a property was a mere mathematical curiosity and that electron pairs that were effectively obeying Bose–Einstein statistics could play a role in explaining superconductivity.[188]

London was involved, in an indirect way, in the discussion about electron pairs. The proposal about the relevance of electron pairs for superconductivity was first made by Ogg (1946) as a result of his experiments on very dilute solutions of alkali metals in liquid ammonia. Ogg had claimed that he could sustain persistent ring currents at surprisingly high temperatures of about 93 K. There was some indirect evidence for the existence of trapped electron pairs in these solutions and Ogg estimated their Bose–Einstein degeneracy temperature and found it to be in the right range. Blatt was convinced that 'Ogg's ideas were discounted so strongly, largely because they were phrased in the language of an experimental chemist, rather than that of a theoretical physicist'.[189] When London was on his way to the Cambridge Conference in Low Temperature Physics in 1946, he visited Tisza at MIT. There was some kind of a meeting and an informal discussion at which London was not present, having already left. 'I thought it would interest you to know that the matter looked somewhat better in the oral presentation by Ogg than in his publications. The general opinion was that there certainly are some experimental effects present and while the theory suggested is still very sketchy, it is possible that its essential features are correct. In particular there seems to be good evidence for the existence of the electron pairs'[190] (London underlined this last sentence and put a question mark in the margin.)

Tisza had talked with Ogg and wrote to London that Ogg was looking for him and had emphasized very much: 'how strongly he was influenced by your ideas on Bose–Einstein condensation'.[190] London replied: 'If I had known that Ogg was in Cambridge, I would have stayed for another day. I wish you could provide me with some detail about what kind of evidence he gave for the existence of the electron pairs.'[191] Tisza wrote: 'Regarding Ogg, he did not bring in a particular point worth mentioning, but the general impression was that while his presentation was rather inadequate, it is still hard to discard his main points.'[192]

The Lorentz Medal

In December 1952, London received an invitation to participate in the celebrations for the centenary of the birthdays of Lorentz and Kamerlingh Onnes at Leiden during June 21–27, 1953.[193] At the beginning of March 1953, London received an official letter from M. W. Woerdeman, Secretary of the Section for Science of the Royal Netherlands Academy of Sciences, informing him that he had been awarded the Lorentz Medal. The decision by the special committee appointed for this purpose was unanimous. The award was given for London's work in homopolar bonding, his 'interpretation of the van der Waals forces on a quantum mechanical basis', the development of the phenomenological theory of superconductivity and his suggestions about the explanation of the behavior of helium below the lambda point.[194]

When the Lorentz Fund was established in 1926, it was decided that part of the revenues from the Fund should be reserved for instituting a gold medal which the Royal Dutch Academy of Sciences would confer every four or five years on a scientist who had distinguished himself in

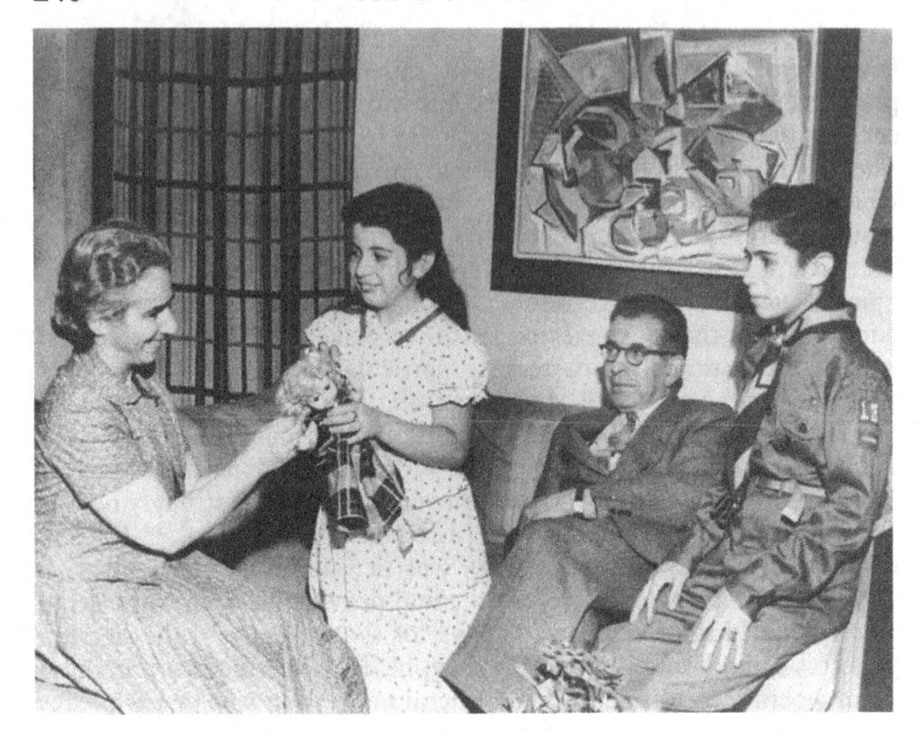

Figure 19. Edith, Rosie, Fritz and Frank London. Photograph published in the *Durham Morning Herald* with an article containing the news about the Lorentz Medal (spring 1953). The painting on the wall is by Edith London. (Courtesy of Edith London.)

the field of theoretical physics. By 1953, the medal had been awarded to only four scientists: Planck, Debye, Sommerfeld and Kramers. London was the first American citizen to receive the medal.

Simon was the first to send his congratulations. Wigner also sent him a congratulatory note saying: 'I know of no more worthy person to whom it could have been awarded'. 'Among the many congratulations I have received', wrote London to Wigner, 'yours is one of the most appreciated ones ... I sometimes greatly doubt whether the high honor which is given to me has been deserved. I am extremely happy that you approve of the decision of the Academy, because I know you and your sense of justice.'[195] The ceremony was planned for June 27, 1953. The *Durham Sun* of March 10 reported the news and there was a leader in the *Durham Morning Herald* on March 12. The Sunday edition of the *New York Times* on March 22 had a short report with a photograph about the 'first American to receive the prestigious Prize'.

London's last visit to Europe was during the summer of 1953. The whole family travelled with the *Maasdam* and reached Southampton on May 20. They took their car with them and drove directly to Harwell. After a night's stay, Edith and Frank went to London so that Frank could attend a boy scout event. They stayed with Heinz and Lucie until May 24 and then Fritz went to Cambridge, where he stayed until June 15. He gave a talk to Lennard-Jones' group at the Department of Chemistry and had many discussions with Pippard about his modified phenomenological theory of superconductivity. Then the whole family went to Leiden. After the ceremony they travelled across Europe, following the Rhine and driving to Venice.

The Lorentz Medal ceremony was attended by the American Consul in Amsterdam and various Dutch dignitaries as well as all the participants at the conference, among them Bohr, Heisenberg, Pauli and Pippard. The President of the Section for Science of the Royal Dutch Academy, A. J. Kluyver, made the introductory speech and R. Kronig, who was the President of the committee appointed to propose the person for the award, talked about London's contributions to science.[196]

Kronig dwelled on the relation between the contributions of London and the fact that almost all the topics in which London made his significant contributions had been initiated by Dutch scientists: the importance of valence in the work of van't Hoff; the weakly attractive intermolecular forces that van der Waals first used; the discovery of superconductivity by Kamerlingh Onnes and the first attempts for its explanation by Rutgers, Casimir and Gorter; and the discovery of the peculiar properties of helium-II by Keesom. Kronig did not mention London's doctoral thesis in philosophy, nor did he talk about London's notion of macroscopic quantum phenomena. Kronig's assessment of London's work on homopolar bonding was that 'the foundations were laid for incorporating theoretical chemistry into physics'.[197] True, of course, but not a formulation London was too eager to hear.

London's response was short, to the point and, uncharacteristically, very personal:

> I am deeply confused by the great honor which you have decided to confer upon me . . . During most of my life I have been so fortunate that I could do the things which my own nature drove me to do. It is embarrassing to earn so much respect for just doing this. Yet, it is a great satisfaction for me to receive this particular sign of recognition, because it tells me that the work which was done, apparently by an internal necessity, has been found to be of some objective value.

All the sentiments expressed in the speech were those of a very private

Fritz London. He did his best to live up to the fact that he was the first American recipient of the medal.

> Science is like a ball game: Someone throws the ball, some other one catches it and carries it a bit further and then it is thrust out of his hands again by someone else. The single individual's merits are difficult to appreciate particularly as long as the game is still going on. I was fortunate to hold the ball sometimes for a while and to carry it a bit further.[198]

Fritz London's speech at the reception given in his honor that evening is, perhaps, his most personal public declaration.

> When the news of the decision of the Netherlands Academy of Sciences became known in the United States, I have often been asked the question 'You are Dutch, aren't you?' As a matter of fact the necessities of life forced me to try *many* countries to become my homeland, four to be exact, but by chance Holland does not belong to them.
>
> However, for me this occasion brings back to me a memory and perhaps, after hearing me, you might decide whether I must not claim to be a Dutchman after all. Here is a personal matter involved a matter which fills me with pride and emotion at the same time. I cannot help thinking today of one of your great men of science who cannot be with us at this occasion and to whom I feel a link of profound devotion and I dare say even of friendship. I must think of the successor of Lorentz, P. Ehrenfest. I have never had the privilege of meeting Lorentz himself. But my various meetings with Ehrenfest had a very profound, I may even say, decisive influence on my life.
>
> I met him first in those days when quantum mechanics was in the *status nascendi* and it seemed as if one had just to stretch out the hands to make a great discovery. I particularly remember a railroad trip from Leipzig to Berlin I took together with Ehrenfest and I shall never forget the fanatical honesty of his character, which I experienced at the occasion for the first time. His insistence in clarity concerned a matter of humanism, of self-education, which, I feel, is so very essential not only for the quest of truth, but for interhuman relations in general. [The following phrase was erased and not included in the published version: 'It has become somehow rare in our times'.]
>
> When the dark age came over my former homeland, when all long established human links seemed to have suddenly died overnight, it was Ehrenfest who came to Berlin, to assure us of the solidarity of our friends abroad, a solidarity which Science perhaps more than any other activity can establish between human beings, beyond any national, beyond even cultural frontiers. Ehrenfest with the acute sensitivity of his soul had immediately grasped what had happened to humanity at that hour, and with a small group of others, similar in mind, he tried to salvage what could be salvaged. He made me come to Leiden in June 1933, just to let me breathe the clear air of a free country and to have

contact with free human beings again. Rightly he assumed that this was needed very urgently. In fact I was in Leiden with Ehrenfest at that time only for a day or so, yet that was long enough to feel revived from death and to be able to take decisions for the new life ahead of me. Humanity took for me a new aspect, a more human one than I experienced in the country of my birth.

This was exactly to the month, perhaps, even to the day, 20 years ago in the end of June 1933. You will understand that I must think of this experience with much affection today. You do not experience your birth, but you may experience your rebirth; and coming back again to this venerable place I remember, that for me Leiden has played the accidental role to be the place of my rebirth. And so I submit to your judgement whether I may consider myself a Dutchman though not by birth, so at least by rebirth.

London was enthusiastic about the ceremony at Leiden. The Dutch had organized a really great and very impressive affair which started at three thirty and ended at eleven thirty in the evening. Many members of the Lorentz family and various dignitaries, including a

Figure 20. Fritz London receiving the Lorentz Medal at Leiden, June 1953. (Courtesy of Edith London.)

representative of the Dutch Queen, were present. Bohr and Pauli arrived on the last day to participate in the celebrations, but Heisenberg left before the end: 'He was most of the time like a lonesome lion'.[199]

Consultancy at Los Alamos and the interview for security clearance

Before leaving for Europe, London was offered a consultancy by the Los Alamos Laboratory's Cryogenic Group. The administrators of the Laboratory had already contacted E. R. Jette, the President of Duke University, who gave the University's approval for such a job. The daily allowance was $50.00 plus travel expenses.[200] The consultancy would have involved discussions of research in progress and of planned research, as well as conducting seminars about liquid helium. More specifically, Los Alamos wanted London because both Los Alamos and Argonne had, at the time, access to tritium, which was being produced by the Atomic Energy Commission in their researches for thermonuclear weapons. Tritium decays to He^3 and the Cryogenics Group at Los Alamos was anxious to search for the superfluid properties of He^3.[201] London agreed to the consultancy, but mentioned that he had developed a heart condition and that his physician objected to his going to high altitudes in winter and that, since he was returning from Europe in mid-September, the best starting date for him would be the spring of 1954.[202] He tentatively planned to go to Los Alamos in mid-June with his family. To be allowed to work at Los Alamos, it was necessary to receive clearance from the Security Division of the Atomic Energy Commission, so he was interviewed at his office in late-October 1952.[203] London had no objection to this interview, provided he could see the transcript afterwards.

He stated his name as Frederick London since, as he mentioned, he did not want to be identified as a German when travelling. He used 'Fritz' only for scientific purposes, since 'in science one has to form a certain identity'. London denied any kind of association with the Communist Party in Germany, England, France and the USA. When asked whether he frequently met people whom he knew were members of the Communist Party, London mentioned Joliot. In different places in the interview, he was asked about Pontecorvo, Sergio Benedetti, Luria, Korwarski, Langevin and Klaus Fuchs. London denied having known Langevin's affiliations with the Communist Party at the time he was in Paris. He praised the way Langevin ran his seminar and noted that he was one of the great physicists of his generation: 'I was very much astonished years later to hear he was a communist'.[204]

He said that his subscription to the leaflets published by the American–Soviet Science Society was because of his interest in the work of the Russians in low temperature physics, who, during the war, were very much advanced. He could find some information in short news items – but gave up his subscription because most of the information was propaganda. When asked whether he thought the friendship between the USA and the Soviet Union would last, London concentrated on issues related to scientific cooperation. He felt that after the war every scientist should do his utmost to foster scientific cooperation. He deeply believed that scientific cooperation was the strongest link between people of different opinions. He admitted that at the present time such a link was not possible, but said: 'I still would say scientific methods are the means of communication which do not allow misunderstanding'.[205]

When asked about his political activities prior to leaving Germany, London stated that his political stand was determined by his strong anti-Nazi feelings. He did not vote for a party, but against Hitler. When pressed to say whether he was a sympathizer of the Social Democratic Party, London answered that his sympathy was with the Democratic Party, which was to the right of the Social Democratic Party. But many people thought that if they voted for the Democrats, it would be a waste and therefore they voted more 'to the left not by reason of conviction but by reasons of necessity of fighting for the ideals which the strongest party opposed'. London was sorry for the split between the socialists and the communists, since this weakened the opposition to the rise of the Nazis. When asked whether he thought the Communist Party in Germany was responsible for the rise of the Nazis, his answer emphasized the missed opportunity of the Left. He was convinced that the only way that war could have been prevented was if there had been a coalition of the Communist Party and the Social Democratic Party.

Later in the interview, London was pressed to state his position about Marxism. He was asked whether he was in sympathy with the socialist experiment at the time he visited the Soviet Union and replied: 'I was interested, I would not deny this. I was not in sympathy'. He admitted having studied – not simply read – Marxism.

Mr Tharp: Do you think it is possible for a person to be a student of Marxism and at the same time not be a victim of the Communist way of life?
Dr London: Yes.
Mr Tharp: You think this is possible?
Dr London: I am an example.
Mr Tharp: Would you explain this, Dr London, briefly.
Dr London: It is not a field of theory as physics. It is not a field in

which thinking can come to sharp cut results. What kind of theory you can realize in political life is not only a question of thinking, but also of historical facts. Some peoples are historically prepared to live according to certain rules, and others live according to other rules, but I was not convinced that a rule was to be prescribed to history.
MR THARP: As Karl Marx attempted to do?
DR LONDON: Yes.
MR THARP: Do you think his theory can become a reality?
DR LONDON: It is a reality in Russia.
MR THARP: His theory is a reality?
DR LONDON: It was at least, very much. It was near reality I would say, but Russia is perhaps the best example that historical elements are involved. Even under the best circumstances the theory is not alone in determining historic developments.[206]

He was asked to explain his attitude towards communism, communism as practiced and the way communism was associated with the Soviet Union. He reminded them that in 1931, when his colleagues in the Soviet Union had made a very attractive offer, he had preferred to stay in Germany. They pressed him to tell them his views about the present state of the Soviet Union. He thought that a lot of things had changed in the meantime. In the thirties, it had been difficult to realise that 'the Stalin regime is different from communism'.[207] But he insisted that in 1931, despite the hard life, the people of the Soviet Union could look at the future with hope.

In his responses to the various questions about the Communist Party of the USA and its affiliations with the Soviet Union, London expressed his opinions about the lack of political education of the American public. Despite the almost obsessive insistence of the interviewers to have London state that the Communist Party of the USA was tied to or controlled by the Soviet Union, London refused to budge. He stressed that one needed much more information to be able to form a scientifically founded opinion and to be led to such a conclusion and that such information could not be found in the newspapers, whose reporting was one-sided and full of errors. His interviewers interjected that these were political issues and could not be decided by newspaper reporting, where errors were quite common. London continued to insist. He could neither speak in favor nor against, with such insufficient information. One of the interviewers asked whether such an attitude was an expression of London's scientific attitude or his character, he replied: 'I cannot separate my science and my character. I avoid such fields of thought in which I can only act biased'.

MR DONOVAN: Can you see the thing that happened in Europe could happen in the United States?

DR LONDON: Yes. I could say the lack of information of the citizens is, of course, a great danger, and the newspapers are here greatly infringing on one side of the information. Mr Eisenhower, many of his talks are very much one-sided of the national interest even in the Presidential election; the information of the citizens of the United States is very one-sided.

MR THARP: Do you yourself have an antipathy toward the political situation in the United States?

DR LONDON: No, not against the political situation, but against the situation of the education of the normal citizen in political issues.[208]

London was asked whether he would consider someone he knew to be a member of the Communist Party as being a subversive. He replied that most probably he would, if he wanted to change the constitution. Then he was asked whether he would consider a student to be a subversive because he was a member of the Party.

DR LONDON: No. I would first probably think he does not know what he is in. I would first think he is an idealist and just looking for something new . . .

MR THARP: Would you form the same conclusion if that person was a professor?

DR LONDON: I would consider that case more serious, but I have seen childish attitudes of professors in Germany.

MR THARP: Some professors are not mature.

DR LONDON: No. I would at least plead for pity. Many people of the United States [have such limited education about] the democratic purpose that one has to be lenient.[209]

London was given the transcribed text and he had a long session with a Professor of Law at Duke before returning the copy to the Atomic Energy Commission. In April 1953, London was informed that his security clearance had been approved by the Atomic Energy Commission.[210]

The last days

Soon after returning to Durham from Europe, London was appointed James Duke Professor and this meant a special distinction and an additional payment of $500 per year (less $200 taxes) from a special fund. But his health was deteriorating, and he had a sudden appendicitis attack and went to hospital for a few days. Nevertheless, the diagnosis was somewhat uncertain and, because of his heart condition, the doctors were unwilling to operate 'which seems to me rather silly'.[211] It was, obviously, 'a dreadful scare. Let us hope that he will in the future be spared such nonsense'.[212]

There arose a possibility of a trip by Heinz to the USA during Christmas 1953 – in '*Chanachten*', which was a Fritz London neologism, a mixture of *Chan*ukah and Weih*nachtnen*. Everyone was very excited about Heinz's impending visit, the first of the close relatives to

Figure 21. Heinz and Fritz London in 1953. (Courtesy of Edith London.)

have visited them during their fourteen years in Durham. 'Hurry up and come before you change your mind', Rosie scribbled on a piece of paper, together with the drawing of her new bicycle.

Fritz and Heinz planned to travel together to a conference held at Houston and then to go to New Orleans.[213] Heinz was to give two lectures at Duke on January 4 and 5. Fritz urged Heinz to spend a day in New York 'the uniquely interesting city in America'[214] before continuing his journey to Durham.[215]

Figure 22. Fritz London in 1953. (Courtesy of Edith London.)

March 1954 was the twenty-fifth anniversary of Fritz's and Edith's wedding. A friend of theirs, Omi, stayed with them for several days and helped Edith during the festive week for their anniversary. One evening, an enthusiastic Rosie and an embarrassed Frank, stuttering through a poem written by Omi for Fritz and Edith, passed the silver paper ornaments they had worn to their parents. The following day they entertained a few friends. They received flowers from Heinz and Lucie and a book about Italy's Rennaisance monuments 'a great reminder of the trip to Ravenna last year'.

Though Fritz had gone through the festive week with no major problems, and had no more sudden pains, he continued to lead a life of complete rest. He had not given any lectures for over four weeks, and did not go out. Two students took care of his lecturing and visited him two or three times a week to discuss various problems. He went to the hospital and had another cardiogram. 'The result one will probably learn next week at the next interview or perhaps never. One has to develop a very thick skin to avoid being constantly and greatly annoyed by doctors, which is known to be poison for heart diseases. We hope that Fritz will be fit for some work again. He is a good, forbearing patient.'[216]

He was feeling much better and, as he wrote to Heinz and Lucie: 'about time too ... Really it is good on such occasions to feel the bond with all the dear ones and to let words and deeds bring us near to each other as you have achieved ... Many thanks again and have a big hug'.[217] And a big hug it was. In less than a week, a few days before this letter reached Heinz and Lucie, they learned of Fritz's death on March 30, 1954.[218] He died at 8.30 am. The previous night, together with their friends the Manasses, they had gone to a cafe in Chapel Hill where Edith had recited German poetry to a group of teachers from the University of North Carolina. It was almost eight years to the day since Fritz had first visited a heart specialist in New York in 1946. The specialist had told Fritz that his heart condition was such that he could not expect to be able to lead an active life for more than about eight years. When he returned home, Fritz told Edith. 'We looked at each other and never discussed it at any other time.'

In his last years, London had become more and more withdrawn. His health had not permitted him to follow the fast rhythms of scientific life after the war. But also, he had felt that he did not need to. Superconductivity, he believed, was in good hands and he was convinced that Fröhlich and Bardeen would solve the problem. There were new proposals about superfluidity which were attempts to bring about a synthesis of his views and those of Landau. He thought that there remained only a few loose ends to complete his program on the study of the macroscopic quantum phenomena. It was a program which, in effect, had its beginnings in the romantic science he had discovered in his high school years and his strong belief in studying 'theories as wholes' which he had pursued so successfully in his thesis in philosophy. Fritz London's view on how to do physics took him from the defense of Goethe's scientific work to the intricacies of the macroscopic quantum phenomena. It was a long and lonesome journey which spanned a short life.

Notes to Chapter 5

1. Private communication by W. Linschitz to K. G.
2. Aleksei Kozhevnikov (1992) published a most illuminating paper about Kapitza's life and role in the former Soviet Union and includes much material from the archives which are kept there.
3. Bohr to John Pellam. The full letter is quoted by Pellam (1989), 315–16.
4. Kapitza to Stalin (April 28, 1938). Some of the correspondence between Kapitza and the Soviet authorities is included in Khalatnikov (1989).
5. Kapitza to Molotov (April 6, 1939).
6. Kozhevnikov reported that in Kapitza's archive there are many letters to politicians, including forty-five to Stalin and about fifty each to Molotov, Malenkov and Mezhlauk. Kapitza's appeals for the release of scientists were not always successful. By writing a letter to Stalin, he had managed to have V. A. Fock, the well known theoretical physicist of Leningrad University who worked on problems of gravitation, released a few days after he was arrested. However, he was unable to do the same for the mathematician Luzin.
7. This incident was mentioned to me by Edith London. Kramers, who was then at Copenhagen, had sent a letter of introduction for Landau to London. Kramers to London (April 21, 1930).
8. Shoenberg, (1989), 227.
9. Kapitza (1941*a*), 581.
10. Kapitza (1940), 24.
11. Kapitza (1941*b*), 638.
12. When Andronikashvilii proposed to title his paper *The temperature dependence of the density of the normal component of helium-II*, Landau disagreed. He suggested the title *Direct observation of two forms of motion in helium-II*. In Andronikashvilii (1989).
13. Brush (1983), 184.
14. Lifshitz (1944), 110.
15. Fritz to Heinz (October 25, 1940).
16. Tyndall to the Royal Society (June 26, 1940. SPSL files).
17. Fritz to Heinz (October 25, 1940). The letter is among the very few letters written in English, because of security reasons.
18. Fritz to Heinz (January 26, 1941).
19. Fritz to Heinz (August 17, 1942).
20. Fritz to Heinz (April 7, 1942).
21. 'I found this in the files and wondered if you would like to see it again. We seldom get such opportunity except for carefully prepared publications' (Philip Handler (School of Medicine, Department of Biochemistry, Duke University) to London (August 31, 1950)). The original letter was sent to Perlzweig from Szent-Györgyi (July 12, 1941).
22. Szent-Györgyi to Perlzweig (July 12, 1941). 'There are various implications of this view which give a very striking interpretation to some of the most mysterious facts of biology; I cannot of course describe them adequately in a few words, but would like to talk sometime with you about them. (See for instance P. Jordan *Anschauliche Quantenmechanik*, last chapter, Springer 1935)'.
23. Pauling (1940), 2643–60.
24. London (1942), 315–16.
25. Fritz to Heinz (August 17, 1942). Fritz's financial situation was not the best it could be: salary $4500; taxes $900; defense bonds $450; annuities $220; Aunt Clara $300. Fritz considered $2500 per year to be the minimum he required to meet their everyday expenses.
26. Fritz to Heinz (August 30, 1943).
27. Fritz to Heinz (July 11, 1944).
28. After short stays in various Government Laboratories, in 1944 Heinz moved to

the Nuffield laboratory at the University of Birmingham to continue work on large scale production of uranium-235. He was horrified when the atom bomb was dropped, but also thought that the Germans would have used it first had they been able to prepare it in time. He continued to work on isotope separation till the mid-1950s, by which time he was already Principal Scientific Officer at Harwell. During these years he developed close relations with Franz Mandl, Myra Ockrent and her future husband, Hans Kronberger. In 1970, Both Mandl and Kronberger visited Heinz in hospital, during the last two weeks of his life. Mandl was in Oxford for a conference and was spotted in the town by Lucie. Kronberger made the trip specially to see Heinz, and, in the hospital room, the two of them talked continuously about physics. 'Heinz showed me how to use physics and he had the most original way of seeing things and working them out – to the bafflement of most onlookers' (Hans Kronberger to Lucie London (October 1970)). Heinz was not particularly enthusiastic about his output, but this did not preoccupy him, since he hated the rat race. He immersed himself in a problem till he could solve it to his own satisfaction, and was always bothered when he had to leave a problem before it was dealt with satisfactorily, because another one had, in the meantime, crept up. Most people who knew him remember his clumsiness and his being very untidy. Many times he had difficulty in expressing himself clearly and his writings are not the easiest to follow. He never felt at home with the English language and never really made peace with it. 'I felt that he kept it at bay as a kind of impersonal revenge against a fate that had made him emigrate and expected him to express and understand highly complex thoughts in any but his mother tongue' says Lucie London (private communication to K. G.).

29. Franz Simon to London (March 19, 1946); London to Simon (March 29, 1946; April 12, 1946).
30. Gross and R. L. Flowers, who was then the President of Duke University, wrote letters to the State Department supporting London's (successful) application to attend the conference.
31. London to Allen (April 17, 1947).
32. See Kozhevnikov (1992). Kapitza was expelled from the directorship by a decision of the Council of Ministers on August 17, 1946. A few weeks earlier he had been dismissed from being head of the *Glavkislorod*, the Chief Department of Oxygen, which oversaw all industries using low temperature methods. Part of the reason for his dismissal was his participation in the Soviet atomic bomb project and his disagreements with the physicist I. V. Kurchatov, who had been the head of the project since 1943, and the police chief Beria, who had assumed the chairmanship of a special committee set up after the atomic bomb was dropped by the Americans in 1945.
33. London to the Passport Division of the State Department (April 2, 1946).
34. 'And then suddenly, he felt me as an ally; it happened at this point only and not till then. So it was only after Peshkov's experiments were in, suddenly he realized that, after all, my idea was OK, and in fact, then everything changed, because he was very much impressed – he was very guilty about it, that he'd attacked me so much, and he could have done me a lot of harm in professional reputation and he went to great lengths, and from then on we were extremely friendly in our contact'. Interview with Laszlo Tisza (see note 10, Chapter 4).
35. London (1947), 1.
36. London, (1946), 4–5.
37. Landau, (1941), 71–90; Landau (1944), 1–3.
38. London to Tisza (May 2, 1946).
39. London to Tisza (May 8, 1946).
40. Tisza to London (June 20, 1946).
41. Tisza to London (July 25, 1946).
42. Tisza to London (November 26, 1946).
43. London (1947), 13.
44. Peshkov (1946), 389–98. In this paper, Peshkov talked of the 'very artificial and unconvincing theoretical speculations' of London and Tisza.
45. Landau (1947), 9–92.
46. London to Tisza (September 10, 1946).
47. Tisza to London (October 10, 1946).
48. Tisza to London (October 17, 1946).
49. London to Tisza (November 21, 1946).
50. London (1947), 8.
51. Tisza to London (November 26, 1946).
52. Tisza (1947), 851.
53. London to Tisza (April 24, 1946); Tisza to London (May 14, 1947). The referee's report of *Physical Review* considered Tisza's paper to be 'midway between a research article and a review' and it was not easy to ascertain whether the results were original. London suggested that Tisza might send his paper to the *Journal of Chemical Physics*, since London

usually reviewed any papers about liquid helium. In any case, he urged Tisza by try publishing it quickly, especially after the attempt by Landau 'to salvage his own theory in view of Peshkov's results'. London to Tisza (June 26, 1947).

54. London to Tisza (September 6, 1947).
55. Tisza to London (September 9, 1947).
56. London to Tisza (September 19, 1947).
57. Tisza to London (September 18, 1947).
58. London to Tisza (September 22, 1947).
59. Tisza to London (October 3, 1947).
60. Pellam & Scott (1949), 869–70; Maurer & Herlin (1949), 948–50.
61. Tisza to London (September 18, 1947); London to Tisza (September 22, 1947); Tisza to London (October 3, 1947).
62. London (1948*b*), 562.
63. Ibid., 565.
64. Heisenberg (1949), 50.
65. Private communication from Sir Brian Pippard to K. G. (May 21, 1991).
66. Pippard (1949).
67. London (1950), 142.
68. Ibid., 146.
69. Ibid., 150. This was before the Fröhlich–Bardeen papers of 1950.
70. Ibid., 150.
71. Among London's notes there is an unfinished review of von Laue's book in Fritz's handwriting. He wrote that the 'experimental physicist will find the book a valuable addition' and concentrated on pointing out the differences between von Laue's and his own approach concerning the question of surface energy. London was quite annoyed with Heinz London when, in 1953, Shoenberg had suggested that Heinz should write a review of von Laue's book for *Nature* and Heinz had accepted, although he never finished it. Fritz to Heinz (January 5, 1953).
72. London to von Laue (July 1, 1947).
73. Von Laue to London (January 13, 1948).
74. London to von Laue (February 20, 1948).
75. Von Laue to London (April 30, 1948).
76. Undated draft letter to von Laue; also London to von Laue (May 27, 1948). See also *Nouvelle Conception* (end of Chapters 8 and 9) and Heinz London (1935). Going from one phase to the other was discussed in Appleyard, Bristow, London & Misener (1939*a*).
77. At the conference, London and Tisza talked about the properties of helium-I and helium-II and He^3, Slater reviewed the theories of superconductivity; and van Vleck discussed the attainment of very low temperatures. 'The two outstanding theorists of low temperature phenomena in this country are Tisza and London. I wish it were as easy to choose the outstanding experimenters. Ten years ago one would have thought first of all Brickwedde (NBS), Giauque (Berkeley), Burton (Toronto)' (Karl Darrow to A. McInnes (January 16, 1948). McInnes papers, Rockefeller University).
78. Tisza (1949*a*) 102–4.
79. Green (1948), 391.
80. Tisza (1948), 102.
81. London to Tisza (June 25, 1948).
82. Von Laue to London (August 8, 1948).
83. Von Laue to London (November 8, 1948).
84. London to von Laue (October 2, 1948); von Laue to London (October 4, 1948).
85. Fritz to Heinz (November 1, 1948).
86. 'The expression used by Heisenberg, "anisotropic form of a tensor" makes no sense. A body can be anisotropic; a tensor can be symmetrical or unsymmetrical ... I cannot understand how Heisenberg could write such non-sense'. Fritz to Heinz (December 27, 1949).
87. Von Laue to London (November 8, 1948).
88. London to von Laue (November 19, 1948).
89. Osborne, Weinstock & Abraham (1949); Sydoriak, Grilly & Hammel (1949).
90. London to Tisza (February 24, 1949).
91. Tisza to London (February 28, 1949).
92. Tisza to London (March 13, 1949).
93. Tisza (1949*b*), 885.
94. London to Tisza (March 17, 1949).
95. London to Tisza (March 10, 1949).
96. Tisza to London (July 26, (1949).
97. Atkins & Osborne (1950), 1081.
98. London to Jacoby (July 2, 1947).
99. The book was to be published in the series titled *The Structure of Matter*, which was first suggested by Franck in 1945 because of the state of the European publishers after the war and because most of the work in scientific research was done in the USA. By the end of 1946, Wiley had decided to go ahead with Maria Goeppert Meyer as the editor of the series.
100. London to D. McPherson (November 11, 1948).
101. W. B. Wiley to London (October 13, 1949).
102. London to Jacoby (October 19, 1949). A similar formulation is found in London's answer in the promotional questionnaire sent to him by Wiley. 'My book takes particular care in explaining why certain equations are chosen as the expression of the basic assumptions of the theory,

while Laue's book merely takes those equations unquestioned and develops their mathematical implications'.

103. Wiley insisted that the additional costs due to these changes amounted to $1000 and asked London to agree to charge this amount against future royalties and they set the price at $5.00 instead of the agreed price of $3.50.
104. He received six free copies and sent one each to Heinz London, von Laue, Shoenberg, Gross and the *News Service*. He asked Wiley to bill him for extra copies to send to Fröhlich, Mott, Pippard and Casimir. Likewise, for the second volume he was planning to send one each to Heinz, Schrödinger, Nielsen, Gross and Hobbs. In the promotional questionnaire sent by Wiley, London provided some revealing details. When asked about the leaders in the field from whom Wiley might request endorsements, London's answer was: Shoenberg, Brickwedde, Squire, Slater and Pauling. When asked what it was about the book that set it apart from other such books, London's response was that superfluids were represented as examples of quantum mechanisms of macroscopic scale.
105. Reviews of the first volume appeared in 1951 in: *American Journal of Science; Chemical Industries Week; Chemistry and Engineering News; Journal of Chemical Education; Journal of the American Chemical Society; Nature; NY New Technical Books; Proceedings of the Physical Society; Record of Chemical Progress; Science; The Philosophical Magazine; US Quarterly Book Review*. Reviews for the second volume appeared in 1955 in: *Journal of the Franklin Institute* and *NY New Technical Books*.
106. Pippard (1951*b*). Paul Marcus's reaction to the book was quite favorable. 'It is more favorable than Pippard's, whose approach to physics is rather different. He is not so responsive to a purely theoretical structure, carefully and elegantly formulated, but takes the empirical, one might say the typical experimentalist's point of view, which emphasizes a mass of phenomenological data which is partially explained or correlated here and there by a theoretical model. This approach seems to flourish in England, and has both its virtues and drawbacks'. Marcus to London (August 29, 1952).
107. But Mendelssohn (1964), in an article which was wholly devoted to the discussion between superconductivity and superfluidity, did not mention London's work.
108. Edith to Heinz and Lucie (September 25, 1948).
109. Fritz to Heinz (March 10, 1949).
110. London to Paul Zilsel (June 1, 1949).
111. Edith to Heinz and Lucie (November 3, 1949).
112. Gorter (1949), 245–250. This issue of *Nuovo Cimento* contains all the other papers of the conference and the discussions.
113. Ibid., 250.
114. London to Zilsel (March 13, 1949). For the MIT Conference in September 1949, in the original program Heisenberg was to give the opening talk, but Slater proposed that London should give it (Tisza to London (July 26, 1949)). However, London did not respond to this proposition. The Londons changed their plans and returned by the *Île de France* on September 5. Cassidy (1991) does not mention Heisenberg attending the conference, even though he attended the International Congress of Mathematicians in Cambridge, Massachusetts (August 30–September 6, 1949).
115. Tisza (1949*c*), 3.
116. Fritz to Heinz (August 14, 1950).
117. Fritz to Heinz (April 23, 1950).
118. Fröhlich (1961), 7.
119. Methods to treat interactions between particles were first devised for quantum field theory in the late 1940s by Fröhlich, Pelzer & Zienau (1950) for a single electron in polar crystals – a problem much simpler than superconductivity. The field theoretical treatment showed that the kinetic energy of the ions attached to a moving electron may be much smaller than the kinetic energy of the electron. 'Clearly such a result was of great interest in connection with the "dictum" that ions, in view of their heavy mass, should be of no importance for superconductivity and the application of appropriate field theoretical methods to metals was called for'. Fröhlich (1961), 7.
120. Three sets of researchers (Allen, 1952; Maxwell, 1950; Serin, Reynolds & Nesbitt, 1950) showed that the shape of the critical field curve for tin was independent of isotopic mass, although the magnitude of the critical field at any temperature did depend on the average isotopic mass.
121. Fröhlich to London (September 23, 1950; September 27, 1950). Fröhlich suggested that London should be invited

to England through a Fulbright program in 1952–1953 and London responded positively in a formal letter. The same day, however, he also sent another, personal, letter to Fröhlich expressing his misgivings. It would, he thought, be nice to renew the contacts with the developments in Europe, but there were certain difficulties. The Fulbright program was 'American propaganda and I do not think naturalized Americans stand much of a chance'. London to Fröhlich (December 26, 1950). Moreover, London would have to spend an additional $3000, for such a trip which would not be covered by Fulbright. Fröhlich did not proceed with the application.

122. Fröhlich to London (October 28, 1950).
123. London proposed that one should try to prove $\bar{p} = 0$, even in the presence of a field, rather than $\boldsymbol{B} = 0$. This could be done for a simple case of a cylinder with a small radius compared to the penetration depth so that one could neglect the magnetic field produced by the current within the superconductor. He pointed out that Fröhlich's strategy was not correct: Fröhlich wanted to prove that $\boldsymbol{B} = 0$, and from that it followed trivially that $\bar{p} = 0$. But, as London noted, in a superconductor one can have $\bar{p} = 0$ even where $\boldsymbol{B}$ is not equal to zero.
124. London to Fröhlich (early November 1950).
125. Fritz to Heinz (November 2, 1950).
126. Bardeen to London (December 9, 1950). Bardeen also wanted to know whether the note on the choice of gauge was an original contribution. 'Do you know of any reference to it in the literature?'
127. London to Bardeen (December 15, 1950).
128. Bardeen to London (January 5, 1951).
129. Pellam to London (November 14, 1950).
130. Pellam to London (March 2, 1951).
131. London to Pellam (March 6, 1951).
132. Von Laue to London (December 30, 1950).
133. Von Laue to London (January 3, 1951).
134. London to von Laue (January 12, 1951).
135. Von Laue to London (February 3, 1951).
136. Von Laue to London (January 25, 1952).
137. London to von Laue (February 1, 1952).
138. Ewald (1960), 148. In his autobiographical essay, von Laue hinted that these attacks may have been due to a great shock he received when he was a young man.
139. Von Laue to Einstein (January 8, 1951).
140. 'I could only suggest (what I believe to be true) that von Laue – that gallant champion of the Jews in Germany – had begun to fall into senile decay – but the pain was not to be alleviated so easily, and Fritz was obviously very deeply upset'. Private communication from Professor Sir Brian Pippard to K. G. (January 4, 1990).
141. Atkins & Osborne (1950), 1078–81.
142. London to Tisza (November 2, 1950).
143. London to Tisza (November 6, 1950). In the April 1950 issue of *Scientific American*, there was a report about the Pellam–Scott experiment and the experimenters were quoted as having said that 'low temperature scientists can now confidently employ Landau's postulates in their research'. London asked Pellam whether he could send a letter to *Scientific American* to clear the issue. He reiterated that the issue between Landau and Tisza–London was not whether the entropy of the superfluid vanished exactly or not, but whether the statistics or the rotons accounted for the presence of the two fluids. And this particular question could only be settled by measurements with the pure isotope He^3 and not by second sound measurements. London to Pellam (March 28, 1950). Pellam had not seen the article, but agreed with London. When they had presented their results they were careful to emphasize 'that the results should cast no doubt on your condensation hypothesis'. In fact, in their letter to the editor they stated that in 'fairness to professor London... [their results] should not necessarily [be] taken as evidence against London's original hypothesis of helium-II as a condensed Bose–Einstein gas', adding that the recent experiments of D. Osborne *et al.*, showing non-superfluidity in He^3, were exceedingly strong evidence for the condensation theory (J. R. Pellam & R. B. Scott (1950), 7; also Pellam to London (March 31, 1950)). London was not convinced of how strong the evidence of second sound measurements were in Landau's favor. It appeared that between 1.4 K and 2.14 K the assumptions of Landau gave much too large values for second sound velocity, whereas Tisza's assumptions fitted *quantitatively* surprisingly well (including the maximum between 1.6 K and 1.7 K). The fact that below 1.2 K the second sound velocity increased with decreasing temperature, in disagreement with Tisza, did not remove the disagreement with Landau above 1.4 K and it was still to be shown that there was a quantitative agreement with

Landau at temperatures below 1.2 K – which London doubted would be possible. 'In my opinion it is asking too much from any two-fluid model to give quantitative predictions for the extreme case where rho-n over rho is less than 0.001'. London to Pellam (April 5, 1950).

144. London to Tisza (November 28, 1950).
145. Tisza to London (December 5, 1950).
146. Tisza to London (August 10, 1951).
147. London to Gorter (May 2, 1951).
148. Edith to Lucie (September 4, 1951).
149. Fritz to Heinz (September 4, 1951).
150. London (1951), 4.
151. London (1954), 60. This was not, of course, true. Tisza, as he himself has related the story and as it is evident from his papers, did not acknowledge in his early papers the possibilities offered by Bose–Einstein condensation and, furthermore, he was not interested in developing a macroscopic theory in London's sense, by a quasi-phenomenological hydrodynamics. Tisza emphasized that he wanted very much to challenge London on the microscopic theory and London, 'correctly, wanted to keep me more macroscopic'. Interview with Tisza (see Chapter 4, note 10).
152. Pauli to London (July 12, 1945).
153. Uhlenbeck to Tisza (January 29, 1942).
154. London (1954), 88.
155. Ibid., 199.
156. Ibid., 200.
157. The first mention in a Russian journal was after 1943 and it was Tisza's articles of 1940 in the *Journal de Physique et Radium*. It was noted that because of the war they were not aware of Tisza's work earlier. But Kapitza (1941*a*, *b*) mentioned both London and Tisza and quoted Tisza's articles in *Nature* and in *Computes Rendues*. Similarly Landau (1941) referred only to Tisza (1938*a*).
158. This is so even in Andronikashvilii (1989). This is a quite remarkable piece of work, where the author devised various ways to avoid hagiographic portraits of the protagonists of Soviet physics during the war. London is mentioned in the very last lines of the long book with a cursory remark about the usefulness of the notion of macroscopic quantum phenomena. Tisza's work was discussed to show that his account of the velocity of the second sound at very low temperatures was wrong and that Landau's theory 'could explain . . . the propagation of second sound'. Although Tisza was the first to predict the phenomenon of second sound, this had hardly been acknowledged in the Soviet literature. In a small booklet which is the supplement to the Russian edition of Keesom's *Helium*, its authors, Lifschitz and Andronikashvilii, stated that it was Tisza who had first introduced the idea of the two fluids and that it was done quite independently of Landau.
159. Fritz to Heinz (November 2, 1950; also April 23, 1950, and August 14, 1950). It was the time of the trial of Klaus Fuchs. 'We got very upset here about the fate of Fröhlich's prospective colleague, which was in all the newspapers. I knew him quite well. We are living in bad times'.
160. Fritz to Heinz (November 3, 1952).
161. Maxwell to London (April 19, 1950).
162. London to Maxwell (April 29, 1950).
163. Fritz to Heinz (April 23, 1950).
164. Fritz to Heinz (November 3, 1949).
165. Fritz to Heinz (January 26, 1950).
166. Fritz to Heinz (December 20, 1952).
167. Interview with Zilsel by S. Heims (March 21, 1988) at Seattle and discussions between K. G. and Zilsel (November 1992).
168. Zilsel developed some of the ideas involved in the two-fluid model. He obtained exact non-linear equations of motion for reversible processes in the two-fluid model from a variational principle of the type introduced by Eckart. Transitions between the two fluids were taken into account and, except for a term giving the effect of these transitions, the derived equations were more or less identical with the equations proposed by Landau. In the limit of small velocities they were reduced to Tisza's equations. In another paper he was able to introduce in mathematical form Landau's phonon and roton spectra within the framework of the Bose–Einstein condensation.
169. London to Franck (January 3, 1946).
170. London to Zilsel (April 27, 1953).
171. Paul Zilsel first went to Israel and then he worked in Canada. He quit physics towards the end of the Vietnam War and opened a bookshop in San Francisco.
172. 'I must say I approached [the book] with fear and trembling. Somewhat to my surprise, I find the book very good. It is clear, concise and modestly written. I have learned a lot from the book and hope that others will too'. Maria G. Mayer to Grimshaw (January 17, 1951).
173. Wiley to London (May 19, 1952); London to Wiley (June 12, 1952).
174. The publishers claimed that they would have a net loss of about $300 even if all

the copies were sold. The book was published in 3000 copies and during the first year 1200 copies were sold.

175. W. H. Grimshaw to London (June 9, 1952).
176. London to Lowdin (May 26, 1952).
177. Grimshaw to London (July 7, 1952).
178. Ibid.
179. London to Marcus (October 27, 1952).
180. Marcus to London (November 12, 1952).
181. London to Grimshaw (March 9, 1954).
182. London to E. F. Taft (March 29, 1954).
183. London to Price (November 30, 1952); Fritz to Heinz (November 3, 1952).
184. At about the same time, Brian Pippard in Cambridge, England, had taken up the same theme and went on to discover microwave superconductivity: he showed that there was a coherence length l, over which superconductivity acted and that this length was inversely proportional to London's penetration depth and it was eventually identified with the distance of the paired electrons of the Bardeen–Cooper–Schrieffer theory of superconductivity.
185. Gordy (1988), 18.
186. In 1950, Pippard had informed London about his experiments which used high-frequency techniques to determine the increase in penetration depth when a steady field was applied to a tin superconductor. Just below T_c the change was about 3% between zero and the critical field, and the effect became smaller at first as the temperature was lowered, so that at 3 K there was a change of much less than 1% at H_c. Below this, a rise occurred, and at 1.7 K the penetration depth changed by 2% at H_c. Heinz London had pointed out to Pippard that when the penetration depth varied with temperature, thermodynamically, it implied that the entropy varied with the field. Pippard told London: 'I do not believe that such a change in entropy density could result in only 2–3% change in lambda, and hence there is evidence that the entropy change is spread over a thicker layer. The Casimir–Gorter model leads to a value of about 10^{-4} cm for the thickness of the layer, and I think this may be taken as the order of magnitude of the range of the superconductor. This agrees quite well with the estimate I made from the sharpness of the resistance transition in zero field'. Pippard to London (April 14, 1950). London was enthusiastic in his answer: 'My congratulations! I am glad that you have studied the question of lambda = lambda(H) which actually amounts to checking the validity of those differential equations. Your theoretical discussion matches your beautiful experiments'. London to Pippard (undated).
187. Private communication from Professor Sir Brian Pippard to K. G. (February 2, 1992).
188. In his paper of 1954, Schafroth considered the pairs were 'metastable' rather than bound states.
189. Blatt (1964), p. 87.
190. Tisza to London (May 2, 1946).
191. London to Tisza (May 8, 1946).
192. Tisza to London (June 20, 1946).
193. Invitation from Gorter, Korringa, van den Handel (December 12, 1952).
194. Woerdeman to London (March 6, 1953).
195. Wigner to London (March 26, 1953); London to Wigner (March 30, 1953).
196. Kronig (1953).
197. Ibid., 65.
198. Ibid., 70.
199. London to Zilsel (end of June or beginning of July, 1953).
200. A. E. Dyhre to London (March 24, 1953).
201. Personal communication from Dr Edward Hammel of Los Alamos National Laboratory to K. G.
202. London to Dyhre (April 2, 1953); E. F. Hammel to London (April 6, 1953); London to Hammel (April 10, 1953).
203. 'Informal Interview with Frederick London' (in possession of Mrs. Edith London) held at the Chemistry Building, Duke University, October 30, 1952, 10 am. Those present were: Robert E. Tharp (Security Division AEC), John D. Donovan (Personnel Division, AEC) and Stanley Ford (Reporter). The report was confidential and was 74 pages long.
204. Ibid., 11–12.
205. Ibid., 7.
206. Ibid., 35–6.
207. Ibid., 62.
208. Ibid., 68–9.
209. Ibid., 70. The only thing London added to the text of the interview was the following. 'I do of course not doubt that Communism is subversive. But I doubt whether every "Communist" knows this. He might believe that he is in merely for "Peace" etc.'
210. Hammel to London (April 6, 1953).
211. Fritz to Heinz (September 25, 1953).
212. Edith to Heinz and Lucie (September 25, 1953).
213. Highly impressed by Feynman's proposals about helium-II, London told the or-

ganisers that he would not be giving the introductory lecture so that Feynman might be included in the program. Then, he was asked by Squire to preside during the session in which Feynman would be presenting his theory and to prepare questions so that he could open the discussion. Squire to London (November 21, 1953).

214. Fritz to Heinz (November 2, 1953).
215. Soon after Fritz died in 1954, Heinz had an operation as a result of a crisis of a duodenal ulcer, and two-thirds of his stomach was removed. The deep grief he felt upon the death of Fritz and his own introverted personality had, undoubtedly, contributed to the deterioration of his health. Throughout the 1960s, Heinz's main occupation was the development of the dilution refrigerator. It was a technique for producing very low temperatures by dilution of He^3 dissolved in He^4. After overcoming a series of technical complications, the machine had been marketed by Oxford Instruments Company, for which he acted as a consultant. The improved machine could reach the millidegree region. In 1959, he was awarded the first Simon Memorial Prize of the Low Temperature Group of the Physical Society and elected a Fellow of the Royal Society in 1961. In the spring of 1966, he developed coronary thrombosis brought on by a mad uphill dash for the morning bus, but he made a quick recovery. During his last illness, he put up a great fight for his life in a very matter-of-fact manner. Lucie London always felt that 'death to Heinz, seemed to be one of the events that happen in life and which must be taken as they come'. He was an almost compulsive smoker, and even though he stopped smoking after his heart attack, soon after he continued his old habit – which, in fact, probably killed him. In 1970, lung cancer was diagnosed, and he died on August 3, 1970. At the time of his death he was three years away from retirement. A few months earlier he had started sensing the unmistakable signs of the new era: the planning authorities did not consider the work of his group as first priority within the new framework of stressing technology. Hints that the group should wind up their projects depressed Heinz, who felt that he was becoming obsolete. (Shoenberg (1971), 441–61. See also the entry in the *Dictionary of Scientific Biographies* by C. W. F. Everitt and W. M. Fairbank.) He used to say that he never, really, expected to live to a very old age, but that he intended to do better than his father and brother – which he did.
216. Edith to Lucie and Heinz (March 17, 1954).
217. Fritz to Heinz (March 23, 1954).
218. Release from the Bureau of Public Information, Duke University, on Wednesday March 31, 1954:

    [March 30] Dr Fritz London, 54, internationally renowned Duke University scientist, died at his home here today at 8.30 am.... He died of a heart attack following an illness of several weeks.

    [University President Hollis Edens] In Dr London was a blend of much of the good of the old world and of the new in scholarship and integrity.

    [Vice-President and Dean, Paul Gross] He had in his scientific thinking that touch of genius which was able to illuminate and clarify some of the most difficult areas of our knowledge of the physical worlds.

    Obituaries were written by de Boer (1954), Fröhlich (1954) and Mendelssohn (1954).

Afterword: background leading to the microscopic theory of superconductivity

John Bardeen

It took almost half a century after the discovery of superconductivity by Kamerlingh Onnes in 1911 before a satisfactory explanation at the microscopic level was given by Robert Schrieffer, Leon Cooper and me in 1957. The theory was the result of many years of effort by many people in both theory and experiment. The first theories were phenomenological: equations were proposed to account for a range of experiments without an understanding of how they could be derived from the equations of motion of the electrons and ions that constitute a superconducting metal.

By far the most important step towards understanding the phenomena was the recognition by Fritz London that both superconductors and superfluid helium are macroscopic quantum systems. Quantum theory was derived to account for the properties of atoms and molecules at the microscopic level. It was Fritz London who first recognized that superconductivity and superfluid flow result from manifestations of quantum phenomena on the scale of large objects.

Perhaps the most striking illustration is that the magnetic flux threading a superconducting ring is an integral multiple of a small flux unit, $hc/2e$. In a footnote in his book published in 1950, London predicted such a relation (with e rather than $2e$ in the denominator). Flux quantization was first observed experimentally eleven years later. Quantization is a direct result of the de Broglie relation between momentum and wave length, $p = h/\lambda$, and the fact that there must be an integral number of wave lengths around the ring, $n\lambda = 2\pi r$, where r is the radius. The same relation was first applied to account for Bohr orbits in the hydrogen atom.

I first learned about superconductivity through reading David Shoenberg's little book, published in 1938. It gives an account of the essentials of what was known, without excessive detail. A unifying theme is the phenomenological theory of Fritz and Heinz London, published a couple of years earlier. The London theory is the subject of the second chapter of the book, following an introductory chapter about Onnes's discovery. Fritz London later pointed out that the London equations would follow from microscopic theory if there were long range order in the momenta of the electrons.

If a simply connected superconductor is placed in a magnetic field, currents flowing in a thin layer near the surface prevent the field from penetrating to the interior. The London theory was proposed to account for the experimental discovery of Meissner and Ochsenfeld in 1933 that in a simply connected body the state with the flux excluded is the unique stable state of a superconductor. If cooled in a magnetic field from the normal state to the superconducting state, the flux in the interior is expelled.

A superconductor may be regarded as a perfect diamagnet such that $B = 0$. Theorists who have thought about superconductivity may be divided into two schools. Those such as Bohr, Heisenberg and Bloch, who first learned about superconductivity before the discovery of the Meissner effect, never got over the feeling that the important thing was to explain the lack of scattering, $E = 0$. Those who first learned about superconductivity after Meissner's discovery felt that the important thing was to understand the nature of a state that would give perfect diamagnetism, $B = 0$. London bridged the two schools. He showed that the London equations that describe a perfect diamagnet could be derived from microscopic theory if there were the long range order in the momentum required to account for lack of effects of scattering on the superconducting current, $E = 0$.

My first attempt to derive a theory, in 1940, was rather naive, but was stimulated by London's approach and the Meissner effect. One can get a large diamagnetism if there is a small energy gap around the Fermi surface of the metal. It was proposed that such a gap could arise from small ionic displacements. If only states below that gap were occupied, the decrease in energy of the electrons hopefully would more than compensate for the energy required to displace the ions. Putting in numbers, it was found that the ionic displacements required an order of magnitude more energy than that gained by the electrons. At first I thought that, under favorable circumstances, this factor could be overcome, but further work indicated that it was unlikely. Only an abstract was published.

This work was interrupted by four years as a civilian scientist with the United States Navy and six years working on semiconductors at Bell Laboratories. In 1950, while still at Bell, I heard about the isotope effect: the dependence of the superconducting transition temperature on isotope mass. The news came in a telephone call from Bernard Serin, leader of the team that made the discovery at Rutgers University. If mass is involved, motion must also be involved, which indicated that interactions between electrons and phonons, the quanta of lattice vibrations, were the key to understanding superconductivity.

I thought immediately of my earlier approach, and tried to revive the theory by getting an energy gap from the electron–phonon interaction rather than static displacements. With use of a variational wave function, I tried to find out under what circumstances one could get an energy gap at the Fermi surfaces. About a week after I sent a note for publication, Herbert Fröhlich visited Bell. He told us about a similar theory he had developed earlier, without knowledge of the isotope effect. He had derived an energy gap from a perturbation expansion of the electron–phonon interaction. It soon was realized that both calculations were incorrect. We both had been using terms from the self-energy of the electrons in the field of phonons. These energies are included in the normal state. When properly taken into account, the energies of states in the vicinity of the Fermi surface are changed without introducing a gap.

At this stage (1951) we knew that it was necessary to take into account the interaction energy between electrons induced by the electron–phonon interaction rather than the self-energies. But we did not know how to describe the quantum states that lead to superconductivity. We tried to sum various terms of the perturbation expansion without success. How could the small energies from the electron–phonon interaction dominate over the much larger screened Coulomb interaction between electrons?

During the early 1950s, many theorists throughout the world were trying to solve the mystery of superconductivity. I benefited from discusssions with Fritz London at scientific meetings. However, his main influence was through his publications, particularly his book on superconductivity.

Unfortunately, London did not live long enough to see how the problem was solved. A brief account of the important steps that led to the microscopic theory will be given to show that he had the correct approach to the problem.

Two generalizations of London's phenomenological theory are important. One is that of Brian Pippard of Cambridge University, who studied the surface impedance at microwave frequencies. To account for his results, he introduced the concept of a coherence distance, ξ_0. According to the London theory, in an appropriate gauge, the supercurrent density is proportional to vector potential, $\mathbf{A}$. The electric field, E, is given by the time derivative of $\mathbf{A}$. In the normal state, in general, the current density is proportional to E, but at microwave frequencies, where the skin depth is small, the field can vary over a free path. The current density at a point is then not given by the local value of the electric field, but by an average of the field over a free path. In order to account for experiments on superconductors, Pippard found that in pure metals, the current density is given by an average of the vector potential over an intrinsic coherence distance, ξ_0, whose magnitude is inversely proportional to the transition temperature.

To account for the thermal properties of the superconductivity states, Gorter and Casimir in 1934 proposed a two-fluid model in which, at finite temperatures, the total current flow could be divided into a normal component, subject to the usual scattering, and a superfluid component. The total current density is then $\rho v \approx \rho_n v_n + \rho_s v_s$, where ρ is the total density $\rho = \rho_n + \rho_s$ and v_n and v_s are the flow velocities of the two components. In a static magnetic field with $E = 0$, the normal current density vanishes. Gorter and Casimir proposed that at finite temperatures, $\rho_n = \rho t^4$ and $\rho_s = \rho(1 - t^4)$, where $t = T/T_c$. The square of the superconducting penetration depth varies inversely with ρ_s.

Later a two-fluid model for superfluid helium was proposed by Tisza. At first the two-fluid model was taken too literally as two independent interpenetrating fluids. Landau showed how the two-fluid model could be derived from flow of the ground state and elementary thermal excitations. The low-lying excitations may be described by a wave-vector $\boldsymbol{k}$ of the magnitude $2\pi/\lambda$. For helium, Landau suggested the phonon–roton spectrum. For superconductivity, the excitations are quasi-particle excitations corresponding to occupied states above, or unoccupied states or holes below, the top of the Fermi sea. They are in one-to-one correspondence with excitations of the normal metal.

It is now recognized that the ground state of a superfluid is a state of macroscopic quantum occupation. For superfluid helium, it corresponds to the ground state of an Einstein–Bose condensate, as suggested by London. For a non-interacting Bose gas, all of the particles would have zero momentum in the rest frame at $T = 0$ K and a finite fraction would remain in this state up to the transition temperature. For an interacting system, the ground-state wave function gives a non-zero probability for occupation of the zero velocity or momentum state. For helium, there is about 10% probability of finding an atom in the zero momentum state at $T = 0$ K. If at finite temperatures the ground state is moving with a velocity v_s, and the excitations come into equilibrium with walls at rest, the net flow is $\rho_s v_s$. If the walls are moving with a velocity of v_n, the net flow is $\rho_n v_n + \rho_s v_s$.

A normal system at finite temperatures is completely defined by the velocity of the walls with which it is in equilibrium, regardless of the relative motion of the observer. In a superfluid, one must also specify the velocity, v_s, of the state of macroscopic occupation. The first to recognize this characteristic feature of superfluids was Fritz London.

In 1950, Ginzburg and Landau suggested that one could define the state of a superconductor by a complex order parameter in the form of a macroscopic wave function, ψ, with amplitude and phase. The square of the amplitude is proportional to the superfluid density and the gradient of the phase to the superfluid velocity. They suggested that near T_c the function ψ can be derived from a non-linear Schrödinger-like equation. In this way one can treat situations in which the amplitude varies rapidly in space, as near the core of a vortex line. The microscopic origin and meaning of the macroscopic wave function remained mysterious. This generalization of the London theory allows one to treat problems in which the superfluid density varies rapidly in space, as near the core of a vortex line.

During London's lifetime, it was not understood what corresponded to the state of macroscopic occupation of a superconductor. We now know that it is the common momentum of pairs of electrons. The ground state of a superconductor is such that there is a certain probability, p_k that a quasi-particle state k in the vicinity of the Fermi surface is occupied and a probability $1 - p_k$ that it is unoccupied. If k is occupied, it is certain that the state of opposite spin and momentum of the paired states k and $-k$ is also occupied. If there is current flow, all paired states have exactly the same net momentum, $\hbar(k_1 + k_2) = 2mv_s$. Thus, v_s defines the net momentum of pairs of electrons. To define the state of the system, one must specify v_s. The long range order of the momentum required by London is that of pairs of electrons.

The low-lying excitations are in one-to-one correspondence with those for a normal metal. They differ in that a finite energy is required to excite a particle from the paired superconducting ground state. If ε_k is the energy of a state k measured from the Fermi energy, the energy of the corresponding state in a superconductor is $(\varepsilon_k^2 + \Delta^2)^{1/2}$. Since quasi-particles are created in pairs from the ground state, the energy gap for creating a pair of excitations is $\varepsilon_g = 2\Delta$.

In our first major paper on the microscopic theory, Cooper, Schrieffer and I suggested that the magnitude of the Ginzburg–Landau order parameter, Ψ, be equated to the gap parameter, Δ. Gorkov later showed that this is indeed the case. Near T_c one can expand the free energy powers of Δ^2. The Ginzburg–Landau theory contains terms to Δ^4.

The microscopic theory confirmed that the earlier phenomenological theories of London, Pippard, Gorter and Casimir, and

Ginzburg and Landau are basically correct. The key to understanding superfluidity is macroscopic occupation of a quantum state: in He^4 it is the momentum of the atoms in the rest frame, in a superconductor it is the common momentum of pairs of electrons. The macroscopic occupation breaks Galilean invariance, since to define the system one must give the velocity of the state of macroscopic occupation, v_s, as well as the velocity, v_n, of the walls that give the rest frame for quasi-particle excitations.

One of Fritz London's major achievements was to make the leap from the microscopic world of atoms and molecules and to show that superfluids are quantum systems on a macroscopic scale.

[Professor John Bardeen (1908–1991) sent this piece in February 1990 to be included as an Afterword.]

Publications by Fritz London

(1922) Die Bedingungen der Moeglichkeit einer Massbestimmung in einer physikalischen Mannigfaltigkeit und das Prinzip der Aehnlichkeit. *Phys. Zs.*, **23**, 1–13.

(1923) Ueber die Bedingungen der Moeglichkeit einer deduktiven Theorie. (Dissertation.) *Jahrb. f. Philosophie und phaenomenologische Forschung*, **VI**, 335–384.

(1924) Ueber die Irreversibilitaet deduktiver Schlussweisen. *Jahrb. der deutschen Mathematischen Vereinigung*, 84–87.

(1925) Ueber die Intensitaeten der Bandenlinien (with H. Hönl). *Zs. f. Phys.*, **33**, 803–809.

(1926*a*) Energiesatz und Rydbergprinzip in der Quantenmechanik. *Zs. f. Phys.*, **36**, 775–777.

(1926*b*) Ueber die Jacobischen Transformationen der Quantenmechanik. *Zs. f. Phys.*, **37**, 915–925.

(1926*c*) Die Zahl der Dispersionselektronen in der Undulationsmechanik. *Zs. f. Phys.*, **39**, 322–326.

(1926*d*) Winkelvariable und Kanonische Transformationen in der Undulationsmechanik. *Zs. f. Phys.*, **40**, 193–210.

(1927*a*) Ueber eine Deutungsmoeglichkeit der Klein'schen fuenfdimensionalen Welt. *Naturwissenschaften*, **15**, 15–16.

(1927*b*) Die Theorie von Weyl und die Quantenmechanik. *Naturwissenschaften*, **15**, 187.

(1927*c*) Quantenmechanische Deutung der Theorie von Weyl. *Zs. f. Phys.*, **42**, 375–389.

(1927*d*) Wechselwirkung neutraler Atome und homöopolare Bindung nach der Quantenmechanik (with W. Heitler). *Zs. f. Phys.*, **44**, 455–472.

(1928*a*) Quantenmechanische Deutung der homöopolaren Valenzzahlen. *Naturwissenschaften*, **16**, 59.

(1928*b*) Zur Quantentheorie der homöopolaren Valenzzahlen. *Zs. f. Phys.*, **46**, 455–477.

(1928*c*) Zur Quantenmechanik der homöopolaren Valenzchemie. *Zs. f. Phys.*, **50**, 24–51.

(1928*d*) Quantentheorie und chemische Bindung. Chapter in the book *Quantentheorie und Chemie*, Hirzel; 59–84.

(1928*e*) Ueber den Mechanismus der homöopolaren Bindung. *Sommerfeld Festschrift*, Hirzel; 104–113.

(1929*a*) Ueber quantenmechanische Energieuebertragung zwischen atomaren Systemen (with H. Kallmann). *Zs. f. physikalische Chemie*, **B2**, 207–243.
(1929*b*) Die Bedeutung der Quantentheorie fuer die Chemie. (Planck Festschrift.) *Naturwissenschaften*, **17**, 516–529.
(1929*c*) Quantenmechanische Deutung der Vorgaenge der Aktivierung. *Zs. f. Elektrochemie*, **35**, 552–555.
(1929*d*) Quantenmechanische Theorie der anomal grossen Wirkungsquerschnitte bei der Energieuebertragung zwischen atomaren Systemen (with H. Kallmann). *Naturwissenschaften*, **17**, 226–227.
(1930*a*) Zur Quantenmechanik der Energieuebertragung (with H. Kallmann). *Zs. f. Phys.*, **60**, 417–419.
(1930*b*) Ueber das Verhaeltnis der Van der Waalsschen Kraefte zu den homöopolaren Bindungskraeften (with Eisenschitz). *Zs. f. Phys.*, **60**, 491–527.
(1930*c*) Zur Theorie und Systematik der Molekularkraefte. *Zs. f. Phys.*, **63**, 245–279.
(1930*d*) Ueber einige Eigenschaften und Anwendungen der Molekularkraefte. *Zs. f. physikalische Chemie*, **B11**, 222–251.
(1930*e*) Ueber die atomtheoretische Deutung der Adsorptionskraefte (with M. Polanyi). *Naturwissenschaften*, **18**, 1099–1100.
(1932*a*) Zur Theorie nicht adiabatisch verlaufender chemischer Prozesse. *Zs. f. Phys.*, **74**, 143–174.
(1932*b*) Théorie électrique des forces entre les atomes et les molécules. *Congrès International d'Electricité*, Paris, 1–18.
(1934) Limitation of the potential theory of the broadening of spectral lines (with H. Kuhn). *Phil. Mag.*, **18**, 983–987.
(1935*a*) The electromagnetic equations of the supraconductor (with H. London). *Proc. Roy. Soc.*, **149**, 71–88.
(1935*b*) Supraleitung und Diamagnetismus (with H. London). *Physica*, **2**, 341–354.
(1935*c*) Zur Theorie der Supraleitung (with H. London and M. v. Laue). *Zs. f. Phys.*, **96**, 359–364.
(1935*d*) Macroscopical interpretation of supraconductivity. *Proc. Roy. Soc.*, **152**, 24–34.
(1936*a*) On condensed helium at absolute zero. *Proc. Roy. Soc.*, **153**, 576–583.
(1936*b*) Zur Theorie magnetischer Felder im Supraleiter. *Physica*, **III**, 450–462.
(1936*c*) Electrodynamics of macroscopic fields in supraconductors. *Nature*, **137**, 991.
(1937*a*) *Une Conception Nouvelle de la Supraconductibilité. Actualités scientifiques et industrielles*. Hermann et Cie, Paris. (Thèse pour le Doctorat ès Sciences Physiques.)
(1937*b*) The general theory of molecular forces. *Trans. Faraday Society*, **189**, XXXIII, 8–26.
(1937*c*) A new conception of supraconductivity. *Nature*, **140**, 793–7.
(1937*d*) On the nature of the supraconducting State. *Phys. Rev.*, **51**, 678–679.
(1937*e*) Supraconductivity in aromatic molecules. *Journ. Chem. Phys.*, **5**, 837–838.
(1937*f*) Théorie quantique du diamagnétisme des combinaisons aromatiques. *Comptes Rendus*, **205**, 28–30.
(1937*g*) Théorie quantique des courants interatomiques dans les combinaisons aromatiques. *Journ. de Physique et le Radium*, **8**, 397–409.
(1937*h*) On a letter of Wick about Supraconductivity. *Phys. Rev.*, **52**, 886.
(1937*i*) La théorie de valence en mécanique quantique. Supra-conductibilité dans les combinasions aromatiques. (Réunion internationale de Physique-Chimie-Biologie, Paris, October) *Actualités Scientifiques et Industrielles*, Hermann et Cie, Paris; 194–202.
(1938*a*) The λ-phenomenon of liquid helium and the Bose–Einstein degeneracy. *Nature*, **141**, 643–4.
(1938*b*) Bemerkungen zu einer Arbeit von Bopp ueber Supraleitung. *Zs. f. Phys.*, **108**, 542–544.
(1938*c*) On the Bose–Einstein condensation. *Physical Review*, **54**, 947–54.

(1939*a*) La Théorie de l'Observation en Mécanique Quantique (with E. Bauer). *Actualités Scientifiques et Industrielles*, Hermann et Cie, Paris.

(1939*b*) Sur les oscillateurs moléculaires dans les molécules aromatiques. *Comptes Rendus*, **208**, 2059–2061.

(1939*c*) The state of liquid helium near absolute zero. *Journ. of Phys. Chem.*, **43**, 49–69.

(1942) On centers of van der Waals attraction. *Journ. of Phys. Chem.*, **46**, 305–317.

(1943*a*) Distance correlations and Bose–Einstein condensation. *Journ. Chem. Phys.*, **11**, 203–213.

(1943*b*) Intermolecular Attraction between Macro Molecules. (Invited paper presented at the Symposium on Surface Chemistry held at the Fiftieth Anniversary of the University of Chicago, September, 1941.) Publication No. 21 of the American Association for the Advancement of Science, 141–149.

(1945) Planck's constant and low temperature transfer. (Invited paper at the occasion of the birthday of Niels Bohr.) *Rev. of Mod. Phys.*, **17**, 310–320.

(1946) On an inequality of quantum hydrodynamics. *Phys. Rev.*, **69**, 254.

(1947) The present state of the theory of liquid helium. (Invited introductory lecture presented at the Conference on Low Temperature Physics, July, 1946, at Cambridge, England.) *Physical Society Cambridge Conference Report*, 1947, pp. 1–18.

(1948*a*) On solutions of He^3 in He^4 (with O. K. Rice). *Phys. Rev.*, **73**, 1188–1193.

(1948*b*) On the problem of the molecular theory of superconductivity. *Phys. Rev.*, **74**, 562–573.

(1948*c*) Heat transfer in liquid helium II by internal convection (with P. R. Zilsel). *Phys. Rev.*, **74**, 1148–1156.

(1949*a*) Sur des mécanisms quantiques à l'échelle macroscopique. (Conférence faite à la Société de Chimie Physique à Paris, 8 Juin 1949.) *Journ. Chimie Physique*, **46**, No. 9–10.

(1949*b*) The rare isotope of helium, He^3; a key to the strange properties of ordinary liquid helium, He^4. *Nature*, **163**, 694.

(1949*c*) Program for the molecular theory of superconductivity. (Invited introductory lecture.) *Proc. Intern. Conf. Low Temperature Physics*, M.I.T., Sept. 1949, pp. 76–83.

(1950) *Superfluids*, Vol. I Wiley and Sons

(1951) Limitations of the two-fluid theory of liquid helium. (Invited introductory lecture.) *Proc. Intern. Conf. on Low Temperature Physics*, Oxford, August 1951, pp. 2–6.

(1952*a*) Zu den 'Bemerkungen zur Theorie der Supraleitung' des Herrn v. Laue. *Annalen der Physik*, **6**, Vol. 10, No. 4–5, 302–304, 314–316.

(1952*b*) Une inégalité de l'hydrodynamique quantique. Extrait du livre *Louis de Broglie, Physicien et Penseur*, Editions Albin Michel, Paris.

(1953) The entropy of liquid He^3 (with Tien Chi Chen). *Phys. Rev.*, **89**, 1038–1040.

(1954) *Superfluids*, Vol. II. Wiley and Sons.

(1955) Alignment of hydrogen molecules by high pressure. *Phys. Rev.*, **102**, 168–171. (Edited version of a posthumous manuscript.)

Bibliography

Abe, Y. (1981) Pauling's revolutionary role in quantum chemistry. *Historia Scientiarum*, **20**, 107–24.
Allen, J. F. (1952) Liquid helium. In Simon (1952).
Allen, J. F. & Jones, H. (1938) New phenomena connected with heat flow in helium II. *Nature*, **141**, 243–4.
Andronikashvilii, E. L. (1946) A direct observation of the two kinds of motion in helium II. *Journal of Physics (USSR)*, **10**, 201–6.
Andronikashvilii, E. L. (1989) *Reflections on Liquid Helium* (translated by R. Berman). American Institute of Physics, New York.
Appleyard, E. T. S., Bristow, J. R., London, H. & Misener, A. D. (1939*a*) Variation of field penetration with temperature in a superconductor. *Nature*, **143**, 433–4.
Appleyard, E. T. S., Bristow, J. R., London, H. & Misener, A. D. (1939*b*) Superconductivity of thin films I. Mercury. *Proceedings of the Royal Society*, **A172**, 540–58.
Arsem, W. C. (1914) A theory of valency and molecular structure. *Journal Of the American Chemical Society*, **36**, 1655–75.
Atkins, K. R. (1959) *Liquid Helium*. Cambridge University Press, Cambridge.
Atkins, K. R. & Osborne, D. V. (1950) The velocity of second sound below 1 K. *Philosophical Magazine*, **41**, 1078–81.
Bantz, D. A. (1980) The structure of discovery: evolution of structural accounts of chemical bonding. In Nickles (1980), 291–329.
Bardeen, J. (1950*a*) Zero point vibrations and superconductivity. *Physical Review*, **79**, 167–8.
Bardeen, J. (1950*b*) Wave functions for superconductivity. *Physical Review*, **80**, 567–74.
Bardeen, J. (1956) On superconductivity. In S. Flugge (ed.) *Encyclopedia of Physics*, volume XV. Springer Verlag, Berlin.
Bardeen, J., Cooper, L. N. & Schrieffer, J. R. (1957) Theory of superconductivity. *Physical Review*, **108**, 1175–200.
Becher, E. (1905) *Der Begriff des Attributes bei Spinoza*. Munich.
Becker, O. (1923) Beitrage zur phanomenologischen Begrundung der Geometrie und ihrer physikalischen Anwendungen. *Jahrbuch für Philosophie und Phanomenologische Forschung*, **6**, 385–560.
Becker, O. (1927) Mathematische Existenz *Jahrbuck für Philosophie und Phanomenologische Forschung*, **8**, 441–809.

Becker, R., Heller, G. & Sauter, F. (1933) Uber die Stromverteilung in einer supraleitenden Kugel. *Zeitschrift für Physik*, **85**, 772–87.

Bernhard, L. (1930) *Akademische Selbstverwaltung in Frankreich und Deutschlald*. Berlin.

Beyerchen, A. S. (1977) *Scientists Under Hitler. Politics and the Physics Community in the Third Reich*. Yale University Press.

Blatt, J. M. (1964) *Theory of Superconductivity*. Academic Press, New York.

Bloch, F. (1928) Uber die Quanten-mechanic der Electronen in Kristallgittern. *Zeitschrift für Physik*, **52**, 555–600.

Bloch, F. (1966) Some remarks on theory of superconductivity. *Physics Today*, **19**, 27–36.

Bogoliubov, N. N. (1962) *The Theory of Superconductivity*. Gordon and Breach, New York.

Born, M. (1930) Zur Quantentheorie der Chemischen Kraften. *Zeitschrift für Physik*, **64**, 729–40.

Born, M. (1931) Chemische Bindung und Quantenmechanik. *Ergebnisse der exakten Naturwissenschaften*, **10**.

Born, M. (1937) The statistical mechanics of condensing systems. *Physica*, **IV**, 215–1156.

Bose, S. N. (1924) Plancks Gesetz und Lichtquantenhypothese. *Zeitschrift für Physik*, **26**, 178–81.

Brush, S. G. (1983) *Statistical physics and the atomic theory of matter, from Boyle and Newton to Landau and Onsager*. Princeton University Press, Princeton.

Buckingham, A. D. (1987) Quantum Chemistry. In Kilminster (1987), 112–17.

Bullivant, K. (1977) *Culture and Society in the Weimar Republic*. Manchester University Press.

Burrau, O. (1927) Berechnung des Energiewertes des Wasserstoffmolekel-Ions (H^+_2). *Det Kig. Videnskabernes Selskab Matematisk-Fysiske Meddelelser*, **7**, 24–38.

Cailletet, L. & Bouty, N. (1885) Sur la conductibilité electrique du mercure et des metaux purs aux basses temperatures. *Séances de la Societé Française de Physique*, **21**, 97–104.

Campbell, N. (1920) *Physics: the Elements*. Cambridge University Press, Cambridge.

Casimir, H. (1973) Superconductivity and superfluidity. In J. Mehra (ed.) *Physicist's Conception of Nature*. Reidel Publishers, Dordrecht, 481–500.

Casimir, H. (1977) Superconductivity. In C. Weiner (ed.) *History of Twentieth century Physics*, Proceedings of the International School of Physics 'Enrico Fermi'. Academic Press, New York.

Cassidy, D. C. (1991) *Uncertainty, the Life and Science of Werner Heisenberg*. Freeman, New York.

Clark, G. L. (1928) Introductory remarks in the symposium on atomic structure and valence. *Chemical Reviews*, **5**, 361–4.

Clusius, K. (1934) Zwei Vorlesungsversuche mit flussingen Wassertoff. *Physikalische Zeitschrift*, **35**, 929–30.

Condon, E. U. (1927) Wave mechanics and the normal state of the hydrogen molecule. *Proceedings of the National Academy of Sciences*, **13**, 466–70.

Coulson, C. A. (1970) Recent developments in valence theory – symposium fifty years of valence. *Pure and Applied Chemistry*, **24**, 257–87.

Crawford, E., Heilbron, J. & Ullrich, R. (1988) *The Nobel Population 1900–1937*. Office for the History of Science and Technology, University of California, Berkeley.

Dahl, P. F. (1993) *Superconductivity, its Historical Roots and Development from Mercury to the Ceramic Oxides*. American Institute of Physics, New York.

Dana, L. I. & Kamerlingh Onnes, H. (1926) Further experiments with liquid helium. BA. Preliminary determinations of the latent heat of vaporisation of liquid helium. *Communications from the Physical Laboratory of the University of Leiden (CPL)*, **179c**, 23–34.

Daunt, J. G. & Mendelssohn, K. (1938) Transfer of helium-II on glass. *Nature*, **141**, 911–12.

Daunt, J. G. & Mendelssohn, K. (1939) The transfer effect in liquid He II. The transfer phenomena. *Proceedings of the Royal Society of London*, **A170**, 423–9.

Daunt, J. G. & Smith, R. S. (1954) The problem of liquid helium – some recent aspects. *Reviews of Modern Physics*, **26**, 172–236.

Deaver, B. S. & Fairbank, W. M. (1961) Experimental evidence for quantized flux in superconducting cylinders. *Physical Review Letters*, **7**, 43–6.

de Boer, J. (1954) Fritz London. *Nederl. Tidjdschr. Natuurkunde*, **20**, 95.

de Broglie, L. (1974) Opening address to the First International Congress of Quantum Chemistry at Menton, France, July 4–10, 1973. In R. Daudel & B. Pullman (eds.) *The World of Quantum Chemistry*. Academic Publishers, New York.

de Haas, W. J. (1933) Supraleiter in Magnetfield. In P. Debye (ed.) *Leipziger Vortrage, Magnetismus*, 59–73. Leipzig.

de Haas, W. J. & Bremmer, H. (1931) Conduction of heat of lead and tin at low temperatures. *CPL*, **214d**.

de Haas, W. J. & Bremmer, H. (1932) Thermal conductivity of indium at low temperatures. *CPL*, **220b**, 5–11.

de Haas, W. J. & Jonker, J. M. (1934) Quantitative Undersuchung ubereinen möglischen Einfluss der Achsenorientierung auf die magnetische Ubergangsfigur. *CPL*, **229c**.

de Haas, W. J. & Voogd, J. (1930) The influence of magnetic fields on superconductors. *Proceedings of the Akademie der Wetenschappen, Amsterdam*, **33**, 262–72.

de Haas, W. J. & Voogd, J. (1931) On the steepness of the transition curve of superconductors. *CPL*, **214c**.

Dewar, J. & Fleming, J. A. (1892) On the electrical resistance of pure metals, alloys, and non-metals at the boiling point of oxygen. *Philosophical Magazine*, **34**, 326–7.

Dewar, J. & Fleming, J. A. (1893) The electrical resistance of metals and alloys at temperatures approaching the absolute zero. *Philosophical Magazine*, **36**, 271–99.

Dewar, J. & Fleming, J. A. (1904) On the electrical resistance thermometry at temperature of boiling hydrogen. *Proceedings of the Royal Society*, **73**, 244–51.

Dingle, R. B. (1952) Theories of helium-II. *Advances in Physics*, **1**, 111–68.

Dirac, P. A. M. (1926) Quantum mechanics and a preliminary investigation of the hydrogen atom. *Proceedings of the Royal Society*, **A110**, 561–79.

Dirac, P. A. M. (1928) The quantum theory of the electron. *Proceedings of the Royal Society*, **A117**, 610–24.

Dirac, P. A. M. (1929) Quantum mechanics of many-electron systems. *Proceedings of the Royal Society*, **A123**, 714–33.

Dresden, M. (1987) *H. A. Kramers, Between Tradition and Revolution*. Springer, Berlin.

Eddington, A. (1928) *The Nature of the Physical World*. Cambridge University Press, Cambridge.

Ehrenfest-Afanassjewa, T. (1916*a*) Der Dimensionsbegriff und der analytische Bau der physikalischen Gleichungen. *Mathematische Annalen*, **77**, 259.

Ehrenfest-Afanassjewa, T. (1916*b*) On Mr. Tolman's 'principle of similitude'. *Physical Review*, **8**, 1–7.

Einstein, A. (1924) Quantentheorie des einatomigen idealen Gases. *Sitzungsberichte der Preussischen Akademie der Wissenschaften, Physikalisch-Matematische Klasse*, 261–67; (1925) 3–16.

Eisenschitz, R. & London, F. (1930) See List of Publications by F. London (1930*b*).

Ewald, P. P. (1960) Max von Laue. *Royal Society of London, Biographical Memoirs*, **6**, 135–56.

Fairbank, J. D., Deaver, B., Everitt, C. & Michelson, P. F. (eds.) (1988) *Near Zero: New Frontiers of Physics*. Freeman, New York.

Feigl, H. (1969) The Weiner Kreis in America. In D. Fleming & B. Bailyn (eds.) *The*

Intellectual Migration, Europe and America 1930–1960. Harvard University Press, 630–73.

Fermi, L. (1971) *Illustrious Immigrants*. University of Chicago Press.

Feynman, R. (1953*a*) Atomic theory of the lambda transition in helium. *Physical Review*, **91**, 1291–301.

Feynman, R. (1953*b*) Atomic theory of liquid helium near absolute zero. *Physical Review*, **94**, 262–70.

Fink, K. (1991) *Goethe's History of Science*. Cambridge University Press, Cambridge.

Fleming, D. and Bailyn, B. (1969) *The Intellectual Migration, Europe and America 1930–1960*. Harvard University Press.

Forman, P. (1971) Weimar culture causality and quantum theory 1918–1927: adaptation by German physicists and mathematicians to a hostile intellectual environment. *Historical Studies in the Physical Sciences*, **3**, 1–115.

Fowler, R. H. (1932) A report on homopolar valency and its quantum mechanical interpretation at the Centenary (1931) meeting of the British Association for the Advancement of Science. Heffer and Sons, Cambridge.

Fox, D. M., Meldrum, M. & Rezak, I. (eds.) (1990) *Nobel Laureates in Medicine or Physiology – A Biographical Dictionary*. Garland Publishers, New York.

Fröhlich, H. (1937) Zur Theorie des lambda Punctes des Heliums *Physica*, **4**, 639–44.

Fröhlich, H. (1950*a*) Theory of the superconducting state I. The ground state at the absolute zero of temperature. *Physical Review*, **79**, 845–56.

Fröhlich, H. (1950*b*) Isotope effect in superconductivity. *Proceedings of the Physical Society (London)*, **A63**, 778.

Fröhlich, H. (1953) Superconductivity and lattice vibrations. *Physica*, **XIX**, 755–64.

Fröhlich, H. (1954) Fritz London. *Nature*, **174**, 63.

Fröhlich, H. (1961) The theory of superconductive state. *Reports on Progress in Physics*, **24**, 1–23.

Fröhlich, H., Pelzer, H. & Zienau, S. (1950) Properties of slow electrons in polar materials. *Philosophical Magazine*, **41**, 221–42.

Fry, H. S. (1928) A pragmatic system of notation for electronic valence conceptions in chemical formulas. *Chemical Reviews*, **5**, 557–68.

Gavroglu, K. & Goudaroulis, Y. (1989) *Methodological Aspects of Low Temperature Physics 1881–1956: Concepts Out of contexts*. Kluwer Academic Publishers, Dordrecht.

Gavroglu, K. & Goudaroulis, Y. (1991) *Through Measurement to Knowledge. The Selected Papers of Heike Kamerlingh Onnes 1856–1926*. Kluwer Academic Publishers, Dordrecht.

Gavroglu, K. & Simoes, A. (1994) The Americans, the Germans and the beginning of quantum chemistry: the confluence of diverging traditions *Historical Studies in the Physical Sciences*, **25**, 1–63.

Gay, P. (1968) *Weimar Culture, The Outsider as Insider*. Harper and Row Publishers, London.

Glaser, H. (1978) *The Cultural Roots of National Socialism*. Croon Helm, London.

Gordy, W. (1988) The nature of a man, William Martin Fairbank. In J. D. Fairbank Jr., B. S. Deaver, C. W. F. Everitt & P. F. Michelson (eds.) *Near Zero: New Frontiers of Physics*. W. H. Freeman, New York.

Gorter, C. J. (1933*a*) Some remarks on the thermodynamics of superconductivity. *Archives du Musée Teyler*, ser. III, **VII**, 378–87.

Gorter, C. J. (1933*b*) Theory of superconductivity. *Nature*, **132**, 931.

Gorter, C. J. (1949) The two-fluid model for helium II. *Nuovo Cimento*, series 9, **6**, 245–50.

Gorter, C. J. (1964) Superconductivity until 1940 in Leiden and as seen from there. *Reviews of Modern Physics*, **36**, 3–7.

Gorter, C. J. (1967) Bad luck in attempts to make scientific discoveries. *Physics Today*, **21**, 84–92.

Gorter, C. J. & Casimir, H. B. G. (1934) On supraconductivity I. *Physica*, **1**, 306–20.

Goudsmit, S. & Uhlenbeck, G. E. (1925) *Naturwissenschaften*, **13**, 953.

Goudsmit, S. & Uhlenbeck, G. E. (1926) Spinning electrons and the structure of spectra. *Nature*, **107**, 264–5.

Green, H. S. (1948) Liquid helium II. *Nature*, **161**, 391.

Griffin, A. (1994) *Excitations in a Bose-Condensed Liquid*. Cambridge University Press, Cambridge.

Hacking, I. (1982) Language, truth and reason. In M. Hollis & S. Lukes (eds.) *Rationality and Relativism*. MIT Press.

Hacking, I. (1993) On styles of reasoning. In K. Gavroglu, J. Christianidis & E. Nikolaidis (eds.) *Trends in the Historiography of Science*. Kluwer Academic Publishers, Dordrecht.

Halperin, S. W. (1946) *Germany Tried Democracy, a Political History of the Reich from 1918 to 1933*. Norton and Company, New York.

Hanle, P. A. (1977) Erwin Schrödinger's reaction to Louis de Broglie's thesis on the quantum theory. *Isis*, **68**, 606–9.

Hanle, P. A. (1979) The Schrödinger–Einstein correspondence and the sources of wave mechanics. *American Journal of Physics*, **47**, 644–9.

Hartshorne Jr, E. Y. (1937) *The German Universities and National Socialism*. George Allen and Unwin, London.

Heilbron, J. H. (1983) The origins of the exclusion principle. *Historical Studies in the Physical Sciences*, **13**, 261–310.

Heilbron, J. H. (1986) *The Dilemmas of an Upright Man: Max Planck as Spokesman for German Science*. University of California Press, Berkeley.

Heisenberg, W. (1932) Contribution to the discussion on the structure of simple molecules. Centenary (1931) meeting of the British Association for the Advancement of Science. Heffer and Sons, Cambridge.

Heisenberg, W. (1947) Zur Electronentheorie der Supraleitung. Abstract of a talk at the Göttingen meeting of German physicists, October 4–6, 1946. *Neue Physikalischen Blatterl*, **2**, 220.

Heisenberg, W. (1949) *The Electron Theory of Superconductivity. Two Lectures*. Cambridge University Press, Cambridge.

Heisenberg, W. & von Laue, M. (1948) Das Barlowsche Rad aus supraleitendem Material. *Zeitschrift für Physik*, **124**, 514–18.

Heitler, W. (1929) Storungsenergie und Austausch beim Mehrkorperproblem. *Zeitschrift für Physik*, **46**, 47–72.

Heitler, W. (1936) *Quantum Theory of Radiation*. Oxford University Press.

Heitler, W. (1945) *Elementary Wave Mechanics*. Oxford University Press.

Heitler, W. (1955) The theory of chemical bond. *Arkiv für Fysik*, **10**, 145–56.

Heitler, W. (1967) Quantum chemistry: the early period. *International Journal of Quantum Chemistry*, **1**, 13–36.

Heitler, W. & London, F. (1927) see list of papers by F. London (1927*d*).

Heitler, W. & Poschl, G. (1934) Ground states of C_2 and O_2 and the theory of valency. *Nature*, **137**, 833–4.

Heitler, W. & Rumer, G. (1931) Quantentheorie der chemischen Bindung für mehratomige Molecule. *Zeitschrift für Physik*, **68**, 12–31.

Heller, E. & Low, F. (eds.) (1933) *Neue Muncherer Philosophische Abhandlungen*. Leipzig.

Hendry, J. (ed.) (1984*a*) *Cambridge Physics in the Thrities*. Adam Hilger, Bristol.

Hendry, J. (1984*b*) *The Creation of Quantum Mechanics and the Bohr–Pauli Dialogue*. Reidel, Dordrecht.

Hirschfelder, J. O., Curtiss, C. F. & Byron Bird, R. (1954) *Molecular Theory of Gases and Liquids*. John Wiley, New York.

Hoch, P. K. (1983) The reception of Central European refugee physicists of the 1930s: USSR, UK, USA. *Annals of Science*, **40**, 217–46.

Hoffmann, D. (1991) Robert Ochsenfeld 90 Jahre. *PTB-Mitteilungen*, **101**, 247–8.
Hund, F. (1929) Chemical binding. *Transactions of the Faraday Society*, **25**, 645–7.
Husserl, E. (1900) *Logische Untersuchungen*. Halle, Germany.
Husserl, E. (1906) *Einleitung in die Logik und Erkenntnistheorie*. Halle, Germany.
Husserl, E. (1929) *Formale und transzendentale Logik*. Halle, Germany.
Jackson, J. (1988) *The Popular Front in France Defending Democracy 1934–1938*. Cambridge University Press.
James, H. M. & Coolidge, A. S. (1933) The ground state of the hydrogen molecule. *Journal of Chemical Physics*, **1**, 825–30.
Jammer, M. (1974) *The Philosophy of Quantum Mechanics*. John Wiley and Sons, New York.
Jammer, M. (1989) *The Conceptual Development of Quantum Mechanics*. American Institute of Physics, New York.
Janouch, F. (1979) 'Lev Landau: his life and work'. CERN yellow reports 79/03, Geneva.
Johnson, M. C. (1929) A method of calculating the numerical equation of state for helium below 6 degrees absolute and of estimating the relative importance of gas degeneracy and interatomic forces. *Proceedings of the Physical Society of London*, **42**, 170–9.
Jurkowitz, E. (1995) 'The Old School and the History of Quantum mechanics'. PhD Dissertation, University of Toronto.
Kahn, B. & Uhlenbeck, G. E. (1938*a*) On the theory of condensation. *Physica*, **IV**, 1155–6.
Kahn, B. & Uhlenbeck, G. E. (1938*b*) On the theory of condensation. *Physica*, **V**, 399–414.
Kamerlingh Onnes, H. (1904) The importance of accurate measurements at very low temperatures. *CPL* (supplement), **9**.
Kamerlingh Onnes, H. (1908) The liquefaction of helium. *CPL*, **108**.
Kamerlingh Onnes, H. (1911) Further experiments with liquid helium D. On the change of the electrical resistance of pure metals at very low temperatures. V. The disappearance of the resistance of mercury. *CPL*, **122b** and *CPL*, **124c**.
Kamerlingh Onnes, H. (1913*a*) Further experiments with liquid helium H. On the electrical resistance of pure metals at very low temperatures. VII. The potential difference necessary for the electric current through mercury below 4.19 K. *CPL*, **133a**, 3–26.
Kamerlingh Onnes, H. (1913*b*) Report on the researches made in the Leiden cryogenics laboratory between the second and third international congress of refrigeration. *CPL* (supplement) **34b**, 37–70.
Kamerlingh Onnes, H. & Tuyn, W. (1922) Further experiments with liquid helium Q. On the electrical resistance of pure metals. X. Measurements concerning the electrical resistance of thalium in the temperature field of liquid helium. *CPL*, **160a**.
Kamerlingh Onnes, H. and Tuyn, W. (1926) Further experiments with liquid helium AA. The disturbance of superconductivity by magnetic fields and currents. The hypothesis of Silsbee. *CPL*, **174a**.
Kapitza, P. (1938) Viscosity of helium below the λ-point. *Nature*, **141**, 74.
Kapitza, P. (1940) Problems of liquid helium. A report at the General Assembly of the USSR Academy of Sciences, 28 December 1940. Reprinted in Kapitza, P. L. (1980) *Experiment, Theory, Practice. Articles and Addresses*. Reidel Publishers, Boston, 12–34.
Kapitza, P. (1941*a*) The study of heat transfer in helium-II. In ter Haar (1965*a*), volume II, 581–624.
Kapitza, P. (1941*b*) Heat transfer and superfluidity of helium-II. In ter Haar (1965*a*), volume II, 625–39.
Kasha, M. & Pullman, B. (eds.) (1962) *Horizons in Biochemistry, Albert Szent-Györgyi Dedicatory Volume*. Academic Press.

Keesom, W. H. (1932) Quelques remarques en rapport avec l'anomalie de la chaleur specifique de l'helium au point lambda. *CPL* (supplement) **71e**, 47–52.
Keesom, W. H. (1942) *Helium*. Elsevier.
Keesom, W. H. & Clusius, K. (1932) Ueber die spezifische Warme des flussingen Heliums. *CPL*, **219e**, 42–58.
Keesom, W. & Keesom, A. P. (1932) On the anomaly in the specific heat of liquid helium. *CPL*, **221d**, 19–26.
Keesom, W. & Keesom, A. P. (1936) On the heat conductivity of liquid helium. *Physica*, **3**, 359–60.
Keesom, W. & MacWood, G. E. (1938) The viscosity of liquid helium. *Physica*, **5**, 737–44.
Keesom, W. H. & van der Ende, J. N. (1930) The specific heat of substances at the temperatures obtainable with the aid of helium-II. Measurements of the atomic heats of lead and of bismuth. *Koninklijke Academie van wetenschappen te Amsterdam, Proceedings of the Section of Sciences*, **33**, 243–54.
Keesom, W. & van der Ende, J. N. (1932) The specific heat of solids obtainable with liquid helium. IV. Measurements of the atomic heats of tin and zinc. *CPL*, **219b**.
Keesom, W. & Wolfke, M. (1927) Two different liquid states of helium. *CPL*, **190b**, 17–22.
Keith, S. T. & Hoch, P. K. (1986) Formation of a research school: theoretical solid state physics at Bristol 1930–1950. *British Journal for the History of Science*, **19**, 19–44.
Khalatnikov, I. M. (ed.) (1989) *Recollections of L. D. Landau* (translated by J. B. Sykes). Pergamon Press, New York.
Kikoin, A. K. & Lasarew, B. G. (1938) Experiments with liquid helium-II. *Nature*, **141**, 912–13.
Kilminster, C. W. (ed.) (1987) *Schrödinger – Centenary Celebration of a Polymath*. Cambridge University Press, Cambridge.
Klein, M. J. (1970) *Paul Ehrenfest*, volume 1, *The making of a Theoretical Physicist*. North-Holland.
Kohler, R. E. (1971) The origin of the G. N. Lewis theory of the shared electron pair bond. *Historical Studies of the Physical Sciences*, **3**, 343–76.
Kohler, R. E. (1975) The Lewis–Langmuir theory of valence and the chemical community, 1920–1928. *Historical Studies in the Physical Sciences*, **6**, 431–68.
Kossel, W. (1916) Ueber Molekulbildung als Frage des Atombaues. *Annalen der Physik*, **49**, 229–362.
Kossel, W. (1920) Die Valenzkrafte im Lichte der neueren physikalischen Forschung. *Z. Elektrochem.*, **26**, 314–23.
Kozhevnikov, Aleksei (1992) Piotr Kapitza and Stalin's government: a study in moral choice. *Historical Studies in the Physical and Biological Sciences*, **22**, 131–63.
Kragh, H. (1990) *Dirac, a Scientific Biography*. Cambridge University Press, Cambridge.
Kronig, R. (1935) *The Optical Basis of Valency*. Cambridge University Press, Cambridge.
Kronig, R. (1953) 'Koninklijke Nederlandske Akademie van Wetenschappen', Zaterdag 27 Juni 1953, Deel LXII, No. 6, 61–71.
Kuhn, T. S., Heilbron, J. L., Forman, P. & Allen, L. (1967) *Sources for the History of Quantum Mechanics*. The American Philosophical Society, Philadelphia.
Kursanov, D. N., Gonikberg, M. G., Dubinin, B., Kabachnik, M. I., Kaveraneva, E. D., Prilezhaeva, E. N., Sokolov, N. D. & Freidlina, R. Kh., (1952) The Present State of the Chemical Structural Theory (translated by I. S. Bengelsdorf). *Journal of Chemical Education*, **78**, 2–13.
Kurti, N. (1958) Franz Eugen Simon. *Biographical Memoirs of the Fellows of the Royal Society*, **4**, 225–56.
Landau, L. (1938) The intermediate state of superconductors. *Nature*, **141**, 688.

Landau, L. D. (1941) The theory of superfluidity of helium II. *Journal of Physics (USSR)*, **5**, 71–90.
Landau, L. D. (1943) On the theory of the intermediate state of superconductors. *Journal of Physics (USSR)*, **7**, 99–107.
Landau, L. D. (1944) On the hydrodynamics of helium II. *Journal of Physics (USSR)*, **8**, 1–3.
Landau, L. D. (1947) On the theory of superfluidity of helium II. *Journal of Physics (USSR)*, **11**, 9–92.
Landau, L. D. (1949) On the theory of superfluidity. *Physical Review*, **75**, 884–85.
Landau, L. D. & Ginzburg, V. L. (1950) On the theory of superconductivity. In ter Haar, D. (1965*b*), 546–68.
Laqueur, W. (1947) *Weimar, A Cultural History 1918–1933*. Putnam's Sons, London.
Lennard-Jones, J. E. (1929) The electronic structure of some diatomic molecules. *Transactions of the Faraday Society*, **25**, 665–86.
Lewis, G. N. (1923) *Valence and the Structure of Atoms and Molecules*. The Chemical Catalog Company, New York.
Lifshitz, E. M. (1944) Radiation of sound in helium II. *Journal of Physics*, **8**, 110–14.
Lifshitz, E. M. & Andronikashvilii, E. L. (1959) *A Supplement to (W. H. Keesom's) 'Helium'* (translated from Russian). Consultants Bureau, Inc, New York.
Lindemann, F. (1932) *The Physical Significance of Quantum Theory*. Clarendon Press, Oxford.
Lindemann, F. (1933) The place of mathematics in the interpretation of the universe. *Philosophy*, **8**, 14–29.
London, H. (1934*a*) 'Uber die Möglichkeit das Auftretens eines Hochfrequenzrestwiderstandes bei Supraleitern.' Doctoral thesis, Breslau University.
London, H. (1934*b*) Production of heat in supraconducters. *Nature*, **133**, 497–8.
London, H. (1935) Phase-equilibrium of supraconductors in a magnetic field. *Proceedings of the Royal Society, London*, **A152**, 650–63.
London, H. (1936) An experimental examination of the electrostatic behaviour of supraconductors. *Proceedings of the Royal Society, London*, **A155**, 102–10.
London, H. (1937) The electromagnetic behaviour of supraconductors. *Proceedings, VII International Congress of Refrigeration*, **1**, 501–7.
London, H. (1938) A ponderomotive effect associated with the flow of heat through liquid helium II. *Nature*, **142**, 612–13.
London, H. (1939) Thermodynamics of the thermomechanical effect of liquid He II. *Proceedings of the Royal Society (London)*, **A171**, 484–96.
London, H. (1960) Superfluid helium. *Year Book of the Physical Society of London*, **10**, 34–48.
Lorentz, H. A. (1924) Application de la théorie des electrons aux proprietes des metaux. *Rapports et Discussions du Quartiéme Conseil de Physique Solvay*. Reprinted in *CPL* (Supplement), **50b**.
Luchtenberg, P. (1929) Erich Becher. *Kantstudien*, XXXIV, 275–90.
Margenau, H. (1939) Van der Waals forces. *Reviews of Modern Physics*, **11**, 1–35.
Margenau, H. (1944) The exclusion principle and its philosophical tradition. *Philosophy of Science*, **11**, 187–208.
Margenau, H. (1978) *Physics and Philosophy: Selected Essays*. Reidel Publishers, Dordrecht.
Margenau, H. & Kestner, N. R. (1969) *Theory of Intermolecular Forces*. Pergamon Press, New York.
Maurer, R. D. & Herlin, M. A. (1949) Second sound velocity in helium II. *Physical Review*, **76**, 948–50.
Maxwell, E. (1950) Isotope effect in the superconductivity of mercury. *Physical Review*, **78**, 477.
Mayer, J. (1937) Statistical mechanics of condensing systems I. *Journal of Chemical Physics*, **5**, 67–73.

Mayer, J. & Ackerman, P. G. (1937) Statistical mechanics of condensing systems II. *Journal of Chemical Physics*, **5**, 74–86.

McCrea, W. H. (1985) How quantum physics came to England. *New Scientist*, 17 October, 58–60.

McLennan, J. C. (1935) Opening address in a discussion on superconductivity and other low temperature phenomena. *Proceedings of the Royal Society of London*, **A152**, 1–8.

McLennan, J. C., Smith, H. D. & Wilhelm, J. O. (1932) The scattering of light by liquid helium. *Philosophical Magazine*, ser. 7, **14**, 161–7.

McWeeny, R. (1970) *Spins in Chemistry*. Academic Press, New York.

Medwick, P. (1988) Douglas Hartree and early computations in quantum mechanics. *Annals in the History of Computing*, **10**, 105–11.

Mehra, J. & Rachenberg, H. (1987) *Historical Development of Quantum Mechanics*. Springer, Berlin.

Meissner, W. (1935) The magnetic effects occurring on transition to the supraconductive state. *Proceedings of the Royal Society (London)*, **A152**, 13–15.

Meissner, W. & Ochsenfeld, R. (1933) Ein neuer Effect bei Eintritt der Supraleitfahigkeit. *Die Naturwissenschaften*, **21**, 787–8.

Mendelssohn, K. (1946) Superconductivity. *Reports of Progress in Physics*, **X**, 358–77.

Mendelssohn, K. (1954) Fritz London. *Naturwissenschaften*, **42**, 617–19.

Mendelssohn, K. (1964) Prewar work on superconductivity as seen from Oxford. *Reviews of Modern Physics*, **36**, 7–12.

Mendelssohn, K. (1969) States of aggregation. *Physics Today*, **23**, 46–51.

Moore, W. (1989) *Schrödinger, Life and Thought*. Cambridge University Press, Cambridge.

Mormann, T. (1991) Husserl's philosophy of science and the semantic approach. *Philosophy of Science*, **58**, 61–83.

Morrell, J. (1992) Research in physics at the Clarendon Laboratory, Oxford 1919–1939. *Historical Studies in the Physical Sciences*, **22**, 263–307.

Moss, R. W. (1988) *Free Radical – Albert Szent-Györgyi and the Battle over Vitamin C*. Paragon House, London.

Mott, Sir Neville (1981) Walter Heitler. *The London Times*, December 11.

Mott, Sir Neville (1982) Walter Heinrich Heitler. *Biographical Memoirs of the Fellows of the Royal Society*, **28**, 141–51.

Mott, Sir Neville (1984) Theory and Experiment circa 1932. In Hendry (1984*a*).

Moyer Hunsberger, I. (1954) Theoretical Chemistry in Russia. *Journal of Chemical Education*, **80**, 504–14.

Mulliken, R. S. (1928*a*) The assignment of quantum numbers for electrons in molecules. Part I. *Physical Review*, **32**, 186–222.

Mulliken, R. S. (1928*b*) The assignment of quantum numbers for electrons in molecules. Part II. *Physical Review*, **32**, 761–772.

Mulliken, R. S. (1929) The assignment of quantum numbers for electrons in molecules. Part III. *Physical Review*, **33**, 731–47.

Mulliken, R. S. (1931) Bonding power of electrons and theory of valence. *Chemical Reviews*, **9**, 347–88.

Mulliken, R. S. (1932*a*) The interpretation of band spectra. *Reviews of Modern Physics*, **4**, 1–86.

Mulliken, R. S. (1932*b*) Electronic structures of polyatomic molecules and valence. II. General considerations. *Physical Review*, **41**, 49–71.

Mulliken, R. S. (1933) Electronic structures of polyatomic molecules and valence. V. *Journal of Chemical Physics*, **1**, 492–503.

Mulliken, R. S. (1935) Electronic structures of polyatomic molecules and valence. VI. On the method of molecular orbitals. *Journal of Chemical Physics*, **3**, 375–8.

Mulliken, R. S. (1965) Molecular scientists and molecular science: some reminiscences. *Journal of Chemical Physics*, **43**, S7.

Mulliken, R. S. (1967) Spectroscopy, molecular orbitals and chemical bonding. *Science*, **157**, 17.

Mulliken, R. S. (1989) *Life of a Scientist*. Springer-Verlag, Berlin.

Nachansohn, D. (1979) *German-Jewish Pioneers of Science 1900–1933*. Springer, Berlin.

Nickles, T. (1980) *Scientific Discovery: Case Studies*. Reidel Publishers, Dordrecht.

Nye, M. J. (1994) *From Chemical Philosophy to Theoretical Chemistry: Dynamics of Matter and Dynamics of Discipline*. University of California Press.

Ogg, R. A. (1946) Bose–Einstein condensation of trapped electron pairs. Phase separation and superconductivity of metalammonia solutions. *Physical Review*, **69**, 243–4.

Osborne, D. W., Weinstock, B. & Abraham, B. M. (1949) Comparison of the flow of isotopically pure liquid He^3 and He^4. *Physical Review*, **75**, 988.

Padoa, A. (1903) Essai d'une théorie algebrique des nombres entiers, précéde d'une introduction logique à une theorie deductive quelconque. *Bibl. d. Congr. d. Phil.* (Paris), **III**.

Pais, A. (1982) *Subtle is the Lord*. Clarendon Press, Oxford.

Pais, A. (1991) *Niels Bohr's Times, in Physics, Philosophy and Polity*. Clarendon Press, Oxford.

Palmer, W. G. (1965) *A Short History of the Concept of Valency to 1930*. Cambridge University Press, Cambridge.

Parson, A. L. (1915) A magneton theory of the structure of the atom. *Smithsonian Miscellaneous Collections*, **65**, 1–80.

Pauli, W. (1927) Zur Quantenmechanik des magnetischen Ellectrons. *Zeitschrift für physik*, **43**, 601–23.

Pauli, W. (1946) Remarks on the history of the exclusion principle. *Science*, **103**, 213–15.

Pauli, W. (1958) *The Theory of Relativity*. Pergamon, London.

Pauling, L. (1928*a*) The application of the quantum mechanics to the structure of the hydrogen molecule. *Chemical Reviews*, **5**, 173–213.

Pauling, L. (1928*b*) The shared-electron chemical bond. *Proceedings of the National Academy of Sciences*, **14**, 359.

Pauling, L. (1931*a*) The nature of the chemical bond I. *Journal of the American Chemical Society*, **53**, 1367.

Pauling, L. (1931*b*) The nature of the chemical bond II. *Journal of the American Chemical Society*, **53**, 3225–37.

Pauling, L. (1931*c*) The nature of the chemical bond. Application obtained from the quantum mechanics and from a theory of paramagnetic susceptibility to the structure of molecules. *Journal of the American Chemical Society*, **53**, 1367–400.

Pauling, L. (1932) The nature of the chemical bond III. *Journal of the American Chemical Society*, **54**, 988–1003.

Pauling, L. (1939) *The Nature of the Chemical Bond*. Cornell University Press.

Pauling, L. (1940) A theory of the structure and process of formation of antibodies. *Science*, **62**, 2643–60.

Pauling, L. (1954) Modern Structural Chemistry. In *Nobel Lectures in Chemistry 1942–1962*. Elsevier Publishing Company, Amsterdam.

Pauling, L. (1960) *The Nature of the Chemical Bond*, third edition. Cornell University Press.

Pauling, L. & Wilson, E. B. (1935) *Introduction to Quantum Mechanics with Applications to Chemistry*. McGraw Hill, New York.

Peierls, R. (1980) Recollections of early solid state physics. *Proceedings of the Royal Society, London*, **A371**, 28–38.

Pellam, J. R. (1949) Investigations of pulsed second sound in liquid helium II. *Physical Review*, **75**, 1183–94.

Pellam, J. R. (1989) Lev Davidovich Landau. In Khalatnikov (1989), 315–16.
Pellam, J. R. & Scott, R. B. (1949) Second sound velocity in paramagnetically cooled liquid helium II. *Physical Review*, **76**, 869–70, 948–50.
Pellam, J. R. & Scott, R. B. (1950) *Scientific American*, **182**, June 1950, 7.
Peshkov, V. (1944) Second sound in helium II. *Journal of Physics*, **VIII**, 381–9.
Peshkov, V. (1946) Determination of the velocity of propagation of the second sound in helium II. *Journal of Physics (USSR)*, **10**, 389–98.
Pfänder, A. (1967) *Phenomenology of Willing and Motivation* (translated with an introduction and supplementary essays by H. Spiegelberg). Northwestern University Press.
Physical Society (1947) 'International Conference on Fundamental Particles and Low Temperatures'. Cavendish Laboratory, Cambridge, 22–27 July 1946. Proceedings published by the Physical Society.
Pippard, A. B. (1949) 'Microwave studies of superconductors'. PhD thesis, supervisor D. Shoenberg. Cambridge University (November 10, 1949).
Pippard, A. B. (1950*a*) Kinetics of the phase transition in superconductors. *Philosophical Magazine*, **xli**, 243–55.
Pippard, A. B. (1950*b*) Field variation of the superconducting penetration depth. *Proceedings of the Royal Society*, **A203**, 210–223.
Pippard, A. B. (1951*a*) The surface energies of superconductors. *Proceedings of the Cambridge Philosophical Society*, **47**, 617–25.
Pippard, A. B. (1951*b*) *Proceedings of the Physical Society*, **64**, Part 9, no. 381A.
Pippard, A. B. (1986) 'Early superconductivity research (except Leiden)'. Paper presented at the *H. Kamerlingh Onnes Symposium on the Origins of Applied Superconductivity*, Maryland.
Pullman, A. & Pullman, B. (1962) From Quantum Chemistry to Quantum Biochemistry. In Kasha & Pullman (1962).
Raman, V. V. & Forman, P. (1969) Why was it Schrödinger who developed de Broglie's ideas? *Historical Studies in the Physical Sciences*, **1**, 291–314.
Ransil, B. J. (ed.) (1989) *Life of a Scientist, Autobiographical Notes by R. S. Mulliken*. Springer-Verlag, Berlin.
Rich, A. & Davidson, N. (eds.) (1968) *Structural Chemistry and Molecular Biology*. Freeman, New York.
Richter, W. (1930) Staatliche Wissenschaftsverwaltung. In Richter, W. & Peters, H. (eds.) *Die Statuten der preussischen Universitäten und Technischen Hochschulen*. Berlin.
Rickert, H. (1962) *Science and History, A Critique of Positivist Epistemology*. Van Nostrand Company, New York.
Rickert, H. (1986) *The Limits of Concept Formation in Natural Science*. Cambridge University Press, Cambridge. (First edition in German, 1896–1902.)
Rider, R. E. (1985) Alarm and opportunity: emigration of mathematicians and physicists to Britain and the United States 1933–1945. *Historical Studies in the Physical Sciences*, **15**, 107–76.
Ringer, F. K. (1969) *The decline of the German Mandarins, the German Academic Community 1890–1933*. Harvard University Press.
Robertson, R. (1923) Opening remarks by the chairman. *Transactions of the Faraday Society*, **19**, 483–4.
Rodebush, W. H. (1928) The electron theory of valence. *Chemical Reviews*, **5**, 509–31.
Ropolyi, L. (ed.) (1990) *Thermodynamics, its History and Philosophy*. Vesprem, Hungary.
Rosen, N. (1931) The normal state of the hydrogen molecule. *Physical Review*, **38**, 2099–144.
Russell, C. (1971) *The History of Valency*. Leicester University Press, Leicester.
Rutgers, A. J. (1934) Note on supraconductivity. *Physica*, **1**, 306–20.

Schafroth, M. R. (1951) A note on Fröhlich's theory of superconductivity. In E. Powers (ed.) *Proceedings of the International Conference of Low Temperature Physics*. Oxford, 112–13.

Schrödinger, E. (1922) Über eine bemerkenswerte Eigenschaft der Quantenbahnen eines einzelnen Elektrons. *Zeitschrift für Physik*, **12**, 13–23.

Schrödinger, E. (1936) Phenomenological theory of Supraconductivity. *Nature*, **137**, 824.

Schweber, S. S. (1990) The young John Slater and the development of quantum chemistry. *Historical Studies in the Physical and Biological Sciences*, **20**, 339–406.

Serin, B., Reynolds, C. A. & Nesbitt, L. B. (1950) Mass dependence of the superconducting transition temperature of mercury. *Physical Review*, **80**, 761–2.

Servos, J. (1990) *Physical Chemistry from Ostwald to Pauling. The Making of a Science in America*. Princeton University Press.

Shimony, A. (1963) Role of observer in quantum theory. *American Journal of Physics*, **31**, 755–77.

Shoenberg, D. (1938) *Superconductivity*. Cambridge University Press, Cambridge.

Shoenberg, D. (1952) *Superconductivity*. Cambridge University Press, Cambridge.

Shoenberg, D. (1971) Heinz London. *Royal Society of London. Biographical Memoirs*, **17**, 441–61.

Shoenberg, D. (1989) Recollections of Landau. In I. M. Kalatnikov (ed.) (1989) *Recollections of L. D. Landau* (translation by J. B. Sykes). Pergamon Press, New York. 227.

Sidgwick, N. V. (1927) *The Electronic Theory of Valency*. Clarendon Press, Oxford.

Sigiura, V. (1927) Quantentheorie und Chemische Bindung. *Zeitschrift für Physik*, **45**, 484–91.

Sigurdsson, S. (1991) 'Hermann Weyl, Mathematics and Physics 1900–1927'. Doctoral dissertation, Department of History of Science, Harvard University.

Simoes, A. (1993) 'Converging trajectories, diverging traditions: chemical bond, valence, quantum mechanics and chemistry 1927–1937.' PhD dissertation, University of Maryland.

Simon, F. E. (1927) Zum Prinzip von der Unerreichbarkeit des absoluten Nullpunktes. *Zeitscrift für Physik*, **41**, 806–9.

Simon, F. E. (1934) Behaviour of condensed helium near absolute zero. *Nature*, **133**, 527.

Simon, F. E. (1952) *Low Temperature Physics, Four Lectures*. Academic Press, New York.

Slater, J. (1931) Directed valence in polyatomic molecules. *Physical Review*, **37**, 481–9.

Slater, J. (1937) The nature of the superconducting state. *Physical Review*, **51**, 195–202.

Slater, J. (1938) The nature of the superconducting state II. *Physical Review*, **52**, 214–22.

Slater, J. & Kirkwood, J. G. (1931) The van der Waals forces in gases. *Physical Review*, **37**, 682–97.

Smith, H. G. & Wilhelm, J. O. (1935) Superconductivity. *Reviews of Modern Physics*, **7**, 238–71.

Sommerfeld, A. (1928) Zur Elekrontheorie der Metalle auf Grund der Fermischen Statistik. *Zeitschrift für Physik*, **47**, 1–32.

Sopka, K. (1988) *Quantum Physics in America*. American Institute of Physics, New York.

Spiegelberg, H. (1973) Is the reduction necessary for phenomenology? Husserl's and Pfander's replies. *Journal of the British Society for Phenomenology*, **4**, 3–15.

Spiegelberg, H. (1982) *The Phenomenological Movement*. Martinus Nijhoff, Dordrecht.

Sturdivant, J. H. (1968) The scientific work of Linus Pauling. In A. Rich & N. Davidson (1968), 4–11.

Sydoriak, S., Grilly, E. & Hammel, E. (1949) Condensation of pure He^3 and its vapour pressures between 1.2° and its critical point. *Physical Review*, series 2, **75**, 303–5.

Szent-Györgyi, A. (1960) *Introduction to Submolecular Biology*. Academic Press, New York.

Tatevskii, V. M. & Shakhparanov, M. I. (1952) About a machistic theory in chemistry and its propagandists. *Journal of Chemical Education*, **78**, 13–14.

Ter Haar, D. (ed.) (1965*a*) *Collected Papers of P. L. Kapitza*. Pergamon Press, New York.

Ter Haar, D. (ed.) (1965*b*) *Collected Papers of L. D. Landau*. Pergamon Press, New York.

Thomson, J. J. (1913) On the structure of the atom. *Philosophical Magazine*, **26**, 757–89.

Thomson, J. J. (1914) The forces between atoms and chemical affinity. *Philosophical Magazine*, **27**, 1655–75.

Thomson, J. J. (1924) Introduction to the session on the electronic theory of valency. *Transactions of the Faraday Society*, **19**, 450.

Tisza, L. (1938*a*) Transport phenomena in helium-II. *Nature*, **141**, 913.

Tisza, L. (1938*b*) Sur la supraconductibilité thermique de l'helium II et le statistique Bose–Einstein. *Comptes Rendus Hebdomadaires des Séances de l'Academie des Sciences* (Paris), **207**, 1035–7.

Tisza, L. (1938*c*) La viscocité de l'helium liquide et la statistique Bose–Einstein. *Comptes Rendus Hebdomadaires des Séances de l'Academie des Sciences* (Paris), 207, 1186–9.

Tisza, L. (1940*a*) Sur la théorie des liquides quantiques. Application a l'helium liquide. I. *Journal de Physique et Radium*, **1**, 164–172.

Tisza, L. (1940*b*) Sur la théorie des liquides quantiques. Application a l'helium liquide. II. *Journal de Physique et Radium*, **1**, 350–8.

Tisza, L. (1947) The theory of liquid helium. *Physical Review*, **72**, 838–54.

Tisza, L. (1948) Helium the unruly liquid. *Physics Today*, **1**, 4–8, 26.

Tisza, L. (1949*a*) Green's theory of liquid helium. *Nature*, **163**, 102–4.

Tisza, L. (1949*b*) On the theory of superfluidity. *Physical Review*, **75**, 885–6.

Tisza, L. (1949*c*) The present state of the helium problem. *Proceedings of the International Conference on the Physics of Very Low Temperatures*. MIT Press, Cambridge, Massachusetts, 1–3.

Tisza, L. (1990) History of the two-fluid model. In Ropolyi (1990).

Tolman, R. C. (1914) The principle of similitude. *Physical Review*, **3**, 244–55.

Tolman, R. C. (1915) The principle of similitude and the principle of dimensional homogeneity. *Physical Review*, **6**, 219–33.

Tolman, R. C. (1916) Note on the homogeneity of physical equations. *Physical Review*, **8**, 8–11.

Uhlenbeck, G. (1927) 'Over statistische methoden in de theorie der quanta'. Doctoral thesis, 's Gravenhage.

Uhlenbeck, G. (1980) Some reminiscences about Einstein's visits to Leiden. In H. Woolf (ed.) *Some Strangeness in Proportion: a Centennial Symposium to Celebrate the Achievements of Albert Einstein*. Addison-Wesley, New York, 524–5.

Van der Waerden, B. L. (1960) Exclusion principle and spin. In M. Fierz & V. F. Weisskopf (ed.) *Theoretical Physics in the Twentieth Century. A Memorial volume to W. Pauli*. Interscience, New York.

Van Vleck, J. (1928) The new quantum mechanics. *Chemical Reviews*, **5**, 467–507.

Van Vleck, J. (1935) The group relation between the Mulliken and Slater–Pauling theories of valence. *Journal of Chemical Physics*, **3**, 803–6.

Van Vleck, J. (1970) Spin, the great indicator of valence behaviour. *Pure and Applied Chemistry*, **24**, 235–55.
Van Vleck, J. & Sherman, A. (1935) The quantum theory of valence. *Reviews of Modern Physics*, 7, 167–228.
Von Laue, M. (1932) Zur Deutung einiger Versuche uber Supraleitung. *Physikalische Zeitschrift*, **33**, 793–5.
Von Laue, M. (1936) Note in *Physicalische Berichte*, **44**, 103.
Von Laue, M. (1938) Zur Thermodynamik der Supraleitung. *Annalen der Physik*, **32**, 253–8.
Von Laue, M. (1947) Theorie der Supraleitung. *Naturwissenschaften*, **34**, 441–2.
Von Laue, M. (1950) *History of Physics*. Academic Press, New York.
Von Laue, M. (1952*a*) Kritische Bemerkungen zur Theorie der Supraleitung. Bemerkungen zur Theorie der Supraleitung II. *Annalen der Physik (Leipzig)*, **10**, 296–301, 305–6.
Von Laue, M. (1952*b*) *Theory of Superconductivity*. Academic Press, New York.
Von Laue, M. & Möglich, F. (1933) Ueber das magnetische Feld in der Umgebung von Supraleitern. Preussischen Akademie der Wissenschaften, *Sitzungberichte*, **16**, 544–65.
Von Neumann, J. (1955) *Mathematical Foundations of Quantum Mechanics* (translated into English by R. T. Beyer). Princeton University Press, Princeton.
Von Wroblewski, S. (1885) Über den gebrauch des siedenden Sauerstoffs Stickstoffs, Kohlenoxyds, sowie der atmospharisharischen Luft als Kaltemittel. *Wiener Sitzungsberichte*, **91**, 667–711.
Wang, S. C. (1928) The problem of the normal hydrogen molecule in the new quantum mechanics. *Physical Review*, **31**, 579–86.
Weinbaum, S. (1933) The normal state of the hydrogen molecule. *Journal of Chemical Physics*, **1**, 593–6.
Weyl, H. (1921) Electricity and Gravitation. *Nature*, **106**, 800–2.
Weyl, H. (1946) Encomium. *Science*, **103**, 216–18.
Wheeler, J. A. & Zurek, W. H. (1983) *Quantum Theory and Measurement*. Princeton University Press.
Wheland, G. (1944) *The Theory of Resonance and its Applications to Organic Chemistry*. John Wiley, New York.
Wilhelm, J. D., Misener, A. D. & Clark, A. R. (1935) The viscocity of liquid helium. *Proceedings of the Royal Society of London*, **A133**, 458–91.
Yang, C. N. (1979) Gauge fields. In S. Kamefuchi *et al.* (eds.) *Proceedings of the International Symposium on the Foundations of Quantum Mechanics*, 1–9.
Yang, C. N. (1983) *Selected Papers 1945–1980, with Commentary*. Freeman, New York.
Yang, C. N. (1986) Hermann Weyl's contributions to physics. In K. Chandrasekharan (ed.) *Hermann Weyl*. Springer, Berlin, 7–21.
Yang, C. N. (1987) Flux quantization, a personal reminiscence. In J. D. Fairbank *et al.* (1987), 255–6.
Zilsel, P. R. (1950) Liquid helium II: the hydrodynamics of the two-fluid model. *Physical Review*, **79**, 309–13.
Zilsel, P. R. (1953) Liquid helium II: Bose–Einstein condensation and the two-fluid model. *Physical Review*, **92**, 1106–12.

Index